ARBITRAGE, CREDIT AND INFORMATIONAL RISKS

PEKING UNIVERSITY SERIES IN MATHEMATICS

Vol. 1: An Introduction to Finsler Geometry
by Xiaohuan Mo (Peking University, China)

Vol. 2: Numerical Methods for Exterior Problems
by Lung-An Ying (Peking University & Xiamen University, China)

Vol. 3: Approaches to the Qualitative Theory of Ordinary Differential Equations: Dynamical Systems and Nonlinear Oscillations
by Tongren Ding (Peking University, China)

Vol. 4: Elliptic, Hyperbolic and Mixed Complex Equations with Parabolic Degeneracy: Including Tricomi–Bers and Tricomi–Frankl–Rassias Problems
by Guo Chun Wen (Peking University, China)

Vol. 5: Arbitrage, Credit and Informational Risks
edited by Caroline Hillairet (Ecole Polytechnique, France), Monique Jeanblanc (Université d'Evry, France) and Ying Jiao (Université Lyon I, France)

Peking University Series in Mathematics — Vol. 5

ARBITRAGE, CREDIT AND INFORMATIONAL RISKS

Editors

Caroline Hillairet
Ecole Polytechnique, France

Monique Jeanblanc
Université d'Evry, France

Ying Jiao
Université Lyon I, France

NEW JERSEY • LONDON • SINGAPORE • BEIJING • SHANGHAI • HONG KONG • TAIPEI • CHENNAI

Published by

World Scientific Publishing Co. Pte. Ltd.
5 Toh Tuck Link, Singapore 596224
USA office: 27 Warren Street, Suite 401-402, Hackensack, NJ 07601
UK office: 57 Shelton Street, Covent Garden, London WC2H 9HE

British Library Cataloguing-in-Publication Data
A catalogue record for this book is available from the British Library.

Peking University Series in Mathematics — Vol. 5
ARBITRAGE, CREDIT AND INFORMATIONAL RISKS

The image on the back cover is from the website of Beijing International Center for Mathematical Research (BICMR), http://www.bicmr.org/

ISBN 978-981-4602-06-8

Printed in Singapore

Learning without thought is labor lost;
thought without learning is perilous

学而不思则罔，思而不学则殆

Confucius

孔子

Preface

This book includes some contributions presented at the workshop "Arbitrage, Credit and Informational Risks". This event has been held at Beijing International Center for Mathematical Research (BICMR) in June 2013 in the framework of the Sino-French Research Program in Mathematics (SFRPM). It was dedicated to the presentation of research results and discussions in the field of financial mathematics around the themes of arbitrage, credit and asymmetric information risks. The book is split into three parts.

In the first part, we collected papers in which two different notions of arbitrages and related concepts are studied. The first one is the well-known condition of No Free Lunch with Vanishing Risk (NFLVR); it is equivalent to the existence of an equivalent local martingale measure under which discounted prices are local martingales. The second one is the condition of No Unbounded Profit with Bounded Risk (NUPBR), which is also known as No Arbitrage of the First Kind; it is equivalent to the existence of a positive local martingale deflator such that prices multiplied by this deflator are local martingales. This part contains the four following papers:

Claudio Fontana studies the stability of NFLVR and NUPBR, as well as other arbitrage conditions, under an absolutely continuous change of probability measure.

Johannes Ruf and Wolfgang J. Runggaldier present a systematic construction of market models that satisfy NUPBR, but not NFLVR, by stating sufficient assumptions on the first hitting time of zero by the inverse of the martingale deflator.

In the context of jump-diffusion market models, Jacopo Mancin and Wolfgang J. Runggaldier construct examples of models that satisfy NUPBR, but not NFLVR.

Working in a progressive enlargement setting, and assuming that NFLVR and NUPBR hold in the reference filtration, Anna Aksamit, Tahir

Choulli, Jun Deng and Monique Jeanblanc give conditions so that NUPBR is satisfied in the enlarged filtration, whereas NFLVR fails to hold.

The second part contains four papers devoted to Credit Risk:

Sébastien Hitier and Ying Zhu study the problem of pricing credit derivatives in a structural model. In structural models, the default time is the hitting time of a given barrier by the firm value. Structural models, although based on sound economic arguments, are difficult to handle for calibration in practice. The authors demonstrate how a structural model can be calibrated and used for risk-neutral pricing of credit derivatives.

Stéphane Crépey proposes and studies a dynamics model for bilateral counterparty risk on credit derivatives, which goes beyond the classic immersion setting.

Shiqi Song develops mathematical background for a dynamic model of a single default, based on the fairly general construction of the conditional law of the default event from its Azéma supermartingale. He establishes a new formula for the semimartingale decomposition of martingales in the reference filtration and gives results related to the existence of the conditional density of the default time.

Laurence Carassus and Simone Scotti apply the error calculus methodology to the problem of an optimal credit allocation under a hidden regime switching model.

The third part presents four contributions in the area of Control Problems and Information Risks:

Ivan Guo and Marek Rutkowski consider a class of recursive multiplayer stopping games in a discrete time setting. Such games have interpretations in economic and financial modelling, for example, as multi-person game options. They prove the existence of an optimal equilibrium and provide an explicit algorithm for the computation of the value of the game.

Monique Jeanblanc and Anthony Réveillac present examples of Backward Stochastic Differential Equations (BSDEs) whose driver is not integrable and degenerates when the terminal time approaches. Such BSDEs may have, depending on the terminal conditions, no solution or an infinite number of solutions.

Caroline Hillairet studies a portfolio optimization problem in a market model characterized by the presence of different prices for the same asset, which arise as a consequence of different information settings.

In the presence of 'shadow costs' of incomplete information, as well as stochastic volatility and jumps in the dynamics of the asset, Sana Mahfoudh and Monique Pontier compare the corresponding cost value process of two different hedging strategies.

Acknowledgement: The workshop "Arbitrage, Credit and Informational Risks" took place at Beijing International Center for Mathematical Research in June 2013 in the framework of Sino-French Research Program in Mathematics. We are grateful for the financial support from National Natural Science Foundation of China, 高等学校数学研究与高等人才培养中心 and several Chinese and French research institutes such as BICMR, Fondation Mathématique Jacques Hadamard, Fondation Sciences Mathématiques de Paris, Institut Fourier at Grenoble and Laboratoire de Probabilité et Modèles Aléatoires at Paris.

During the preparation of this volume, Marek Rutkowski provided us with valuable help and many suggestions, we would like to warmly thank him. We also benefited a lot from the support of the World Scientific Press, notably from the editors Lai Fun Kwong, Dena Li, Hongbing Shi and Ji Zhang.

Last but not least, we thank all the participants of the workshop for their contributions, making this event a great success:

Anna Aksamit, Université d'Evry Val d'Essonne,
Stefan Ankirchner, Universität Bonn,
Giuseppe Benedetti, Université Paris Dauphine,
Christophette Blanchet-Scalliet, Ecole Centrale de Lyon,
Lijun Bo, Xidian University,
Giorgia Callegaro, Università degli Studi di Padova,
Luciano Campi, Université Paris Nord,
Laurence Carassus, Université de Reims Champagne Ardenne,
Sébastien Choukroun, Université Paris Diderot - Paris 7,
Stéphane Crépey, Université d'Evry Val d'Essonne,
Albina Danilova, London School of Economics,
Cristina Di Girolami, Università G. D'Annunzio di Pescara,
Nicole El Karoui, Université Pierre et Marie Curie - Paris 6,
Laure Elie, Université Paris Diderot - Paris 7,
Claudio Fontana, Inria Paris - Rocquencourt,
Noufel Frikha, Université Paris-Diderot -Paris 7,
Zorana Grbac, Universität Berlin,

Caroline Hillairet, Ecole Polytechnique,
Sébastien Hitier, BNP Paribas Hong Kong,
Chau Ngoc Huy, Université Paris Diderot - Paris 7,
Monique Jeanblanc, Université d'Evry Val d'Essonne,
Ying Jiao, Université Paris Diderot - Paris 7,
Wanmo Kang, Korea Advanced Institute of Science and Technology,
Céline Labart, Université de Savoie,
Sophie Laruelle, Ecole Centrale Paris,
Jérôme Lelong, Ecole Nationale Supérieure d'Informatique et Mathématiques Appliquées de Grenoble,
Marta Leniec, Uppsala Universitet,
Libo Li, Ritsumeikan University,
Shanqiu Li, Université Paris Diderot - Paris 7,
Chunhua Ma, Nankai University,
Gilles Pagès, Université Pierre et Marie Curie - Paris 6,
Xianhua Peng, Hong Kong University of Science and Technology,
Monique Pontier, Université Paul Sabatier - Toulouse 3,
Johannes Ruf, University of Oxford,
Wolfgang Runggaldier, Università degli Studi di Padova,
Marek Rutkowski, University of Sydney,
Abass Sagna, Ecole Nationale Supérieure d'Informatique pour l'Industrie et l'Entreprise,
Simone Scotti, Université Paris Diderot - Paris 7,
Shiqi Song, Université d'Evry Val d'Essonne,
Laura Vinckenbosch, INRIA Nancy,
Duo Wang, Peking University,
Lan Wu, Peking University,
Jianming Xia, Chinese Academy of Science,
Dewen Xiong, Shanghai Jiao Tong University,
Jingping Yang, Peking University,
Kai Zhang, Jilin University.

Contents

PART 1
Arbitrage

No-arbitrage Conditions and Absolutely Continuous Changes of Measure

Claudio Fontana*

Abstract. We study the stability of several no-arbitrage conditions with respect to absolutely continuous, but not necessarily equivalent, changes of measure. We first consider models based on continuous semimartingales and show that no-arbitrage conditions weaker than NA and NFLVR are always stable. Then, in the context of general semimartingale models, we show that an absolutely continuous change of measure does never introduce arbitrages of the first kind as long as the change of measure density process can reach zero only continuously.

1. Introduction

The notion of no-arbitrage is of paramount importance in financial economics and mathematical finance. The search for an adequate no-arbitrage condition, economically meaningful and mathematically tractable, has a rather long history and we refer to [37] for an overview of the main developments. An important step was marked by the papers [6, 9], where the authors established the equivalence between the existence of an Equivalent (Local-/σ-) Martingale Measure and the validity of the No Free Lunch with Vanishing Risk (NFLVR) condition.

More recently, motivated mainly by developments in Stochastic Portfolio Theory (see e.g. [13] for an overview) and in the Benchmark Approach to Quantitative Finance (see e.g. [33]), an increasing attention has been paid to no-arbitrage concepts weaker than the classical NFLVR condition, as documented by the recent papers [16, 18, 24–26, 35, 39] and [41] (see also [29] for an earlier contribution to this strand of literature and [15, 17] for a unifying analysis of the different no-arbitrage conditions appearing in the context of continuous semimartingale models). In particular, it has been demonstrated that the full strength of NFLVR may not be needed in

*Université d'Évry Val d'Essonne, Laboratoire Analyse et Probabilités.
E-mail: `claudio.fontana@univ-evry.fr`.

order to solve portfolio optimisation problems as well as to perform pricing and hedging.

In quantitative finance, the typical approach starts with the specification of a model under a given probability measure $\mathbb{P}$, under the assumption that arbitrage profits (according to one of the many possible definitions) cannot be realised. One may then wonder whether such a no-arbitrage assumption is robust with respect to a change of measure from $\mathbb{P}$ to an absolutely continuous, but not necessarily equivalent, measure $\mathbb{Q}$. The present paper aims at answering this question.

In the mathematical finance literature, it has already been shown that absolutely continuous, but not equivalent, changes of measure may lead to arbitrage opportunities (in the classical sense of [6]), see e.g. [7, 31] and, more recently, [36]. However, this still leaves open the question of whether no-arbitrage conditions weaker than NA and NFLVR can be altered by absolutely continuous changes of measure. As we are going to show in Section 3, the first (but actually not very surprising) main result of this paper is that this is never the case in models based on continuous semimartingales.

Among the no-arbitrage conditions that are weaker than NFLVR, the condition of No Arbitrage of the First Kind (NA1), or, equivalently, No Unbounded Profit with Bounded Risk (see [24] and [25]), plays a particularly important role. Indeed, it has been shown that pricing and hedging can be satisfactorily performed as long as NA1 holds and, moreover, the same condition is the minimal one in order to solve portfolio optimisation problems in general semimartingale models. This motivates the study of the stability of NA1 with respect to absolutely continuous, but not necessarily equivalent, changes of measure in general semimartingale models. As already observed in [24, Remark 4.13], the NA1 condition may not be preserved by an absolutely continuous change of measure when jumps are present. In that regard, the second main result of this paper shows that the loss of NA1 after a change of measure is intimately linked to the possibility of the density process jumping to zero (see Section 5).

The remainder of the paper is structured as follows. Section 2 introduces the modeling framework for the first part of the paper, based on continuous semimartingales, and presents five different notions of arbitrage. In the continuous semimartingale setting, Section 3 shows that weak no-arbitrage conditions are invariant with respect to absolutely continuous changes of measure, while Section 4 shows by means of a counterexample that this is not the case for the classical NA and NFLVR conditions. Finally, Section 5 studies the stability of the NA1 condition in general semimartingale models.

2. Continuous Financial Markets and No-arbitrage Conditions

Let $(\Omega, \mathcal{F}, \mathbb{P})$ be a given probability space endowed with a right-continuous filtration $\mathbb{F} = (\mathcal{F}_t)_{0 \leq t \leq T}$, where $T \in (0, \infty)$ represents a fixed time horizon. We denote by $\mathcal{M}(\mathbb{P})$ the family of all (uniformly integrable) $\mathbb{P}$-martingales and by $\mathcal{M}_{\text{loc}}(\mathbb{P})$ the family of all $\mathbb{P}$-local martingales. Without loss of generality, we assume that all elements of $\mathcal{M}_{\text{loc}}$ are càdlàg and we denote by $\mathcal{M}^c(\mathbb{P})$ and $\mathcal{M}^c_{\text{loc}}(\mathbb{P})$ the families of all continuous processes belonging to $\mathcal{M}(\mathbb{P})$ and $\mathcal{M}_{\text{loc}}(\mathbb{P})$, respectively.

We consider an abstract financial market with d risky assets, whose discounted prices (with respect to a given reference security) are represented by the $\mathbb{R}^d$-valued *continuous* semimartingale $S = (S_t)_{0 \leq t \leq T}$, with $S_t = (S^1_t, \ldots, S^d_t)^\top$, with $^\top$ denoting transposition. Being S a special semimartingale, its unique canonical decomposition can be written as $S = S_0 + A + M$, where A is a continuous $\mathbb{R}^d$-valued predictable process of finite variation and M is an $\mathbb{R}^d$-valued process in $\mathcal{M}^c_{\text{loc}}(\mathbb{P})$ with $M_0 = A_0 = 0$. By Proposition II.2.9 of [21], we can write, for all $i, j = 1, \ldots, d$,

$$A^i = \int a^i dB \qquad \text{and} \qquad \langle S^i, S^j \rangle = \langle M^i, M^j \rangle = \int c^{ij} dB, \tag{2.1}$$

for some continuous real-valued predictable strictly increasing process B and where $a = (a^1, \ldots, a^d)^\top$ and $c = \big((c^{i1})_{1 \leq i \leq d}, \ldots, (c^{id})_{1 \leq i \leq d}\big)$ are predictable processes taking values in $\mathbb{R}^d$ and in the cone of symmetric non-negative $d \times d$ matrices, respectively. We do not necessarily assume that S takes values in the positive orthant of $\mathbb{R}^d$. For all $t \in [0, T]$, let us denote by c^+_t the Moore-Penrose pseudoinverse of the matrix c_t. The proof of Proposition 2.1 of [11] (see also [10], Lemma 4.3) shows that the process $c^+ = (c^+_t)_{0 \leq t \leq T}$ is predictable and, hence, the process a can be represented as

$$a = c\lambda + \nu, \tag{2.2}$$

where $\lambda = (\lambda_t)_{0 \leq t \leq T}$ is defined by $\lambda_t := c^+_t a_t$, for all $t \in [0, T]$, and $\nu = (\nu_t)_{0 \leq t \leq T}$ is an $\mathbb{R}^d$-valued predictable process with $\nu_t \in \text{Ker}(c_t) := \{x \in \mathbb{R}^d : c_t x = 0\}$, for all $t \in [0, T]$.

Let us now introduce the notion of *admissible strategy*, assuming a frictionless financial market. Let $L(S; \mathbb{P}) := L^2_{\text{loc}}(M; \mathbb{P}) \cap L^0(A; \mathbb{P})$, where $L^2_{\text{loc}}(M; \mathbb{P})$ and $L^0(A; \mathbb{P})$ are the sets of all $\mathbb{R}^d$-valued predictable processes H such that $\int_0^T H_t^\top d\langle M, M \rangle_t H_t < \infty$ $\mathbb{P}$-a.s. and $\int_0^T |H_t^\top dA_t| < \infty$ $\mathbb{P}$-a.s., respectively. It is well-known that $L(S;\mathbb{P})$ is the largest set of predictable integrands with respect to the continuous semimartingale S under the probability measure $\mathbb{P}$. For $H \in L(S;\mathbb{P})$, we denote by $H \cdot S$ the stochastic integral process $\left(\int_0^t H_u\, dS_u\right)_{0\leq t\leq T}$, which has to be understood as a vector stochastic integral (see, e.g., [21, Section III.6]). We are now in position to formulate the following classical definition.

Definition 2.1. Let $a \in \mathbb{R}_+$. An element $H \in L(S;\mathbb{P})$ is said to be an *aadmissible strategy* if $H_0 = 0$ and $(H \cdot S)_t \geq -a$ $\mathbb{P}$-a.s. for all $t \in [0,T]$. An element $H \in L(S;\mathbb{P})$ is said to be an *admissible strategy* if it is an a-admissible strategy for some $a \in \mathbb{R}_+$.

For $a \in \mathbb{R}_+$, we denote by $\mathcal{A}_a(\mathbb{P})$ the set of all a-admissible strategies and by $\mathcal{A}(\mathbb{P})$ the set of all admissible strategies, i.e., $\mathcal{A}(\mathbb{P}) = \bigcup_{a\in\mathbb{R}_+}\mathcal{A}_a(\mathbb{P})$. As usual, H_t^i represents the number of units of asset i held in the portfolio at time t. For $(x,H)\in \mathbb{R}\times \mathcal{A}(\mathbb{P})$, we define the *gains from trading* process $G(H) = \left(G_t(H)\right)_{0\leq t\leq T}$ by $G_t(H) := (H \cdot S)_t$, for all $t \in [0,T]$, and the *portfolio value* process $V(x,H) := x + G(H)$. This corresponds to consider portfolios generated by *self-financing* admissible strategies.

We now introduce five different notions of arbitrage that will be considered in the present paper.

Definition 2.2. (i) A strategy $H \in \mathcal{A}_0(\mathbb{P})$ is said to yield an *increasing profit* if the process $G(H)$ is $\mathbb{P}$-a.s. non-decreasing and $\mathbb{P}\left(G_T(H) > 0\right) > 0$. If there exists no such strategy we say that the *No Increasing Profit (NIP)* condition holds;
(ii) a strategy $H \in \mathcal{A}_0(\mathbb{P})$ is said to yield a *strong arbitrage opportunity* if $\mathbb{P}\left(G_T(H) > 0\right) > 0$. If there exists no such strategy we say that the *No Strong Arbitrage (NSA)* condition holds;
(iii) a non-negative random variable ξ is said to yield an *arbitrage of the first kind* if $\mathbb{P}(\xi > 0) > 0$ and for every $v \in (0,\infty)$ there exists an element $H^v \in \mathcal{A}_v(\mathbb{P})$ such that $V_T(v,H^v) \geq \xi$ $\mathbb{P}$-a.s. If there exists no such random variable we say that the *No Arbitrage of the First Kind (NA1)* condition holds;
(iv) a strategy $H \in \mathcal{A}(\mathbb{P})$ is said to yield an *arbitrage opportunity* if $G_T(H) \geq 0$ $\mathbb{P}$-a.s. and $\mathbb{P}\left(G_T(H) > 0\right) > 0$. If there exists no such strategy we say that the *No Arbitrage (NA)* condition holds;
(v) a sequence $\{H^n\}_{n\in\mathbb{N}} \subset \mathcal{A}(\mathbb{P})$ is said to yield a *Free Lunch with Vanishing Risk* if there exist a positive constant $\varepsilon > 0$ and an increasing sequence $\{\delta_n\}_{n\in\mathbb{N}}$ with $0 \leq \delta_n \nearrow 1$ such that $\mathbb{P}\left(G_T(H^n) > -1 + \delta_n\right) = 1$ and

$\mathbb{P}(G_T(H^n) > \varepsilon) \geq \varepsilon$. If there exists no such sequence we say that the *No Free Lunch with Vanishing Risk (NFLVR)* condition holds.

The NIP condition corresponds to the notion of No Unbounded Increasing Profit introduced in [24] and represents the weakest condition among those listed above.† The NSA condition corresponds to the notion of absence of arbitrage opportunities adopted in Section 3 of [29] as well as to the NA$^+$ condition studied in [40]. The notion of arbitrage of the first kind goes back to [19], while the formulation adopted in Definition 2.2 is due to [25]. In particular, the NA1 condition is equivalent to the notion of No Unbounded Profit with Bounded Risk of [24] (see [25, Proposition 1]), which in turn corresponds to the BK condition considered in [23]. Finally, the NA and NFLVR conditions are classical (see, e.g., [6]).

We close this section by recalling the probabilistic characterisations of the conditions introduced in Definition 2.2, referring to [15] for a more detailed analysis of the no-arbitrage properties of financial models based on continuous semimartingales. As a preliminary, let the *mean-variance trade-off process* $\widehat{K} = (\widehat{K}_t)_{0 \leq t \leq T}$ be defined as (see, e.g., [38])

$$\widehat{K}_t := \int_0^t \lambda_u^\top d\langle M, M\rangle_u \, \lambda_u = \int_0^t a_u^\top c_u^+ a_u \, dB_u, \qquad \text{for all } t \in [0,T]. \quad (2.3)$$

Let us also define $\widehat{K}_s^t := \widehat{K}_t - \widehat{K}_s$, for all $s, t \in [0,T]$ with $s \leq t$, and the (possibly infinite valued) stopping time

$$\sigma := \inf\{t \in [0,T] : \widehat{K}_t^{t+h} = \infty, \forall h \in (0, T-t]\}, \quad (2.4)$$

with the usual convention $\inf \emptyset = \infty$. The next proposition provides necessary and sufficient conditions for the validity of the no-arbitrage conditions introduced in Definition 2.2, collecting several important results obtained by [6, 8, 23, 24, 40] and [25]. In the current formulation, a proof can be found in [15].

Proposition 2.3. *The following hold:*
(i)*NIP holds if and only if* $\nu = 0$ $\mathbb{P} \otimes B$*-a.e.;*
(ii) *NSA holds if and only if* $\nu = 0$ $\mathbb{P} \otimes B$*-a.e. and* $\sigma = \infty$ $\mathbb{P}$*-a.s.;*
(iii) *NA1 holds if and only if* $\nu = 0$ $\mathbb{P} \otimes B$*-a.e. and* $\widehat{K}_T < \infty$ $\mathbb{P}$*-a.s.;*
(iv) *NFLVR holds if and only if NA1 and NA hold;*

†Note that if H yields an increasing profit in the sense of Definition 2.2-(i), it holds that $H^n := nH \in \mathcal{A}_0(\mathbb{P})$ and $G(H^n) \geq G(H)$, for every $n \in \mathbb{N}$. This means that the increasing profit generated by H can be scaled to arbitrarily large levels of wealth, thus explaining the adjective unbounded.

(v) *NFLVR holds if and only if there exists* $\mathbb{Q} \sim \mathbb{P}$ *such that* $S \in \mathcal{M}^c_{\text{loc}}(\mathbb{Q})$;
where ν, σ *and* $\widehat{K}$ *are defined in* (2.2), (2.4) *and* (2.3), *respectively.*

3. Absolutely Continuous Changes of Measure

Let $\mathbb{Q}$ be a probability measure on $(\Omega, \mathcal{F})$ with $\mathbb{Q} \ll \mathbb{P}$, but not necessarily equivalent to $\mathbb{P}$. It is well known (see, e.g., [21, Section III.3]) that there exists a unique (up to $\mathbb{P}$- and $\mathbb{Q}$-indistinguishability) non-negative $\mathbb{P}$-martingale $Z = (Z_t)_{0 \le t \le T}$ such that $Z_t = d\mathbb{Q}|_{\mathcal{F}_t}/d\mathbb{P}|_{\mathcal{F}_t}$, for all $t \in [0,T]$ (note however that Z is not necessarily continuous). Moreover, we also have $\mathbb{Q}\big(Z_t > 0 \text{ and } Z_{t-} > 0 \text{ for all } t \in [0,T]\big) = 1$. The Girsanov-Lenglart theory of absolutely continuous changes of measure (see [28]) allows us to compute the canonical decomposition of S with respect to $\mathbb{Q}$.

Lemma 3.1. *Let* $\mathbb{Q}$ *be a probability measure on* $(\Omega, \mathcal{F})$ *such that* $\mathbb{Q} \ll \mathbb{P}$. *The canonical decomposition of* S *with respect to* $\mathbb{Q}$ *is given by* $S = S_0 + \bar{A} + \bar{M}$, *where*

$$
\begin{aligned}
\bar{A} &:= \int \big(c\,\bar{\lambda} + \bar{\nu}\big)\, dB\,, \quad \text{with } \bar{\lambda} := \lambda + \theta/Z_- \text{ and } \bar{\nu} := \nu\,, \\
\bar{M} &:= M - \int \frac{c\,\theta}{Z_-}\, dB \in \mathcal{M}^c_{\text{loc}}(\mathbb{Q})\,,
\end{aligned}
\tag{3.1}
$$

where the $\mathbb{R}^d$*-valued predictable process* $\theta = (\theta_t)_{0 \le t \le T} \in L^2_{\text{loc}}(M;\mathbb{P})$ *can be chosen such that* $\theta = c^+ d\langle M, Z\rangle/dB$ $\mathbb{P} \otimes B$*-a.e.*

Proof. Theorem 1 of [28] implies that S is a continuous semimartingale with respect to $\mathbb{Q}$ and, hence, it admits a unique canonical decomposition under $\mathbb{Q}$. Since S is continuous, the predictable quadratic variation $\langle M, Z\rangle$ always exists and the process $\bar{M} := M - \int Z_-^{-1} d\langle M, Z\rangle$ belongs to $\mathcal{M}^c_{\text{loc}}(\mathbb{Q})$, see Theorem 2 of [28]. By applying the Galtchouk-Kunita-Watanabe decomposition to Z with respect to M (see [1]), we get the existence of a process $\theta \in L^2_{\text{loc}}(M;\mathbb{P})$ such that $d\langle M, Z\rangle_t = d\langle M, M\rangle_t \theta_t = c_t\, \theta_t\, dB_t$. Together with (2.1)-(2.2), this gives the canonical decomposition (3.1) of S with respect to the measure $\mathbb{Q}$. Finally, define the process $\tilde{\theta} := c^+ c\, \theta$, so that $\tilde{\theta} = c^+ d\langle M, Z\rangle/dB$ holds $\mathbb{P} \otimes B$-a.e. Since c^+ is predictable and using the properties of the Moore-Penrose pseudoinverse, it is easy to check that $\tilde{\theta}$ belongs to $L^2_{\text{loc}}(M;\mathbb{P})$ and satisfies $c\,\tilde{\theta} = c\,\theta$. □

Similarly as in the previous section, let the process $\widehat{K}^{\mathbb{Q}} = (\widehat{K}^{\mathbb{Q}}_t)_{0\leq t\leq T}$ be defined by

$$\widehat{K}^{\mathbb{Q}}_t =: \int_0^t \bar{\lambda}^\top_u \, d\langle \bar{M}, \bar{M}\rangle_u \, \bar{\lambda}_u \,, \qquad \text{for all } t \in [0,T] \,,$$

where $\bar{\lambda}$ and $\bar{M}$ are as in Lemma 3.1. Furthermore, let us define the (possibly infinite valued) stopping time $\sigma^{\mathbb{Q}}$ by

$$\sigma^{\mathbb{Q}} := \inf\{t \in [0,T] : \widehat{K}^{\mathbb{Q}}_{t+h} - \widehat{K}^{\mathbb{Q}}_t = \infty, \forall h \in (0, T-t]\} \,.$$

The next theorem is the main result of the first part of the present paper and shows that *weak* no-arbitrage conditions (i.e., NIP, NSA and NA1), as opposed to *strong* no-arbitrage conditions (i.e., NA and NFLVR), are always stable with respect to absolutely continuous changes of measure and not only to equivalent changes of measure. The proof is a rather direct consequence of Proposition 2.3 and Lemma 3.1, but we prefer to give full details for the reader's convenience.

Theorem 3.2. *Let $\mathbb{Q}$ be a probability measure on $(\Omega, \mathcal{F})$ such that $\mathbb{Q} \ll \mathbb{P}$. Then the following hold:*
(i) *if NIP holds with respect to $\mathbb{P}$, then NIP holds with respect to $\mathbb{Q}$ as well;*
(ii) *if NSA holds with respect to $\mathbb{P}$, then NSA holds with respect to $\mathbb{Q}$ as well;*
(iii) *if NA1 holds with respect to $\mathbb{P}$, then NA1 holds with respect to $\mathbb{Q}$ as well.*

Proof. If NIP holds with respect to $\mathbb{P}$, Lemma 3.1 and part *(i)* of Proposition 2.3 imply that $\bar{\nu} = \nu = 0$ $\mathbb{P}\otimes B$-a.e. and also $\mathbb{Q}\otimes B$-a.e., since $\mathbb{Q} \ll \mathbb{P}$. The first claim then follows again from part *(i)* of Proposition 2.3 (now applied under $\mathbb{Q}$). In order to prove parts *(ii)*-*(iii)*, observe that, for every $t \in [0,T]$:

$$\widehat{K}^{\mathbb{Q}}_t = \int_0^t \left(\lambda_u + \frac{\theta_u}{Z_{u-}}\right)^\top c_u \left(\lambda_u + \frac{\theta_u}{Z_{u-}}\right) dB_u \leq \widehat{K}_t + \int_0^t \frac{1}{Z^2_{u-}} \, \theta^\top_u \, c_u \, \theta_u \, dB_u \,, \tag{3.2}$$

using the Cauchy-Schwarz inequality together with (2.3). Note that, due to Lemma 3.1, we have $\theta \in L^2_{\text{loc}}(M;\mathbb{P}) \subseteq L^2_{\text{loc}}(\bar{M};\mathbb{Q})$, where the last inclusion follows from the assumption that $\mathbb{Q} \ll \mathbb{P}$. Since the process Z_- is $\mathbb{Q}$-a.s. strictly positive, adapted and left-continuous, hence locally bounded, the second term on the right-hand side of (3.2) is $\mathbb{Q}$-a.s. finite for every $t \in [0,T]$. In particular, this implies that $\sigma^{\mathbb{Q}} \geq \sigma$ $\mathbb{Q}$-a.s. By part *(ii)* of

Proposition 2.3, if NSA holds with respect to $\mathbb{P}$, then $\sigma = \infty$ $\mathbb{P}$-a.s. and also $\mathbb{Q}$-a.s., since $\mathbb{Q} \ll \mathbb{P}$, thus giving $\sigma^{\mathbb{Q}} = \infty$ $\mathbb{Q}$-a.s. Then, again part *(ii)* of Proposition 2.3 implies that NSA holds with respect to $\mathbb{Q}$. Similarly, due to part *(iii)* of Proposition 2.3, if NA1 holds with respect to $\mathbb{P}$ then $\widehat{K}_T < \infty$ $\mathbb{P}$-a.s. and also $\mathbb{Q}$-a.s., since $\mathbb{Q} \ll \mathbb{P}$. By (3.2), this implies that $\widehat{K}_T^{\mathbb{Q}} < \infty$ $\mathbb{Q}$-a.s. and, hence, again by part *(iii)* of Proposition 2.3, NA1 holds with respect to $\mathbb{Q}$. □

Remark 3.3. The condition $\widehat{K}_T < \infty$ $\mathbb{P}$-a.s., which characterises NA1, is called *finiteness condition* in [10]. In Lemma 4.5 of that paper, it is shown that the finiteness condition is stable under equivalent changes of measure. A result analogous to part *(iii)* of Theorem 3.2 (albeit with a different terminology) is also established in Proposition 2.7 and Remark 2.10 of [5].

We close this section by pointing out that, if we restrict our attention to *equivalent* changes of measure, rather than only absolutely continuous changes of measure, then also the NA and NFLVR conditions are stable. Indeed, suppose that NA holds with respect to $\mathbb{P}$ and let $\mathbb{Q} \sim \mathbb{P}$. Arguing by contradiction, suppose that there exists a strategy $H \in \mathcal{A}(\mathbb{Q})$ which realises an arbitrage opportunity under $\mathbb{Q}$, in the sense of part (iv) of Definition 2.2. Then, since $\mathbb{P} \ll \mathbb{Q}$, Proposition III.6.24 of [21] shows that $H \in L(S;\mathbb{P})$ and the stochastic integral $H \cdot S$ viewed with respect to $\mathbb{P}$ coincides with the stochastic integral with respect to $\mathbb{Q}$. Moreover, since $\mathbb{Q} \sim \mathbb{P}$, we also have $H \in \mathcal{A}(\mathbb{P})$ as well as $G_T(H) \geq 0$ $\mathbb{P}$-a.s. and $\mathbb{P}\big(G_T(H) > 0\big) > 0$, thus contradicting the validity of NA with respect to $\mathbb{P}$. An analogous reasoning allows to show the stability of NFLVR with respect to equivalent changes of measure. Note also that this argument does not rely on the continuity of S.

4. The NA and NFLVR Conditions: a Counterexample

As shown in Theorem 3.2, in the context of continuous semimartingale models, the *weak* NIP, NSA and NA1 conditions are always stable with respect to absolutely continuous changes of measure. We now show that, in general, the classical *strong* NA and NFLVR conditions are not robust with respect to absolutely continuous changes of measure, even for continuous processes. We proceed to illustrate this phenomenon by means of a counterexample, which has been already developed in [7] in relation to strict local martingales (i.e., local martingales which fail to be martingales, see, e.g., [12]).

Let $W = (W_t)_{0 \leq t \leq T}$ be a real-valued Brownian motion starting from $W_0 = 1$ and let $\tau := \inf\{t \in [0,T] : W_t = 0\} \wedge T$. We define S as the stopped process $S := W^\tau$ and let the filtration $\mathbb{F}$ be the $\mathbb{P}$-augmented natural filtration of S, with $\mathcal{F} = \mathcal{F}_T$. Since $S \in \mathcal{M}(\mathbb{P})$, NFLVR trivially holds with respect to $\mathbb{P}$, in view of part *(v)* of Proposition 2.3. We then define a probability measure $\mathbb{Q} \ll \mathbb{P}$ by $d\mathbb{Q}/d\mathbb{P} := S_T = W_\tau$. Clearly, $\mathbb{P} \ll \mathbb{Q}$ does not hold, since $\mathbb{P}(S_T = 0) > 0$, so that $\mathbb{Q}$ and $\mathbb{P}$ fail to be equivalent. Note also that S represents the density process of $\mathbb{Q}$ with respect to $\mathbb{P}$.

Proposition 4.1. *In the context of the present section, the process S allows for arbitrage opportunities with respect to the probability measure $\mathbb{Q}$ in the filtration $\mathbb{F}$.*

Proof. By Theorem 2 of [28], the process $N := S - \int S^{-1} d\langle S \rangle$ belongs to $\mathcal{M}^c_{\text{loc}}(\mathbb{Q})$. Observe that $\langle N \rangle_t = \langle S \rangle_t = t \wedge \tau$ for all $t \in [0,T]$. Furthermore, noting that $\mathbb{Q}(S_T = 0) = \mathbb{E}_\mathbb{P}\left[\mathbf{1}_{\{S_T=0\}} S_T\right] = 0$, it holds that $\tau = T$ $\mathbb{Q}$-a.s. and hence $\langle N \rangle_t = t$ $\mathbb{Q}$-a.s. for all $t \in [0,T]$. Lévy's characterisation theorem then implies that N is a $\mathbb{Q}$-Brownian motion starting at $N_0 = 1$. Let us now denote by $\mathbb{G}$ the $\mathbb{Q}$-augmented natural filtration of N (or, equivalently, of S), which coincides with $\mathbb{F}$ augmented by the subsets of $\{S_T = 0\}$. The process S satisfies $dS_t = S_t^{-1} dt + dN_t$. Hence, with respect to the measure $\mathbb{Q}$ and the filtration $\mathbb{G}$, the process S is a three-dimensional Bessel process (see, e.g., [34, Section XI.1]) and the corollary on page 361 of [7] implies that S admits arbitrage opportunities with respect to $\mathbb{Q}$ in the filtration $\mathbb{G}$. This means that there exists a $\mathbb{G}$-predictable admissible strategy $H \in L(S;\mathbb{Q})$ such that $G_T(H) \geq 0$ $\mathbb{Q}$-a.s. and $\mathbb{Q}\big(G_T(H) > 0\big) > 0$. As remarked on page 360 of [7], there also exists an $\mathbb{F}$-predictable process K that is $\mathbb{Q}$-indistinguishable from H, so that $K \cdot S \equiv H \cdot S$, where both stochastic integrals are considered with respect to $(\mathbb{Q},\mathbb{G})$. But then, since S is $\mathbb{F}$-adapted, Proposition III.6.25 of [21] shows that the stochastic integral $K \cdot S$ viewed with respect to $(\mathbb{Q},\mathbb{G})$ coincides with the stochastic integral viewed with respect to $(\mathbb{Q},\mathbb{F})$. We have thus proved that K realises an arbitrage opportunity with respect to $\mathbb{Q}$ in the filtration $\mathbb{F}$. □

Actually, the counterexample given in Proposition 4.1 can be generalised to a whole class of models for which NA (and, hence, NFLVR as well) is destroyed by an absolutely continuous, but not equivalent, change of measure. The following result corresponds essentially to Theorem 3 of [7] (see also Proposition 2.8 of [31] as well as Theorem 1 of [36] for an extension to incomplete markets).

Proposition 4.2. *Let $S = (S_t)_{0 \le t \le T}$ be a non-negative real-valued process in $\mathcal{M}(\mathbb{P})$ with the predictable representation property with respect to $\mathbb{P}$, with $S_0 = 1$ $\mathbb{P}$-a.s. Define the probability measure $\mathbb{Q} \ll \mathbb{P}$ by $d\mathbb{Q}/d\mathbb{P} := S_T$. If $\mathbb{P}(S_T = 0) > 0$ then S does not satisfy NA with respect to $\mathbb{Q}$.*

Remark 4.3. We want to point out that the results of the present section do not actually need the completeness of the filtration. Indeed, the predictable representation property can be also established in right-continuous but not complete filtrations, see e.g. Theorem III.4.33 of [21], while the failure of NA under $\mathbb{Q}$ in the proof of Proposition 4.1 can be proven as on page 59 of [16].

5. Stability of NA1 in General Semimartingale Models

Among the different notions of no-arbitrage considered in Definition 2.2, the NA1 condition is of particular interest. Indeed, as shown in Proposition 4.19 of [24] (see also [4]), NA1 is the minimal condition that allows for a meaningful solution of portfolio optimisation problems in general semimartingale models. This section aims at investigating whether the NA1 condition is stable with respect to absolutely continuous changes of measure in financial models based on a *general* (i.e., not necessarily continuous or locally bounded) $\mathbb{R}^d$-valued semimartingale $S = (S_t)_{0 \le t \le T}$.[‡]

The following result, due to [41], characterises NA1 in a general semimartingale setting. We recall that an $\mathbb{R}^d$-valued semimartingale S is called a *σ-martingale* if there exists an increasing sequence $\{\Sigma_n\}_{n \in \mathbb{N}}$ of predictable sets such that $\bigcup_{n \in \mathbb{N}} \Sigma_n = \Omega \times [0, T]$ and $\mathbf{1}_{\Sigma_n} \cdot S^i \in \mathcal{M}(\mathbb{P})$, for all $n \in \mathbb{N}$ and $i = 1, \ldots, d$ (see [21, Section III.6e]). We denote by $\mathcal{M}_\sigma(\mathbb{P})$ the family of all σ-martingales with respect to $\mathbb{P}$.

Proposition 5.1. *NA1 holds if and only if there exists a* strict martingale density (SMD), *i.e., a strictly positive real-valued process $L = (L_t)_{0 \le t \le T}$ belonging to $\mathcal{M}_{loc}(\mathbb{P})$ such that $\mathbb{E}_\mathbb{P}[L_0] < \infty$ and $LS \in \mathcal{M}_\sigma(\mathbb{P})$.*

Let now $\mathbb{Q}$ be a probability measure on $(\Omega, \mathcal{F})$ with $\mathbb{Q} \ll \mathbb{P}$. Until the end of this section, we denote by $Z = (Z_t)_{0 \le t \le T}$ the density process of $\mathbb{Q}$

[‡]As already explained in Remark 4.13 in [24], the NA1 condition may no longer be preserved by an absolutely continuous change of measure if jumps are present, unlike the continuous case considered so far.

with respect to $\mathbb{P}$ and define the following stopping times:

$$\begin{aligned} \tau &:= \inf\{t \in [0,T] : Z_{t-} = 0 \text{ or } Z_t = 0\}\,, \\ \tau_n &:= \inf\{t \in [0,T] : Z_t < 1/n\} \wedge T\,, \qquad \text{for all } n \in \mathbb{N}. \end{aligned} \tag{5.1}$$

A key insight, first exploited by [14, 30] and more recently by [2, 27] among others, consists in looking at the process $1/Z$ under the measure $\mathbb{Q}$.

Lemma 5.2. *Let $\mathbb{Q}$ be a probability measure on $(\Omega, \mathcal{F})$ such that $\mathbb{Q} \ll \mathbb{P}$. Then the following hold:*
(i) *the process $1/Z$ is a strictly positive $\mathbb{Q}$-supermartingale;*
(ii) *for every $M = (M_t)_{0 \le t \le T} \in \mathcal{M}(\mathbb{P})$ the process M/Z belongs to $\mathcal{M}_{\text{loc}}(\mathbb{Q})$ if and only if $\mathbb{P}(\tau > \tau_n) = 1$, for all $n \in \mathbb{N}$.*

Proof. The first claim follows from simple computations (see also [2, Theorem 2.1]). In order to prove the second claim, suppose that $\mathbb{P}(\tau > \tau_n) = 1$, for all $n \in \mathbb{N}$. Then, for every $M \in \mathcal{M}(\mathbb{P})$, we have $M/Z \in \mathcal{M}_{\text{loc}}(\mathbb{Q})$, see e.g. part *(iii)* of Proposition 2.3 of [2]. Conversely, taking $M \equiv 1$, if $1/Z \in \mathcal{M}_{\text{loc}}(\mathbb{Q})$, Theorem 2.1 of [2] implies that $\mathbb{P}(\tau > \tau_n) = 1$ for all $n \in \mathbb{N}$. □

The next theorem is the main result of this section and shows that the NA1 condition is stable with respect to an absolutely continuous change of measure if the corresponding density process Z does not jump to zero. Note also that the proof is constructive, in the sense that it exhibits an explicit SMD for S with respect to the measure $\mathbb{Q}$.

Theorem 5.3. *Let $\mathbb{Q}$ be a probability measure on $(\Omega, \mathcal{F})$ such that $\mathbb{Q} \ll \mathbb{P}$. If NA1 holds with respect to $\mathbb{P}$ and $\mathbb{P}(\tau > \tau_n) = 1$, for all $n \in \mathbb{N}$, then NA1 holds with respect to $\mathbb{Q}$.*

Proof. By Proposition 5.1, there exists a real-valued strictly positive process $L \in \mathcal{M}_{\text{loc}}(\mathbb{P})$ such that $\mathbb{E}_{\mathbb{P}}[L_0] < \infty$ and $LS \in \mathcal{M}_\sigma(\mathbb{P})$. Let $\{\varrho_n\}_{n \in \mathbb{N}}$ be a $\mathbb{P}$-localizing sequence for L. Then, due to part *(ii)* of Lemma 5.2, it holds that $L^{\varrho_n}/Z \in \mathcal{M}_{\text{loc}}(\mathbb{Q})$. Since $\mathbb{Q} \ll \mathbb{P}$, this implies that $L/Z \in \mathcal{M}_{\text{loc}}(\mathbb{Q})$. By Definition III.6.33 of [21], there exists an increasing sequence of predictable sets $\{\Sigma_n\}_{n \in \mathbb{N}}$ such that $\bigcup_{n \in \mathbb{N}} \Sigma_n = \Omega \times [0,T]$ and $\mathbf{1}_{\Sigma_n} \cdot LS^i \in \mathcal{M}(\mathbb{P})$, for every $i = 1, \ldots, d$. Again by part *(ii)* of Lemma 5.2, we have $(\mathbf{1}_{\Sigma_n} \cdot LS^i)/Z \in \mathcal{M}_{\text{loc}}(\mathbb{Q})$ and an application of the product rule leads to $\mathbf{1}_{\Sigma_n} \cdot (LS^i/Z) \in \mathcal{M}_{\text{loc}}(\mathbb{Q})$, for every $i = 1, \ldots, d$. Since $\mathbb{Q} \ll \mathbb{P}$ and the class $\mathcal{M}_\sigma(\mathbb{Q})$ is stable by localization (see, e.g., [21, Proposition III.6.34]), this shows that $LS^i/Z \in \mathcal{M}_\sigma(\mathbb{Q})$, for all $i = 1, \ldots, d$. Since we also have $\mathbb{E}_{\mathbb{Q}}[L_0/Z_0] \leq \mathbb{E}_{\mathbb{P}}[L_0] < \infty$, we have thus shown that L/Z is an SMD under $\mathbb{Q}$. Proposition 5.1 then implies that NA1 holds with respect to $\mathbb{Q}$. □

In general, the condition $\mathbb{P}(\tau > \tau_n) = 1$, for all $n \in \mathbb{N}$, cannot be weakened, as shown in the following result (see also Example 5.5 and Proposition 5.7 below).

Proposition 5.4. *Let $S = (S_t)_{0 \leq t \leq T}$ be a non-negative real-valued process in $\mathcal{M}(\mathbb{P})$ with the predictable representation property with respect to $\mathbb{P}$, with $S_0 = 1$ $\mathbb{P}$-a.s. Define the probability measure $\mathbb{Q} \ll \mathbb{P}$ by $d\mathbb{Q}/d\mathbb{P} := S_T$ and $\tau := \inf\{t \in [0,T] : S_{t-} = 0 \text{ or } S_t = 0\}$. If $\mathbb{P}(\{\tau \leq T\} \cap \{S_{\tau-} > 0\}) > 0$ then S does not satisfy NA1 with respect to $\mathbb{Q}$.*

Proof. Arguing by contradiction, suppose that S satisfies NA1 under $\mathbb{Q}$. Then, due to Proposition 5.1, there exists an SMD L with respect to $\mathbb{Q}$, so that $LS \in \mathcal{M}_{\text{loc}}(\mathbb{Q})$, since a non-negative σ-martingale is a local martingale (see, e.g., [21, Proposition III.6.35]). Due to the lemma on page 67 of [28], there exists a sequence of stopping times $\{\varrho_n\}_{n \in \mathbb{N}}$ with $\varrho_n \nearrow \infty$ $\mathbb{Q}$-a.s. as $n \to \infty$ such that $(LS)^{\varrho_n} \in \mathcal{M}_{\text{loc}}(\mathbb{P})$ and $(LS^2)^{\varrho_n} \in \mathcal{M}_{\text{loc}}(\mathbb{P})$. In view of Theorem 11.2 of [20], since $S \in \mathcal{M}(\mathbb{P})$ enjoys the predictable representation property with respect to $\mathbb{P}$, this implies that $(LS)^{\varrho_n}$ is $\mathbb{P}$-a.s. trivial, for all $n \in \mathbb{N}$. Since $\mathbb{Q} \ll \mathbb{P}$ and $\varrho_n \nearrow \infty$ $\mathbb{Q}$-a.s. as $n \to \infty$, this implies that $L = 1/S$ $\mathbb{Q}$-a.s. However, due to part *(ii)* of Lemma 5.2 (see also [32, Example 4.1]), the condition $\mathbb{P}(\{\tau \leq T\} \cap \{S_{\tau-} > 0\}) > 0$ implies that $1/S$ is a $\mathbb{Q}$-supermartingale which does not belong to $\mathcal{M}_{\text{loc}}(\mathbb{Q})$, thus contradicting the hypothesis that L is an SMD with respect to $\mathbb{Q}$. □

Example 5.5. We now present a simple example where the assumptions of Proposition 5.4 are satisfied (compare also with [32], Example 4.2). Let $(\Omega, \mathcal{F}, \mathbb{P})$ be a given probability space supporting a standard exponential random variable $\xi : \Omega \to \mathbb{R}_+$, so that $\mathbb{P}(\xi > t) = e^{-t}$, for all $t \in \mathbb{R}_+$, and let $\mathbb{F}$ be the right-continuous filtration generated by the process $(\mathbf{1}_{\{\xi \leq t\}})_{0 \leq t \leq T}$. Define then the process $S = (S_t)_{0 \leq t \leq T}$ by $S_t := \mathbf{1}_{\{\xi > t\}} e^t$, for $t \in [0,T]$. As follows from Propositions 7.2.3.2 and 7.2.5.1 of [22], the process S belongs to $\mathcal{M}(\mathbb{P})$ and enjoys the predictable representation property in $\mathbb{F}$. Clearly, using the notation of Proposition 5.4, it holds that $\tau = \xi$ and, moreover:

$$\mathbb{P}(\{\tau \leq T\} \cap \{S_{\tau-} > 0\}) = \mathbb{P}(\xi \leq T) = 1 - e^{-T} > 0.$$

Proposition 5.4 then shows that S fails to satisfy NA1 under $\mathbb{Q}$. Indeed, since $\mathbb{Q}(\xi > T) = 1$, the process S is perceived as a strictly increasing process under $\mathbb{Q}$ and, therefore, it allows for increasing profits (in the sense of part (i) of Definition 2.2).

It is interesting to compare Proposition 5.4 with the result of Proposition 4.2. Indeed, both propositions show that, if one starts from a model where $S \in \mathcal{M}(\mathbb{P})$ (and, hence, NFLVR trivially holds with respect to $\mathbb{P}$) and S enjoys the predictable representation property, then arbitrage profits may arise after an absolutely continuous, but not equivalent, change of measure from $\mathbb{P}$ to $\mathbb{Q}$. More specifically, Proposition 4.2 shows that, if $d\mathbb{Q} = S_T\, d\mathbb{P}$ fails to define a measure $\mathbb{Q} \sim \mathbb{P}$, then the process S allows for arbitrage opportunities (i.e., NA fails) when viewed under $\mathbb{Q}$. Moreover, due to Proposition 5.4, if the process S can jump to zero, then S also allows for arbitrages of the first kind when viewed under $\mathbb{Q}$ (compare also with Example 5.5).

Remark 5.6. In the recent paper [36], the authors provide a systematic procedure for constructing market models that satisfy NA1 but allow for arbitrage opportunities (i.e., NA fails to hold). Their approach is intimately linked to our results: indeed, they start from a market model where S is a non-negative $\mathbb{R}^d$-valued martingale and then pass to an absolutely continuous, but not equivalent, probability measure whose density process is allowed to reach zero only continuously. Hence, the change of measure adopted in [36] has the effect of disrupting the NA component of NFLVR, while, in view of Theorem 5.3, the NA1 component is preserved.

The next proposition represents a converse result to Theorem 5.3 and shows that, if the condition $\mathbb{P}(\tau > \tau_n) = 1$ for all $n \in \mathbb{N}$ fails to hold, then one can find a semimartingale S satisfying NFLVR with respect to $\mathbb{P}$ but allowing for increasing profits (and, hence, arbitrages of the first kind) under $\mathbb{Q}$. The proof exploits an idea already used in the proof of Theorem 3.16 of [3].

Proposition 5.7. *Let $\mathbb{Q}$ be a probability measure on $(\Omega, \mathcal{F})$ with $\mathbb{Q} \ll \mathbb{P}$ and suppose that $\mathbb{P}(\tau = \tau_n) > 0$ for some $n \in \mathbb{N}$. Then there exists a process $S \in \mathcal{M}(\mathbb{P})$ which allows for increasing profits with respect to $\mathbb{Q}$.*

Proof. Let $B = (\mathbf{1}_{[\![\tau,T]\!]})^p$ be the predictable compensator (see, e.g., [21, Theorem I.3.17]) of the increasing process $\mathbf{1}_{[\![\tau,T]\!]}$ and $S := -(\mathbf{1}_{[\![\tau,T]\!]} - B)$, so that $S \in \mathcal{M}(\mathbb{P})$. Since $\tau = \infty$ $\mathbb{Q}$-a.s., it holds that $S = B$ $\mathbb{Q}$-a.s. The predictable process $H := \mathbf{1}_{(\!(0,T]\!]}$ satisfies $G(H) = H \cdot S = B - B_0 \geq 0$ $\mathbb{Q}$-a.s., since B is $\mathbb{P}$-a.s. increasing and $\mathbb{Q} \ll \mathbb{P}$. Moreover, using the properties of predictable compensators (see [20, Proposition 1.47]) together with the $\mathbb{P}$-martingale property of Z, we get

$$\begin{aligned}\mathbb{E}_{\mathbb{Q}}\big[G_T(H)\big] &= \mathbb{E}\big[Z_T B_T - Z_0 B_0\big] = \mathbb{E}\left[\int_0^T Z_{t-}\, dB_t\right] \\ &= \mathbb{E}\left[\int_0^T Z_{t-}\, d\mathbf{1}_{[\![\tau,T]\!]}\right] = \mathbb{E}\left[Z_{\tau-}\mathbf{1}_{\{\tau\leq T\}}\right] > 0,\end{aligned}$$

where the last inequality follows from the fact that $\mathbb{P}(\tau = \tau_n) > 0$ for some $n \in \mathbb{N}$, meaning that the process Z has a strictly positive probability of jumping to zero. □

Summing up, in the context of general $\mathbb{R}^d$-valued semimartingale models, Lemma 5.2, Theorem 5.3 and Proposition 5.4 together yield the equivalence between the three following statements, where $\mathbb{Q} \ll \mathbb{P}$ and according to the notation introduced in (5.1):

(a) *for every $\mathbb{R}^d$-valued semimartingale $S = (S_t)_{0\leq t\leq T}$ satisfying NA1 with respect to $\mathbb{P}$, NA1 holds with respect to $\mathbb{Q}$ as well;*
(b) $\mathbb{P}(\tau > \tau_n) = 1$, *for all $n \in \mathbb{N}$;*
(c) $1/Z \in \mathcal{M}_{\mathrm{loc}}(\mathbb{Q})$.

Acknowledgements

This research was supported by a Marie Curie Intra European Fellowship within the 7th European Community Framework Programme under grant agreement PIEF-GA-2012-332345. The author is thankful to Johannes Ruf for useful comments on an earlier version of the paper.

References

[1] Ansel, J.P. and Stricker, C. (1993). Décomposition de Kunita-Watanabe, *Séminaire de Probabilités* **XXVII**, Vol. 1557, Lecture Notes in Mathematics, pp. 30–32, Springer, Berlin - Heidelberg.

[2] Carr, P., Fisher, T. and Ruf, J. (2013). On the hedging of options on exploding exchange rates, *Finance Stoch.* Volume 18, Issue 1 (2014), Page 115-144.

[3] Choulli, T., Aksamit, A., Deng, J. and Jeanblanc, M. (2013). Non-arbitrage up to random horizon and after honest times for semimartingale models. Preprint, available at `http://arxiv.org/abs/1310.1142`.

[4] Choulli, T., Deng, J. and Ma, J. (2013). How non-arbitrage, viability and numéraire portfolio are related. Preprint, available at `http://arxiv.org/abs/1211.4598`.

[5] Choulli, T. and Stricker, C. (1996). Deux applications de la décomposition de Galtchouk-Kunita-Watanabe, *Séminaire de Probabilités* **XXX**, Vol. 1626, Lecture Notes in Mathematics, pp. 12–23, Springer, Berlin - Heidelberg.

[6] Delbaen, F. and Schachermayer, W. (1994). A general version of the fundamental theorem of asset pricing, *Math. Ann.* **300**, 463–520.

[7] Delbaen, F. and Schachermayer, W. (1995a). Arbitrage possibilities in Bessel processes and their relations to local martingales, *Prob. Theory Rel. Fields* **102**, 357–366.

[8] Delbaen, F. and Schachermayer, W. (1995b). The existence of absolutely continuous local martingale measures, *Ann. Appl. Prob.* **5**, 926–945.

[9] Delbaen, F. and Schachermayer, W. (1998). The fundamental theorem of asset pricing for unbounded stochastic processes, *Math. Ann.* **312**, 215–250.

[10] Delbaen, F. and Shirakawa, H. (1996). A note on the no-arbitrage condition for international financial markets, *Financial Engineering and the Japanese Markets* **3**, 239–251.

[11] Dzhaparidze, K. and Spreij, P. (1993). On correlation calculus for multivariate martingales, *Stoch. Proc. Appl.* **46**, 283–299.

[12] Elworthy, K.D., Li, X.M. and Yor, M. (1999). The importance of strictly local martingales; applications to radial Ornstein-Uhlenbeck processes, *Prob. Theory Rel. Fields* **115**, 325–355.

[13] Fernholz, R. and Karatzas, I. (2009). Stochastic portfolio theory: an overview. In eds. A. Bensoussan and Q. Zhang, *Mathematical Modeling and Numerical Methods in Finance*, Vol. XV, *Handbook of Numerical Analysis*, pp. 89–167, North-Holland, Oxford.

[14] Föllmer, H. (1972). The exit measure of a supermartingale, *Z. Wahrscheinlichkeit.* **21**, 154–166.

[15] Fontana, C. (2013). Weak and strong no-arbitrage conditions for continuous financial markets. Preprint, available at `http://arxiv.org/abs/1302.7192`.

[16] Fontana, C. and Runggaldier, W.J. (2013). Diffusion-based models for financial markets without martingale measures. In eds. F. Biagini, A. Richter and H. Schlesinger, *Risk Measures and Attitudes*, EAA Series, pp. 45–81, Springer, London.

[17] Hulley, H. (2009). *Strict Local Martingales in Continuous Financial Market Models*, PhD thesis, University of Technology Sydney.

[18] Hulley, H. and Schweizer, M. (2010). M^6 - on minimal market models and minimal martingale measures. In eds. C. Chiarella and A. Novikov, *Contemporary Quantitative Finance: Essays in Honour of Eckhard Platen*, pp. 35–51, Springer, Berlin - Heidelberg.

[19] Ingersoll, J.E. (1987). *Theory of Financial Decision Making*, Rowman & Littlefield, Savage (MD).

[20] Jacod, J. (1979). *Calcul Stochastique et Problèmes de Martingales*, Vol. 714, Lecture Notes in Mathematics, Springer, Berlin - Heidelberg.

[21] Jacod, J. and Shiryaev, A.N. (2003). *Limit Theorems for Stochastic Processes*, 2nd edn., Springer, Berlin - Heidelberg - New York.

[22] Jeanblanc, M., Yor, M. and Chesney, M. (2009). *Mathematical Methods for Financial Markets*, Springer, London.

[23] Kabanov, Y. (1997). On the FTAP of Kreps-Delbaen-Schachermayer. In eds. Y. Kabanov, B.L. Rozovskii and A.N. Shiryaev, *Statistics and Control of Stochastic Processes: The Liptser Festschrift*, pp. 191–203, World Scientific, Singapore.

[24] Karatzas, I. and Kardaras, K. (2007). The numeraire portfolio in semimartingale financial models, *Financ. Stoch.* **11**, 447–493.

[25] Kardaras, C. (2010). Finitely additive probabilities and the fundamental theorem of asset pricing. In eds. C. Chiarella and A. Novikov, *Contemporary Quantitative Finance: Essays in Honour of Eckhard Platen*, pp. 19–34, Springer, Berlin - Heidelberg.

[26] Kardaras, C. (2012). Market viability via absence of arbitrages of the first kind, *Financ. Stoch.* **16**, 651–667.

[27] Kardaras, C., Kreher, D. and Nikeghbali, A. (2013). Strict local martingales, bubbles. Preprint, available at `http://arxiv.org/abs/1108.4177`.

[28] Lenglart, E. (1977). Transformation des martingales locales par changement absolument continu de probabilités, *Z. Wahrscheinlichkeit.* **39**, 65–70.

[29] Loewenstein, M. and Willard, G.A. (2000). Local martingales, arbitrage, and viability, *Econ. Theory* **16**, 135–161.

[30] Meyer, P.A. (1972), La mesure de H. Föllmer en théorie des surmartingales, *Séminaire de Probabilités* **VI**, Vol. 258, Lecture Notes in Mathematics, pp. 118–129, Springer, Berlin - Heidelberg

[31] Osterrieder, J.R. and Rheinländer, T. (2006). Arbitrage opportunities in diverse markets via a non-equivalent measure change, *Ann. Financ.* **2**, 287–301.

[32] Perkowski, N. and Ruf, J. (2013). Supermartingales as Radon-Nikodym densities and related measure extensions. Preprint, available at `http://arxiv.org/abs/1309.4623`.

[33] Platen, E. and Heath, D. (2006). *A Benchmark Approach to Quantitative Finance*, Springer, Berlin - Heidelberg.

[34] Revuz, D. and Yor, M. (1999). *Continuous Martingales and Brownian Motion*, 3rd edn., Springer, Berlin - Heidelberg.

[35] Ruf, J. (2013). Hedging under arbitrage, *Math. Financ.* **23**, 297–317.

[36] Ruf, J. and Runggaldier, W.J. (2013). A systematic approach to constructing market models with arbitrage, in this volume.

[37] Schachermayer, W. (2010). The fundamental theorem of asset pricing. In ed. R. Cont, *Encyclopedia of Quantitative Finance*, pp. 792–801, Wiley, Chichester.

[38] Schweizer, M. (1995). On the minimal martingale measure and the Föllmer-Schweizer decomposition, *Stoch. Anal. Appl.* **13**, 573–599.

[39] Song, S. (2013). An alternative proof of a result of Takaoka. Preprint, available at `http://arxiv.org/abs/1306.1062`.

[40] Strasser, E. (2005). Characterization of arbitrage-free markets, *Ann. Appl. Prob.* **15**, 116–124.

[41] Takaoka, K. (2013). A note on the condition of no unbounded profit with bounded risk, to appear in: *Finance Stoch.*

A Systematic Approach to Constructing Market Models with Arbitrage

Johannes Ruf*, Wolfgang J. Runggaldier†

Abstract. This short note provides a systematic construction of market models without unbounded profits but with arbitrage opportunities.

1. Introduction

One of the fundamental notions in modern mathematical finance is that of absence of arbitrage. In fact, without no-arbitrage conditions one cannot meaningfully solve problems of pricing, hedging or portfolio optimization. A fundamental step in the historical development of the no-arbitrage theory was made by Harrison, Kreps, and Pliska in a series of papers, see [12, 25], and [13]. In [2, 4] (see also [19]) the authors proved the equivalence between the economic notion of No Free Lunch With Vanishing Risk (NFLVR) and the mathematical concept of an Equivalent Local Martingale Measure (ELMM) in full generality.

The more recent Stochastic Portfolio Theory (see, for example, [6] or the survey in [7]), which is a more descriptive rather than normative theory, shows that the behavior in real markets corresponds to weaker notions of no-arbitrage than NFLVR. Somehow in parallel to this theory, the so-called Benchmark Approach to quantitative finance (see [29] or the textbook [30]) was introduced with the aim of showing that pricing and hedging can also be performed without relying on the existence of an ELMM.

Various weaker notions of no-arbitrage have henceforth been studied as well as their consequences on asset pricing and portfolio optimization; for a recent unifying analysis of the whole spectrum of no-arbitrage conditions for continuous financial market models, see [9]. A crucial concept in

*Oxford-Man Institute of Quantitative Finance and Mathematical Institute, University of Oxford, United Kingdom, E-Mail: johannes.ruf@oxford-man.ox.ac.uk
†Department of Mathematics, University of Padova, Italy, E-Mail: runggal@math.unipd.it

this development is the notion of an Equivalent Local Martingale Deflator (ELMD), which is the counterpart of the density process for the case when an ELMM exists. Like the density process, the ELMD is a local martingale, but it may fail to be a martingale. In parallel with this theory, also a theory of asset price bubbles was developed where, under an ELMM, discounted asset prices are strict local martingales but we do not touch this issue here, and refer instead to [17, 18].

Along the development of the weaker notions of no-arbitrage, a crucial step was made in the paper [20] where the authors show that a condition, which they call No Unbounded Profit With Bounded Risk (NUPBR), is the minimal condition for which portfolio optimization can be meaningfully performed. For the corresponding hedging problem, we refer to [33, 34]. The notion of NUPBR has also appeared under the name of No Arbitrage of the First Kind (NA1); see [16] and [23]. A related problem of interest, but that we do not deal with here, is that of the robustness of the concept of arbitrage, an attempt in this direction is made in [11].

As a consequence of the above, the interest arose in finding market models that fall between NFLVR and NUPBR. Such models would then allow for classical arbitrage, but make it still possible to perform pricing and hedging as well as portfolio optimization. A classical example for continuous market models appears already in [3]. The relevance of Bessel processes in this context is also stressed in [30]; see also the note [32] and the survey [14]. At this point, one might then wonder whether there are other financially significant models, beyond those based on Bessel processes, that satisfy NUPBR but not NFLVR and whether there could be a systematic procedure to generate such models. Equivalently, whether there is a procedure to generate strict local martingales and the present paper is an attempt in this direction.

Our approach is inspired by a recent renewed interest (see, for example, [1, 5, 24, 28, 35], in the so-called Föllmer exit measure of a strictly positive local martingale, constructed in [8], as it was already initiated for continuous processes in [3] (see Theorem 1 there). Another, but related point of view to look at our approach is to interpret the expectation process (as a function of time) of a nonnegative local martingale as the distribution function of a certain random variable, namely the time of explosion of a process that is related to the nonnegative martingale; this point of view is inspired by [26] and further explored in [21].

Our approach is closely related to the method suggested in [27], where diverse markets are constructed through an absolutely continuous but not equivalent change of measure. Parallel to this work, [31] has developed an approach via a shrinkage of the underlying filtration to systematically obtain strict local martingales. Using such an insight to construct strict local martingales might yield a further method to obtain models that satisfy NUPBR, but not NFLVR. We shall not pursue this direction and leave it open for future research. On the contrary, [10] construct models that satisfy NUPBR (at least, up to a certain time), but not NFLVR, via a filtration enlargement.

2. The Model and Preliminary Notions

Given a finite time horizon $T < \infty$, consider a market with d assets, namely a pair $\left(\Omega, \mathcal{F}, (\mathcal{F}(t))_{t\in[0,T]}, \mathbb{P}\right), S$ of a filtered probability space and a d-dimensional vector S of nonnegative semimartingales $(S_i)_{i=1,\dots,d}$ with $S_i = (S_i(t))_{t\geq 0}$. We assume that $\mathcal{F}(0)$ is trivial and that the filtration $(\mathcal{F}(t))_{t\in[0,T]}$ is right-continuous. Each component of the process S represents the price of one of d assets that we assume already discounted with respect to the money market account; that is, we assume the short rate of interest to be zero. Agents invest in this market according to a self-financing strategy $H = (H(t))_{t\in[0,T]}$ and we shall denote by

$$V^{x,H} = (V^{x,H}(t))_{t\in[0,T]} = x+((H\cdot S)_t)_{t\in[0,T]} = x+\left(\int_0^t H(u)\mathrm{d}S(u)\right)_{t\in[0,T]}$$

the value process corresponding to the strategy H with initial value $V^{x,H}(0) = x$.

Definition 2.1. Let $\alpha > 0$ be a positive number. An S−integrable predictable process H is called *α–admissible* if $H(0) = 0$ and the process $V^{0,H}$ satisfies $V^{0,H}(t) \geq -\alpha$ for all $t \in [0,T]$ almost surely. The strategy H is called *admissible* if it is α–admissible for some $\alpha > 0$.

Definition 2.2. An *arbitrage strategy* H is an admissible strategy for which $\mathbb{P}(V^{0,H}(T) \geq 0) = 1$ and $\mathbb{P}(V^{0,H}(T) > 0) > 0$. We call it a *strong arbitrage strategy* if $\mathbb{P}(V^{0,H}(T) > 0) = 1$.

We also recall that the classical notion of absence of arbitrage, namely *No Free Lunch with Vanishing Risk (NFLVR)*, is equivalent to the existence of a probability measure $\mathbb{Q}$, equivalent to $\mathbb{P}$, under which the price processes are local martingales (as we assume the prices to be nonnegative).

Among the more recent weaker notions of absence of arbitrage we recall the following:

Definition 2.3. An $\mathcal{F}(T)$–measurable random variable ξ is called an *Arbitrage of the First Kind* if $\mathbb{P}(\xi \geq 0) = 1$, $\mathbb{P}(\xi > 0) > 0$, and for all $x > 0$ there exists an x–admissible strategy H such that $V^{x,H}(T) \geq \xi$. We shall say that the market admits *No Arbitrage of the First Kind (NA1)*, if there is no arbitrage of the first kind in the market.

Definition 2.4. There is *No Unbounded Profit with Bounded Risk (NUPBR)* if the set

$$\mathcal{K}_1 = \left\{V^{0,H}(T) \mid H = (H(t))_{t\in[0,T]} \text{ is a 1–admissible strategy for } S\right\}$$

is bounded in L^0, that is, if

$$\lim_{c\uparrow\infty} \sup_{W\in\mathcal{K}_1} \mathbb{P}(W > c) = 0$$

holds.

It can be shown that NA1 and NUPBR are equivalent (see [22]) and that NFLVR implies NUPBR, but there is no equivalence between the latter two notions (see [2] or [20]). A market satisfying NA1 (or, equivalently, NUPBR) is also called (weakly) *viable* and it can be shown that market viability in the sense of NA1 (NUPBR) is a minimal condition to meaningfully solve problems of pricing, hedging and portfolio optimization; see [20].

The last notion to be recalled is that of an *Equivalent Local Martingale Deflator (ELMD)*, which generalizes the notion of the density process for an ELMM:

Definition 2.5. An *Equivalent Local Martingale Deflator (ELMD)* is a nonnegative local martingale Z, not necessarily a martingale, such that $Z(0) = 1$ and $\mathbb{P}(Z(T) > 0) = 1$, and the price processes, when multiplied by Z, become local martingales.

The following result has only recently been proven in full generality; see [23, 37], and [36]:

Proposition 2.6. *A market satisfies NUPBR if and only if the set of equivalent local martingale deflators is not empty.*

The goal of this note is now to provide a systematic way to construct a market that satisfies NUPBR, but not NFLVR.

3. Main Result

In this section we formulate two assumptions under which we can construct a market that satisfies NA1 (equivalently NUPBR) but not NFLVR. In the following Section 4, we then present some examples for markets in which those assumptions are satisfied, namely for which NUPBR thus holds, but not NFLVR.

Based on [3], as it is said there, we now turn things upside down. While in the previous section we had started from a probability space $(\Omega, \mathcal{F}, (\mathcal{F}(t))_{t\in[0,T]}, \mathbb{P})$, on which the d asset price processes S_i are semimartingales, now we consider a filtered probability space $(\Omega, \mathcal{F}, (\mathcal{F}(t))_{t\in[0,T]}, \mathbb{Q})$, on which the d processes S_i are $\mathbb{Q}-$local martingales. On this same probability space we then consider a further nonnegative $\mathbb{Q}$–martingale $Y = (Y(t))_{t\in[0,T]}$ with $Y(0) = 1$. Let the stopping time τ denote the first hitting time of 0 by the $\mathbb{Q}$–martingale Y. We shall assume that Y has positive probability to hit zero, but it only hits zero continuously; to wit,

$$\mathbb{Q}(Y(T) = 0) = \mathbb{Q}(\tau \leq T) > 0; \quad \mathbb{Q}(\{Y(\tau-) > 0\} \cap \{\tau \leq T\}) = 0. \tag{3.1}$$

Since Y was assumed to be $\mathbb{Q}$–martingale we also have $\mathbb{Q}(\tau \leq T) < 1$.

Since Y is a $\mathbb{Q}$–martingale it generates a probability measure $\mathbb{P}$ (it corresponds to the $\mathbb{P}$ from the previous section) via the Radon-Nikodym derivative $\mathrm{d}\mathbb{P}/\mathrm{d}\mathbb{Q} = Y(T)$; the probability measure $\mathbb{P}$ is absolutely continuous with respect to $\mathbb{Q}$, but not equivalent to $\mathbb{Q}$.

Lemma 3.1. *Under Assumption* (3.1), *the process* $1/Y$ *is a nonnegative* $\mathbb{P}-$*strict local martingale with* $\mathbb{P}(1/Y(T) > 0) = 1$.

This statement of Lemma 3.1 follows directly from simple computations; see, for example, Theorem 2.1 in [1].

We introduce the following basic assumption:

$$\begin{gathered}\text{There exists } x \in (0,1) \text{ and an admissible strategy}\\ H = (H(t))_{t\in[0,T]} \text{ s.t. } V^{x,H}(T) \geq \mathbf{1}_{\{Y(T)>0\}}.\end{gathered} \tag{3.2}$$

Note that the market $(\Omega, \mathcal{F}, (\mathcal{F}(t))_{t\in[0,T]}, \mathbb{Q}), S$ of the last subsection satisfies NFLVR. Thus, Assumption (3.2) is equivalent to the assumption that the minimal superreplication price of the contingent claim $\mathbf{1}_{\{Y(T)>0\}}$ is smaller than 1 in this market; to wit,

$$\sup_{R\in\mathcal{M}} \mathbb{E}^R[\mathbf{1}_{\{Y(T)>0\}}] < 1,$$

where $\mathcal{M}$ denotes the set of all probability measures that are equivalent to $\mathbb{Q}$ and under which S_i is a local martingale for each $i = 1, \ldots, d$.

We now state and prove the main result of this note:

Theorem 3.2. *Under the setup of this section and under Assumptions* (3.1) *and* (3.2), *the market* $(\Omega, \mathcal{F}, (\mathcal{F}(t))_{t\in[0,T]}, \mathbb{P}), S$ *satisfies NUPBR but does not satisfy NFLVR. Moreover, any predictable process* H *that satisfies the condition in Assumption 3.2 is a strong arbitrage strategy in the market* $(\Omega, \mathcal{F}, (\mathcal{F}(t))_{t\in[0,T]}, \mathbb{P}), S$.

Proof. Note that the process H from Assumption (3.2) is an admissible trading strategy under $\mathbb{Q}$, and thus, under $\mathbb{P}$, too. Since $\mathbb{P}(\mathbf{1}_{\{Y(T)>0\}} = 1) = 1$ and since $x < 1$, the strategy H is a strong arbitrage strategy under $\mathbb{P}$. Thus, the market $(\Omega, \mathcal{F}, (\mathcal{F}(t))_{t\in[0,T]}, \mathbb{P}), S$ does not satisfy NFLVR.

To see that the market satisfies NUPBR, note that, by Lemma 3.1, the process $1/Y$ is a $\mathbb{P}$–local martingale and that S_i/Y is also a $\mathbb{P}$–local martingale for each $i = 1, \ldots, d$ by a generalized Bayes' formula; see, for example, Proposition 2.3(iii) in [1]. Thus, a local martingale deflator exists and, by Proposition 2.6, the market satisfies NUPBR. □

Basically any market $(\Omega, \mathcal{F}, (\mathcal{F}(t))_{t\in[0,T]}, \mathbb{P}), S$ that satisfies NUPBR but not NFLVR implies the existence of a probability measure $\mathbb{Q}$ and of a $\mathbb{Q}$–local martingale Y that satisfies (3.1) and such that $\mathrm{d}\mathbb{P}/\mathrm{d}\mathbb{Q} = Y(T)$; see [3, 34], and [15]. In this sense, Theorem 3.2 provides the reverse direction and therefore may be considered as a systematic construction of markets satisfying NUPBR but not NFLVR.

4. Examples

In the setup of the previous section, we now discuss several examples for markets in which Assumptions (3.1) and (3.2) hold. Theorem 3.2 then proves that all those markets satisfy, under $\mathbb{P}$, NUPBR but not NFLVR.

Example 4.1. Let Y denote a nonnegative $\mathbb{Q}$–martingale that satisfies $Y(0) = 1$ and Assumption (3.1). Let S_1 be the right-continuous modification of the process $(E^{\mathbb{Q}}[1_{\{Y(T)>0\}}|\mathcal{F}(t)])_{t\in[0,T]}$ and let S_i for $i = 2, \ldots, d$ denote any $\mathbb{Q}$–local martingale. Then, clearly Assumption (3.2) is satisfied and Theorem 3.2 may be applied. □

Example 4.2. As a second example, worked out in detail in the dissertation of Chau Ngoc Huy, consider the case when Y is a compensated

$\mathbb{Q}$–Poisson process with intensity $\lambda \geq 1/T$ started in one, stopped when hitting zero or when it first jumps. Set $S_1 = Y$ and let S_i for $i = 2, \ldots, d$ denote any $\mathbb{Q}$–local martingale. Clearly, Assumption (3.1) (with $\mathbb{Q}(Y(T) = 0) = \exp(-1)$) and Assumption (3.2) (with $x = 1 - \exp(-1)$) hold, and thus, Theorem 3.2 applies. $\square$

Example 4.3. Let us now slightly generalize the previous Poisson setup of Example 4.2. Towards this end, we fix $\lambda \geq 1/T$ and let $F_{\min} \leq 1 \leq F_{\max}$ denote two strictly positive reals such that

$$x := \frac{F_{\max}}{F_{\min}} \left(1 - \exp\left(-\frac{1}{F_{\max}}\right)\right) < 1.$$

Furthermore, we choose an arbitrary distribution function F with support $[F_{\min}, F_{\max}]$ and expectation 1. Now, we let Y denote a compensated compound Poisson process with intensity λ and jump distribution F (under the probability measure $\mathbb{Q}$), started in one and stopped when hitting zero or at its first jump. As before, we set $S_1 = Y$ and let S_i for $i = 2, \ldots, d$ denote any $\mathbb{Q}$–local martingale, without making any further assumptions on them. Again, Assumption (3.1) is clearly satisfied.

As the nonnegative $\mathbb{Q}$–local martingale Y might have jumps of different sizes, the process S does not necessarily have the predictable martingale representation property. However, we may check Assumption (3.2) by hand. To make headway, we fix τ as the first hitting time of zero by the compensated compound Poisson process Y and let ρ denote its first jump time. Note that $\tau \wedge \rho \leq 1/\lambda \leq T$ since at time $1/\lambda$ the compound Poisson process Y has either made a jump or hit zero. We define the S–integrable predictable process $H = (H_1, \ldots, H_d)$ by $H_i \equiv 0$ for all $i = 2, \ldots, d$, and,

$$H_1(t) = \frac{\exp\left(-\frac{1-\lambda t}{F_{\max}}\right)}{F_{\min}} \mathbf{1}_{\{t \leq \rho \wedge \tau\}}$$

for all $t \in [0, T]$. Then, with x as defined above, we obtain

$$\begin{aligned} V^{x,H}(T) = V^{x,H}(\tau \wedge \rho) &= x + \int_0^{\tau \wedge \rho} \frac{\exp\left(-\frac{1-\lambda t}{F_{\max}}\right)}{F_{\min}} \mathrm{d}Y(t) \\ &= x - \lambda \int_0^{\tau \wedge \rho} \frac{\exp\left(-\frac{1-\lambda t}{F_{\max}}\right)}{F_{\min}} \mathrm{d}t + \frac{\exp\left(-\frac{1-\lambda \rho}{F_{\max}}\right)}{F_{\min}} \Delta Y(\rho) \mathbf{1}_{\{\rho \leq \tau\}} \end{aligned}$$

$$\begin{aligned}
&\geq \frac{F_{\max}}{F_{\min}}\left(1-\exp\left(-\frac{1-\lambda(\tau\wedge\rho)}{F_{\max}}\right)\right)+\exp\left(-\frac{1-\lambda\rho}{F_{\max}}\right)\mathbf{1}_{\{\rho\leq\tau\}}\\
&=\left(\frac{F_{\max}}{F_{\min}}\left(1-\exp\left(-\frac{1-\lambda\rho}{F_{\max}}\right)\right)+\exp\left(-\frac{1-\lambda\rho}{F_{\max}}\right)\right)\mathbf{1}_{\{\rho\leq\tau\}}\\
&\geq \mathbf{1}_{\{\rho\leq\tau\}}=\mathbf{1}_{\{Y(T)>0\}},
\end{aligned}$$

where the first inequality follows from the observation that

$$\Delta Y(\rho)/F_{\min}\mathbf{1}_{\{\rho\leq\tau\}}\geq \mathbf{1}_{\{\rho\leq\tau\}},$$

the equality just afterwards follows from the observation that $\lambda\tau = 1$ on the event $\{\tau \leq \rho\}$, and the last inequality follows from the two facts that $F_{\max}/F_{\min} \geq 1$ and that the inequality $\lambda\rho \leq 1$ holds on the event $\{\rho \leq \tau\}$. Thus, Assumption (3.2) is satisfied and Theorem 3.2 may be applied. □

Remark 4.4. Note that Assumption (3.2) is always satisfied if the process S has the predictable martingale representation property under $\mathbb{Q}$. To wit, Assumption (3.2) is always satisfied if the market $(\Omega, \mathcal{F}, (\mathcal{F}(t))_{t\in[0,T]}, \mathbb{Q}), S$ is complete. □

Example 4.5. Consider the filtration $(\mathcal{F}(t))_{t\in[0,T]}$ to be generated by a d-dimensional $\mathbb{Q}$–Brownian motion $B = (B_i)_{i=1,\dots,d}$ with $B_i = (B_i(t))_{t\in[0,T]}$. Let $\sigma = (\sigma(t))_{t\in[0,T]}$ denote a progressively measurable matrix-valued process of dimension $d \times d$ such that $\sigma(t)$ is $\mathbb{Q}$–almost surely invertible for *Lebesgue*-almost every $t \in [0,T]$ and let $S(\cdot) = \int_0^\cdot \sigma(t)\mathrm{d}B(t)$. Assume that the process σ is chosen so that the price process S is strictly positive. Let Y be a nonnegative local martingale hitting zero with positive probability, for example, the process $1 + B_1$ stopped when hitting zero. Then, the process S has the predictable martingale representation property under $\mathbb{Q}$, and the construction of Section 3 generates a market that satisfies NUPBR but not NFLVR; see Remark 4.4. We also refer to Theorem 3 in [3], where a similar setup is discussed. □

5. Acknowledgments

This project started at the Sino-French Workshop at the Beijing International Center for Mathematical Research in June 2013. We are deeply indebted to Ying Jiao, Caroline Hillairet, and Peter Tankov for organising this wonderful meeting and we are very grateful to all participants of the workshop for stimulating discussions on the subject matter of this note. We thank Claudio Fontana and Chau Ngoc Huy for many helpful comments on an earlier version of this note.

References

[1] Carr, P., Fisher, T. and Ruf, J. (2013). On the hedging of options on exploding exchange rates, *Finance and Stochastics*, Volume 18, Issue 1 (2014), Page 115-144.

[2] Delbaen, F. and Schachermayer, W. (1994). A general version of the Fundamental Theorem of Asset Pricing, *Mathematische Annalen* **300**, 3, pp. 463–520.

[3] Delbaen, F. and Schachermayer, W. (1995). Arbitrage possibilities in Bessel processes and their relations to local martingales, *Probability Theory and Related Fields* **102**, 3, pp. 357–366.

[4] Delbaen, F. and Schachermayer, W. (1998). The Fundamental Theorem of Asset Pricing for unbounded stochastic processes, *Mathematische Annalen* **312**, 2, pp. 215–250.

[5] Fernholz, D. and Karatzas, I. (2010). On optimal arbitrage, *Annals of Applied Probability* **20**, 4, pp. 1179–1204.

[6] Fernholz, E. R. (2002). *Stochastic Portfolio Theory*, Springer.

[7] Fernholz, E. R. and Karatzas, I. (2009). Stochastic Portfolio Theory: an overview, in A. Bensoussan. ed., *Handbook of Numerical Analysis*, Vol. Mathematical Modeling and Numerical Methods in Finance, Elsevier.

[8] Föllmer, H. (1972). The exit measure of a supermartingale, *Zeitschrift für Wahrscheinlichkeitstheorie und Verwandte Gebiete* **21**, pp. 154–166.

[9] Fontana, C. (2013). Weak and strong no-arbitrage conditions for continuous financial markets, Preprint, arXiv:1302.7192.

[10] Fontana, C., Jeanblanc, M. and Song, S. (2013). On arbitrages arising from honest times, *Finance and Stochastics*,forthcoming.

[11] Guasoni, P. and Rásony, M. (2011). Fragility of arbitrage and bubbles in diffusion models, Preprint, http://ssrn.com/abstract=1856223.

[12] Harrison, J. M. and Kreps, D. M. (1979). Martingales and arbitrage in multiperiod securities markets, *Journal of Economic Theory* **20**, 3, pp. 381–408.

[13] Harrison, J. M. and Pliska, S. (1981). Martingales and stochastic integrals in the theory of continuous trading, *Stochastic Processes and Their Applications* **11**, 3, pp. 215–260.

[14] Hulley, H. (2010). The economic plausibility of strict local martingales in financial modelling, in C. Chiarella and A. Novikov. eds., *Contemporary Mathematical Finance*, pp. 53–75, Springer.

[15] Imkeller, P. and Perkowski, N. (2013). The existence of dominating local martingale measures, Preprint, arXiv:1111.3885.

[16] Ingersoll, J. E. (1987). *Theory of Financial Decision Making*, Rowman & Littlefield Publishers.

[17] Jarrow, R., Protter, P. and Shimbo, K. (2007). Asset price bubbles in complete markets, in M. C. Fu, R. A. Jarrow, J.-Y. J. Yen and R. J. Elliott. eds., *Advances in Mathematical Finance*, Vol. in honor of Dilip Madan, pp. 97–121, Birkhäuser.

[18] Jarrow, R., Protter, P. and Shimbo, K. (2010). Asset price bubbles in incomplete markets, *Mathematical Finance* **20**, 2, pp. 145–185.

[19] Kabanov, Y. M. (1997). On the FTAP of Kreps-Delbaen-Schachermayer, *Statistics and Control of Stochastic Processes (Moscow, 1995/1996)* , pp. 191–203.

[20] Karatzas, I. and Kardaras, C. (2007). The numéraire portfolio in semimartingale financial models, *Finance and Stochastics* **11**, 4, pp. 447–493.

[21] Karatzas, I. and Ruf, J. (2013). Distribution of the time to explosion for one-dimensional diffusions, Preprint, available on http://www.oxford-man.ox.ac.uk/~jruf/papers/Distribution of Time to Explosion.pdf.

[22] Kardaras, C. (2010). Finitely additive probabilities and the Fundamental Theorem of Asset Pricing, in C. Chiarella and A. Novikov. eds., *Contemporary Mathematical Finance*, Springer.

[23] Kardaras, C. (2012). Market viability via absence of arbitrage of the first kind, *Finance and Stochastics* **16**, 4, pp. 651–667.

[24] Kardaras, C., Kreher, D. and Nikeghbali, A. (2012). Strict local martingales and bubbles, Preprint, arXiv:1108.4177.

[25] Kreps, D. M. (1981). Arbitrage and equilibrium in economies with infinitely many commodities, *Journal of Mathematical Economics* **8**, 1, pp. 15–35.

[26] McKean, H. P. (1969). *Stochastic Integrals*, Academic Press, New York.

[27] Osterrieder, J. and Rheinländer, T. (2006). Arbitrage opportunities in diverse markets via a non-equivalent measure change, *Annals of Finance* **2**, 3, pp. 287–301.

[28] Perkowski, N. and Ruf, J. (2013). Supermartingales as Radon-Nikodym densities and related measure extensions, Preprint, arXiv:1309.4623.

[29] Platen, E. (2006). A benchmark approach to finance, *Mathematical Finance* **16**, 1, pp. 131–151.

[30] Platen, E. and Heath, D. (2006). *A Benchmark Approach to Quantitative Finance*, Springer.

[31] Protter, P. (2013). Strict local martingales with jumps, Preprint, arXiv:1307.2436.

[32] Ruf, J. (2010). Optimal trading strategies and the Bessel process, in *Actuarial and Financial Mathematics Conference. Interplay between Finance and Insurance*, pp. 81–86.

[33] Ruf, J. (2011). *Optimal Trading Strategies Under Arbitrage*, Ph.D. thesis, Columbia University, New York, USA, retrieved from http://academiccommons.columbia.edu/catalog/ac:131477.

[34] Ruf, J. (2013a). Hedging under arbitrage, *Mathematical Finance* **23**, 2, pp. 297–317.

[35] Ruf, J. (2013b). A new proof for the conditions of Novikov and Kazamaki, *Stochastic Processes and Their Applications* **123**, pp. 404–421.

[36] Song, S. (2013). An alternative proof of a result of Takaoka, Preprint, arXiv:1306.1062.

[37] Takaoka, K. (2013). A note on the condition of no unbounded profit with bounded risk, *Finance and Stochastics*, forthcoming.

On the Existence of Martingale Measures in Jump Diffusion Market Models

Jacopo Mancin*, Wolfgang J. Runggaldier†

Abstract. In the context of jump-diffusion market models we construct examples that satisfy the weaker no-arbitrage condition of NA1 (NUPBR), but not NFLVR. We show that in these examples the only candidate for the density process of an equivalent local martingale measure is a supermartingale that is not a martingale, not even a local martingale. This candidate is given by the supermartingale deflator resulting from the inverse of the discounted growth optimal portfolio. In particular, we consider an example with constraints on the portfolio that go beyond the standard ones for admissibility.

1. Introduction

The First Fundamental Theorem of Asset Pricing states the equivalence of the classical no-arbitrage concept of *No-Free-Lunch-With-Vanishing-Risk* (NFLVR) and the existence of an *equivalent* $\sigma-$*martingale measure* (EσMM) (see [8]). Since in this paper we shall consider only nonnegative price processes, we may restrict ourselves to the sub-class of *equivalent local martingale measures* (ELMM), which can be characterized by their density processes (see [6]). In real markets some forms of arbitrage however exist that are incompatible with FLVR. This is taken into account in the recent Stochastic Portfolio Theory (see the survey in [9]), where the NFLVR condition is not imposed as a normative assumption and it is shown that some arbitrage opportunities may arise naturally in a real market; furthermore, one of the roles of portfolio optimization is in fact also that of exploiting possible arbitrages. On the other hand the full strength of NFLVR is not necessarily needed to solve fundamental problems of valuation, hedging

*Department of Mathematics, LMU University Munich, Germany, e-mail: mancin.jacopo@gmail.com
†Dipartimento di Matematica Pura ed Applicata, Universitá di Padova, Italy, e-mail: runggal@math.unipd.it

and portfolio optimization. In addition, the notion of NFLVR is not robust with respect to changes in the numeraire or the reference filtration and it is not easy to check it in real markets. In parallel, there is also the so-called *Benchmark approach to Quantitative Finance* (see e.g. [20]) that aims at developing a theory of valuation that does not rely on the existence of an ELMM. For the hedging problem in market models that do not admit an ELMM we refer to [22].

Weaker forms of no-arbitrage were thus introduced recently and it turns out that the equivalent notions of no-arbitrage given by the *No-Arbitrage of the First Kind* (NA1) (see [15] , see also [18] for an earlier version of this notion) and *No Unbounded Profit with Bounded Risk* (NUPBR) (see [13]) represent the minimal condition that still allows one to meaningfully solve problems of pricing, hedging and portfolio optimization. It is thus of interest to develop economically significant market models, which satisfy NA1 (NUPBR), but not NFLVR. As long as one remains within continuous market models, there may not exist many models that satisfy NA1 (NUPBR) but not NFLVR. The classical examples are based on Bessel processes (see e.g. [7, 11, 20]). As hinted at in [4] , discontinuous market models may offer many more possibilities. To this effect one may however point out that in [14] the author shows that for exponential Lévy models the various no-arbitrage notions weaker than NFLVR are all equivalent to NFLVR. This still leaves open the possibility of investigating the case of jump-diffusion models, which is the setting that we shall consider in this paper.

Working under NA1 (NUPBR) we cannot rely upon the existence of an ELMM, nor upon the corresponding density process. In fact, if there exists some form of arbitrage beyond NFLVR, then a possible candidate density of an EσMM (ELMM) turns out to be a strict local martingale. The density process can however be generalized to the notion of an *Equivalent Supermartingale Deflator* (ESMD) which, considering a finite horizon $[0, T]$, is a process D_t with $D_0 = 1$, $D_t \geq 0$, $D_T > 0$ $P - a.s.$ and such that $D\bar{V}$ is a supermartingale for all discounted value processes $\bar{V}_t$ of a self-financing admissible portfolio strategy (D_t is thus itself a supermartingale). Notice that, if there exists an *Equivalent Supermartingale Measure* (ESMM), namely a measure $Q \sim P$ under which all discounted self-financing portfolio processes are supermartingales, then the process $D_t := (dQ/dP)_{|\mathcal{F}_t}$ is an ESMD that is actually a martingale. An ESMD is however not necessarily a density process, it may not even be a (local) martingale. Instead of approaching directly the problem of constructing jump-diffusion market models that satisfy NA1 (NUPBR) but not NFLVR, in this paper we do it indirectly: for specific jump-diffusion market models that satisfy NA1 (NUPBR) we construct an ESMD that is not a martingale, not even a local martingale. We shall then show that for these models the so-constructed ESMD is the only candidate for the density process of an ELMM (ESMM). If, thus, this only candidate is a supermartingale that is not a martingale, then there cannot exist an ELMM (ESMM). Since, furthermore, for these cases we shall show that the physical measure P cannot be an ELMM, the property of NFLVR fails to hold.

A basic tool to obtain an ESMD is via the *Growth Optimal Portfolio* (GOP), which is a portfolio that outperforms any other self-financing portfolio in the sense that the ratio between the two processes is a supermartingale. For continuous markets it can in fact be shown that, under local square integrability of the market price of risk process, if $\bar{V}_t^*$ denotes the discounted value process of the GOP, then $\hat{Z}_t := 1/\bar{V}_t^*$ is an ESMD (see e.g. [10]). Furthermore, always in continuous markets and under local square integrability of the market price of risk, the undiscounted GOP process V_t^* is a *Numeraire Portfolio* in the sense that self-financing portfolio values, expressed in units of V_t^*, are supermartingales under the physical measure (in the continuous case they are actually positive local martingales so that NFLVR holds in the V_t^*–discounted market and P itself is an ELMM). This is however not true in general: the inverse of the discounted GOP may fail to be even a local martingale (see e.g. Example 5.1bis in [17]). In particular, this may happen when jumps are present (see e.g. Example 6 in [1], as well as [4]). On the other hand, also for general semimartingale models one can show (see [13]) the equivalence of

i) Existence of a numeraire portfolio.
ii) Existence of an ESMD.
iii) Validity of NA1 (NUPBR).

Since, when the GOP exists, this GOP is a numeraire portfolio even if the inverse of its discounted value is not a local martingale (see e.g. [12]), by the previous equivalence there exists then an ESMD. In this case, namely when the inverse of the discounted GOP is not a local martingale, we then also have that the physical measure is not an ELMM when using the GOP as numeraire. In general this does not exclude that NFLVR may nevertheless hold (see [3] , see also [24]). However, in the jump-diffusion case we shall show that the process $\hat{Z}_t$ is the only candidate for the density process of an ELMM (ESMM). Therefore, in this case the failure of the inverse of the discounted GOP to be a (local) martingale excludes the possibility of

NFLVR.

In our quest for examples in the jump-diffusion setting, where the GOP exists and thus NA1 (NUPBR) holds, but the inverse of the discounted GOP may fail to be a local martingale, we shall start by studying the existence and the properties of the GOP, thereby focusing on the characteristics of the market price of risk vector and showing that the existence of an ELMM (ESMM) depends strictly on the relationship between the components of this vector and the corresponding jump intensities. We then prove that the inverse of the discounted GOP defines the only candidate to be the density process of an ELMM and provide examples in which the discounted GOP not only fails to be the inverse of a martingale density process, but is a supermartingale that is not a local martingale. The examples concern both the case when there are no constraints on the portfolio strategies, beyond admissibility, as well as when there are.

The outline of the paper is as follows: In section 2 we describe our jump-diffusion market model and the admissible strategies. Section 3 is devoted to the existence and the properties of the GOP as well as its relation with ESMDs. Examples where the discounted GOP fails to be the inverse of a (super)martingale density process are then discussed in section 4. The model described in section 2 as well as the contents of subsection 3.1 are based on the first part of Chapter 14 in [20] (see also [5]). We recall them here to make the presentation self-contained thereby giving also some additional detail that will be needed later in the paper.

2. The Jump Diffusion Market Model

Let there be given a complete probability space $(\Omega, \mathcal{F}, P)$, with a filtration $\mathcal{F} = (\mathcal{F}_t)_{t\geq 0}$ satisfying the usual conditions of right-continuity and completeness. We consider a market containing $d \in \mathbb{N}$ sources of uncertainty. Continuous uncertainty is represented by an m-dimensional standard Wiener process $W = \{W_t = (W_t^1, \ldots, W_t^m)^\top,\ t \in [0,\infty)\}$. Event driven uncertainty on the other hand is modeled by an $(d\text{-}m)$-variate point process, identified by the $\mathcal{F}$-adapted counting process $N = \{N_t = (N_t^1, \ldots, N_t^{d-m})^\top,\ t \in [0,\infty)\}$, whose intensity $\lambda = \{\lambda_t = (\lambda_t^1, \ldots, \lambda_t^{d-m})^\top,\ t \in [0,\infty)\}$ is a given, predictable and strictly positive process satisfying $\lambda_t^k > 0$ and $\int_0^t \lambda_s^k ds < \infty$ almost surely for all $t \in [0,\infty)$ and $k \in \{1, 2, \ldots, d-m\}$. We shall denote by T_n the jump times of N_t. For each univariate point process we can define a corresponding jump martingale $M^k = \{M_t^k,\ t \in [0,\infty)\}$ which, following [20], we define via its stochastic differential

$$dM_t^k = \frac{dN_t^k - \lambda_t^k dt}{\sqrt{\lambda_t^k}} \tag{2.1}$$

for all $t \in [0, \infty)$ and $k \in \{1, 2, \ldots, d - m\}$. We assume that W and N are independent, generate all the uncertainty in the model and that pairwise the N^ks do not jump at the same time.

The financial market consists of $d + 1$ securities S^j, for $j = 0, 1, \ldots, d$, that model the evolution of wealth due to the ownership of primary securities, with all income and dividends reinvested. As usual, the first account S_t^0 is assumed to be locally risk free, which means that it is of finite variation and the solution of the differential equation

$$dS_t^0 = S_t^0 r_t dt,$$

for $t \in [0, \infty)$, with $S_0^0 = 1$. The remaining assets S^j, for $j = 1, \ldots, d$, are supposed to be risky and to be the solution to the jump diffusion SDE

$$dS_t^j = S_{t-}^j \left(a_t^j dt + \sum_{k=1}^{m} b_t^{j,k} dW_t^k + \sum_{k=m+1}^{d} b_t^{j,k} dM_t^{k-m} \right) \qquad S_0^j > 0, \tag{2.2}$$

for $t \in [0, \infty)$, where the short rate process r, the appreciation rate processes a^j, the generalized volatility processes $b^{j,k}$ and the intensity processes λ^k are almost surely finite and predictable. Assuming that these processes are such that a unique strong solution to the system of SDEs (2.2) exists, we obtain the following explicit expression for every $j = 1, \ldots, d$

$$S_t^j = S_0^j \exp\left\{ \int_0^t \left(a_s^j - \frac{1}{2} \sum_{k=1}^{m} (b_s^{j,k})^2 \right) ds + \sum_{k=1}^{m} \int_0^t b_s^{j,k} dW_s^k \right\} \cdot$$
$$\cdot \prod_{k=m+1}^{d} \left[\exp\left\{ - \int_0^t b_s^{j,k} \sqrt{\lambda_s^{k-m}} ds \right\} \prod_{n=1}^{N_t^{k-m}} \left(\frac{b_{T_n}^{j,k}}{\sqrt{\lambda_{T_n}^{k-m}}} + 1 \right) \right]$$

To ensure non-negativity for each primary security account, and exclude jumps that would lead to negative values for S_t^j, we need to make the following assumption:

Assumption 2.1. The condition

$$b_t^{j,k} \geq -\sqrt{\lambda_t^{k-m}}$$

holds for all $t \in [0, \infty)$, $j \in \{1, 2, \ldots, d\}$ and $k \in \{m + 1, \ldots, d\}$.

In what follows b_t will denote the *generalized volatility* matrix $[b_t^{j,k}]_{j,k=1}^d$ for all $t \in [0, \infty)$ and we make the further assumption:

Assumption 2.2. The generalized volatility matrix b_t is invertible for Lebesgue-almost-every $t \in [0, \infty)$.

The condition stated in Assumption 2.2 means that no primary security account can be formed as a portfolio of other primary security accounts, i.e. the market does not contain redundant assets.

We can now introduce the *market price of risk* vector

$$\theta_t = (\theta_t^1, \dots, \theta_t^d)^\top = b_t^{-1}[a_t - r_t \mathbf{1}] \tag{2.3}$$

for $t \in [0, \infty)$. Here $a_t = (a_t^1, \dots, a_t^d)^\top$ is the *appreciation rate* vector and $\mathbf{1} = (1, \dots, 1)^\top$ is the *unit* vector. Using (2.3), we obtain $a_t = b_t\theta_t + r_t\mathbf{1}$ so that we can rewrite the SDE (2.2) in the form

$$dS_t^j = S_{t-}^j \left(r_t dt + \sum_{k=1}^m b_t^{j,k}(\theta_t^k dt + dW_t^k) + \sum_{k=m+1}^d b_t^{j,k}(\theta_t^k dt + dM_t^{k-m}) \right) \tag{2.4}$$

for $t \in [0, \infty)$ and $j \in \{1, \dots, d\}$. For $k \in \{1, 2, \dots, m\}$, the quantity θ_t^k denotes the market price of risk with respect to the k-th Wiener process W^k. In a similar way, if $k \in \{m+1, \dots, d\}$, then θ_t^k can be interpreted as the market price of the $(k-m)$-th event risk with respect to the counting process N^{k-m}.

The vector process $S = \{S_t = (S_t^0, \dots, S_t^d)^\top, \ t \in [0, \infty)\}$ characterizes the evolution of all primary security accounts. In order to rigorously describe the activity of trading in the financial market we now recall the concept of *trading strategy.* We emphasize that we only consider self-financing trading strategies which generate *positive portfolio processes.*

Definition 2.3. (*Admissible strategies*)

- An admissible trading strategy with initial value s is a $\mathbb{R}^{d+1}$-valued predictable stochastic process $\delta = \{\delta_t = (\delta_t^0, \dots, \delta_t^d)^\top, \ t \in [0, \infty)\}$, where δ_t^j denotes the number of units of the j-th primary security account held at time $t \in [0, \infty)$ in the portfolio, and it is such that the Itô integral $\int_0^T \delta_t^j dS_t^j$ is well-defined for any $T > 0$ and $j \in \{0, 1, \dots, d\}$. Furthermore, the value of the corresponding portfolio process at time t, which as in [20] we denote by $S_t^{s,\delta} = \sum_{j=0}^d \delta_t^j S_t^j$ with $S_0^{s,\delta} = s > 0$, is nonnegative for any $t \in [0, \infty)$. We let

$S_t^\delta := S_t^{1,\delta}$ so that $S_t^{s,\delta} = s\,S_t^\delta$. Since the strategy is self-financing, we have also $dS_t^{s,\delta} = \sum_{j=0}^d \delta_t^j dS_t^j$.

- For any admissible trading strategy δ the value of the corresponding discounted portfolio process at time t is defined as $\bar{S}_t^{s,\delta} = S_t^{s,\delta}/S_t^0$.

For a given strategy δ with strictly positive portfolio process $S^{s,\delta}$ we denote as usual by $\pi_{\delta,t}^j$ the fraction of wealth invested in the j-th primary security account at time t, that is $\pi_{\delta,t}^j = \delta_t^j \left(S_{t-}^j / S_{t-}^{s,\delta}\right)$ for $t \in [0,\infty)$ and $j \in \{0,1,\dots,d\}$. In terms of the vector of fractions $\pi_{\delta,t} = (\pi_{\delta,t}^0, \dots, \pi_{\delta,t}^d)^\top$ we obtain from (2.4), the self-financing property, and taking (2.1) into account, the following SDE for $S_t^{s,\delta}$

$$
\begin{aligned}
dS_t^{s,\delta} &= S_{t-}^{s,\delta} \\
&\cdot \left\{ \left(r_t + \sum_{k=1}^m \sum_{j=1}^d \pi_{\delta,t-}^j b_t^{j,k} \theta_t^k + \sum_{k=m+1}^d \sum_{j=1}^d \pi_{\delta,t-}^j b_t^{j,k} \left(\theta_t^k - \sqrt{\lambda_t^{k-m}} \right) \right) dt \right. \\
&\left. + \sum_{k=1}^m \sum_{j=1}^d \pi_{\delta,t}^j b_t^{j,k} dW_t^k + \sum_{k=m+1}^d \sum_{j=1}^d \pi_{\delta,t}^j \frac{b_t^{j,k}}{\sqrt{\lambda_t^{k-m}}} dN_t^{k-m} \right\}
\end{aligned}
\tag{2.5}
$$

Therefore, the value at time t of the portfolio $S^{s,\delta}$ is

$$
\begin{aligned}
S_t^{s,\delta} &= s \exp\left\{ \int_0^t \left(r_s + \sum_{k=1}^d \sum_{j=1}^d \pi_{\delta,s}^j b_s^{j,k} \theta_s^k - \frac{1}{2} \sum_{k=1}^m \left(\sum_{j=1}^d \pi_{\delta,s}^j b_s^{j,k} \right)^2 \right) ds \right. \\
&\left. + \sum_{k=1}^m \int_0^t \sum_{j=1}^d \left(\pi_{\delta,s}^j b_s^{j,k} \right) dW_s^k \right\} \prod_{k=m+1}^d \left[\exp\left\{ - \int_0^t \sum_{j=1}^d \pi_{\delta,s}^j b_s^{j,k} \sqrt{\lambda_s^{k-m}} ds \right\} \right. \\
&\left. \cdot \prod_{n=1}^{N_t^{k-m}} \left(\frac{\sum_{j=1}^d \pi_{\delta,T_n}^j b_{T_n}^{j,k}}{\sqrt{\lambda_{T_n}^{k-m}}} + 1 \right) \right]
\end{aligned}
\tag{2.6}
$$

From the last term on the right hand side in (2.6) it is immediately seen that a portfolio process remains strictly positive if and only if

$$
\sum_{j=1}^d \pi_{\delta,t}^j b_t^{j,k} > -\sqrt{\lambda_t^{k-m}} \quad \text{a.s.} \tag{2.7}
$$

for all $k \in \{m+1,\ldots,d\}$ and $t \in [0,\infty)$. This condition is guaranteed by Assumption 2.1.

3. The Growth Optimal Portfolio (GOP)

3.1. *Derivation of the GOP and its dynamics*

Definition 3.1. For an admissible trading strategy δ, leading to a strictly positive portfolio process, the **growth rate process** $g^\delta = (g_t^\delta)_{t\geq 0}$ is defined as the drift term in the SDE satisfied by the process $\log S^\delta = (\log S_t^\delta)_{t\geq 0}$. An admissible trading strategy δ^* (and the corresponding portfolio process S^{δ^*}) is said to be growth-optimal if $g_t^{\delta^*} \geq g_t^\delta$ P-a.s. for all $t \in [0,\infty)$ for any admissible trading strategy δ. We shall use the acronym *GOP* to denote the growth-optimal portfolio.

By applying Itô's formula and suitably adding and subtracting terms (see [19]) one immediately obtains the general expression for g_t^δ, namely

Lemma 3.2. *For any admissible trading strategy δ, the SDE satisfied by $\log S^\delta$ is*

$$d\log S_t^\delta = g_t^\delta dt$$
$$+\sum_{k=1}^{m}\sum_{j=1}^{d}\pi_{\delta,t}^j b_t^{j,k} dW_t^k + \sum_{k=m+1}^{d} \log\left(1+\sum_{j=1}^{d}\pi_{\delta,t}^j \frac{b_t^{j,k}}{\sqrt{\lambda_t^{k-m}}}\right)\sqrt{\lambda_t^{k-m}}\, dM_t^{k-m} \tag{3.1}$$

where g_t^δ is the growth rate given by

$$g_t^\delta = r_t + \sum_{k=1}^{m}\left[\sum_{j=1}^{d}\pi_{\delta,t}^j b_t^{j,k}\theta_t^k - \frac{1}{2}\left(\sum_{j=1}^{d}\pi_{\delta,t}^j b_t^{j,k}\right)^2\right]$$
$$+\sum_{k=m+1}^{d}\left[\sum_{j=1}^{d}\pi_{\delta,t}^j b_t^{j,k}\left(\theta_t^k - \sqrt{\lambda_t^{k-m}}\right) + \log\left(1+\sum_{j=1}^{d}\pi_{\delta,t}^j\frac{b_t^{j,k}}{\sqrt{\lambda_t^{k-m}}}\right)\lambda_t^{k-m}\right] \tag{3.2}$$

for $t \in [0,\infty)$.

In order to obtain the GOP dynamics we now maximize separately the two sums on the right hand side of (3.2) with respect to the *portfolio volatilities* $c_t^k := \sum_{j=1}^d \pi_{\delta,t}^j b_t^{j,k}$ for $k \in \{1, \ldots, d\}$. Note that for the first sum a unique maximum exists, because it is a negative definite quadratic form with respect to the portfolio volatilities. In order to guarantee the existence and the uniqueness of a maximum also in the second sum, we have to impose the following condition

Assumption 3.3. The intensities and the market price of event risk components satisfy

$$\sqrt{\lambda_t^{k-m}} > \theta_t^k,$$

for all $t \in [0, \infty)$ and $k \in \{m+1, \ldots, d\}$.

This is because the first derivative of

$$c_t^k \left(\theta_t^k - \sqrt{\lambda_t^{k-m}}\right) + \log\left(1 + \frac{c_t^k}{\sqrt{\lambda_t^{k-m}}}\right)\lambda_t^{k-m} \tag{3.3}$$

with respect to c_t^k, which is $\left(\theta_t^k - \sqrt{\lambda_t^{k-m}}\right) + \left(\lambda_t^{k-m} / \left(\sqrt{\lambda_t^{k-m}} + c_t^k\right)\right)$, is positive for all $c_t^k > -\sqrt{\lambda_t^{k-m}}$ if Assumption 3.3 does not hold. These, by virtue of (2.7), are precisely all the possible values of the *event driven volatility* $\sum_{j=1}^d \pi_{\delta,t}^j b_t^{j,k}$, $k \in \{m+1, \ldots, d\}$. Therefore if Assumption 3.3 fails to hold there will not exist an optimal growth rate, since (3.3) tends to infinity as $c_t^k \to \infty$, for any $k \in \{m+1, \ldots, d\}$.

This condition allows us to introduce the predictable vector process $c_t^* = (c_t^{*1}, \ldots, c_t^{*d})^\top$ which describes the optimal generalized portfolio volatilities. For the components with $k \in \{1, \ldots, m\}$, we get from the first order condition that identifies the maximum growth rate the following

$$\theta_t^k - c_t^k = 0 \Longleftrightarrow c_t^k = \theta_t^k$$

For the last $(d-m)$ components we get, again from the first order condition, noticing that the function of c_t^k in (3.3) is strictly concave and that we must have $c_t^k > -\sqrt{\lambda_t^{k-m}}$,

$$\left(\theta_t^k - \sqrt{\lambda_t^{k-m}}\right) + \frac{\lambda_t^{k-m}}{\sqrt{\lambda_t^{k-m}} + c_t^k} = 0 \Longleftrightarrow c_t^k = \frac{\theta_t^k}{1 - \theta_t^k (\lambda_t^{k-m})^{-\frac{1}{2}}}$$

Therefore the vector c_t^* has the following representation

$$c_t^{*k} = \begin{cases} \theta_t^k, & \text{for} \quad k \in \{1, 2, \ldots, m\} \\ \frac{\theta_t^k}{1 - \theta_t^k (\lambda_t^{k-m})^{-\frac{1}{2}}}, & \text{for} \quad k \in \{m+1, \ldots, d\} \end{cases} \tag{3.4}$$

for $t \in [0,\infty)$. Note that a very large jump intensity with $\lambda_t^{k-m} \gg 1$ or $\theta_t^k/\sqrt{\lambda_t^{k-m}} \ll 1$ causes the corresponding component c_t^{*k} to approach the market price of jump risk θ_t^k asymptotically for given $t \in [0,\infty)$ and $k \in \{m+1,\dots,d\}$. In this case the structure of the components $c_t^{*k} \approx \theta_t^k$ for $k \in \{m+1,\dots,d\}$ is similar to those obtained with respect to the Wiener processes. Intuitively this is because, when jumps occur more and more frequently, almost continuously, the jump martingales M^k become nearly indistinguishable from the continuous ones.

The above considerations lead immediately to the following (see also Corollary 14.1.5 in [20])

Lemma 3.4. *Under the Assumptions 2.1, 2.2 and 3.3 the fractions*

$$\pi_{\delta_*,t} = (\pi^1_{\delta_*,t},\dots,\pi^d_{\delta_*,t}) = (c_t^{*\top} b_t^{-1})^\top \tag{3.5}$$

determine uniquely the GOP and the corresponding portfolio process $S^{\delta_*} = \{S_t^{\delta_*},\ t\in[0,\infty)\}$ satisfies the SDE

$$\begin{aligned} dS_t^{\delta_*} = S_{t-}^{\delta_*}\Bigg(& r_t dt + \sum_{k=1}^m \theta_t^k(\theta_t^k dt + dW_t^k) \\ & + \sum_{k=m+1}^d \frac{\theta_t^k}{1-\theta_t^k(\lambda_t^{k-m})^{-\frac12}}(\theta_t^k dt + dM_t^{k-m})\Bigg) \end{aligned} \tag{3.6}$$

for $t \in [0,\infty)$, with $S_0^{\delta_*} > 0$. Note that Assumption 3.3 guarantees that the portfolio process S^{δ_*} is strictly positive.

By (3.2), (3.4) and (3.5) we obtain the optimal growth rate of the GOP in the form

$$\begin{aligned} g_t^{\delta_*} &= r_t + \sum_{k=1}^m \left[\left(\theta_t^k\right)^2 - \frac12\left(\theta_t^k\right)^2\right] + \sum_{k=m+1}^d \left[\frac{\theta_t^k}{1-\theta_t^k(\lambda_t^{k-m})^{-\frac12}}\left(\theta_t^k - \sqrt{\lambda_t^{k-m}}\right)\right. \\ &\quad \left. + \log\left(1 + \frac{\theta_t^k}{1-\theta_t^k(\lambda_t^{k-m})^{-\frac12}}\frac{1}{\sqrt{\lambda_t^{k-m}}}\right)\lambda_t^{k-m}\right] \\ &= r_t + \frac12\sum_{k=1}^m\left(\theta_t^k\right)^2 + \sum_{k=m+1}^d \lambda_t^{k-m}\left(\log\left(1+\frac{\theta_t^k}{\sqrt{\lambda_t^{k-m}}-\theta_t^k}\right) - \frac{\theta_t^k}{\sqrt{\lambda_t^{k-m}}}\right) \end{aligned}$$

for $t \in [0,\infty)$.

3.2. *GOP and Martingale Deflators*

In what follows we consider a fixed finite time horizon $T < \infty$ and investigate whether the model introduced above represents a viable financial market, in particular we shall check whether properly defined arbitrage opportunities are excluded. First we recall the following

Definition 3.5. An *equivalent local martingale measure* (ELMM) is a measure $Q \sim P$ such that all price processes, expressed in units of the risk-free asset (i.e. discounted prices), are local martingales. Analogously, $Q \sim P$ is an *equivalent supermartingale measure* (ESMM) if the price processes, expressed in units of the risk-free asset, are supermartingales.

From [15] we also recall the

Definition 3.6. An $\mathcal{F}_T$-measurable nonnegative random variable ξ is called *arbitrage of the first kind* if $P(\xi > 0) > 0$ and, for all initial values $s \in (0, \infty)$, there exists an admissible trading strategy δ such that $\bar{S}_T^{s,\delta} \geq \xi$ P-a.s. We say that the financial market is viable if there are no arbitrages of the first kind, i.e. the condition NA1 holds.

We next show that a sufficient condition for the absence of arbitrages of the first kind is the existence of a *supermartingale deflator* which we describe as

Definition 3.7. An equivalent supermartingale deflator (ESMD) is a real-valued nonnegative adapted process $D = (D_t)_{0 \leq t \leq T}$ with $D_0 = 1$ and $D_T > 0$ P-a.s. and such that the process $D\bar{S}^\delta = (D_t \bar{S}_t^\delta)_{0 \leq t \leq T}$ is a supermartingale for every admissible trading strategy δ. We denote by $\mathcal{D}$ the set of all supermartingale deflators. (By taking $\delta \equiv (1, 0, \cdots, 0)$, one has that D is itself a supermartingale).

Proposition 3.8. *If $\mathcal{D} \neq \emptyset$ then there cannot exist arbitrages of the first kind.*

Proof. (adapted from [10]). Let $D \in \mathcal{D}$ and suppose that there exists a random variable ξ yielding an arbitrage of the first kind. Then, for every $n \in \mathbb{N}$, there exists an admissible trading strategy δ^n such that, $\bar{S}_T^{1/n,\delta^n} \geq \xi$ P-a.s. For every $n \in \mathbb{N}$, the process $D\bar{S}^{1/n,\delta^n} = (D_t \bar{S}_t^{1/n,\delta^n})_{0 \leq t \leq T}$ is a supermartingale. So, for every $n \in \mathbb{N}$ one has $E[D_T \xi] \leq E[D_T \bar{S}_T^{1/n,\delta^n}] \leq E[D_0 \bar{S}_0^{1/n,\delta^n}] = 1/n$. Letting $n \to \infty$ gives $E[D_T \xi] = 0$ and hence $D_T \xi = 0$ P-a.s. Since, due to Definition 3.7, we have $D_T > 0$ P-a.s. this implies that $\xi = 0$ P-a.s., which contradicts the assumption that ξ is an arbitrage of the first kind. □

In the remaining part of this section we will derive a fundamental property of the GOP. Dealing with GOP-denominated portfolio processes, following [20] we first introduce the following notation.

Definition 3.9. For any portfolio process S^δ, the process $\hat{S}^\delta = \left(\hat{S}^\delta_t\right)_{0 \le t \le T}$, defined as $\hat{S}^\delta_t := S^\delta_t / S^{\delta_*}_t$ for $t \in [0,T]$, is called *benchmarked portfolio process.*

Remark 3.10. We shall often refer to the inverse of the discounted GOP as $\hat{Z}_t := 1/\bar{S}^{\delta_*}_t$.

Proposition 3.11. *Under Assumptions 2.1, 2.2 and 3.3 the discounted GOP process* $\bar{S}^{\delta_*} = \{\bar{S}^{\delta_*}_t,\ t \in [0,T]\}$ *is the inverse of a supermartingale deflator* D*. Equivalently,* $\hat{Z}_t$ *is a supermartingale deflator.*

Proof. It suffices to show that every benchmarked portfolio is a local martingale that, being nonnegative, is then a supermartingale by Fatou's lemma. According to the product formula (see Corollary II-2 in [21]) we first have that

$$d\left(\frac{S^\delta_t}{S^{\delta_*}_t}\right) = d\left(\frac{\bar{S}^\delta_t}{\bar{S}^{\delta_*}_t}\right) = \frac{d\bar{S}^\delta_t}{\bar{S}^{\delta_*}_{t-}} + d\left(\frac{1}{\bar{S}^{\delta_*}_{t-}}\right)\bar{S}^\delta_t + d\left[\bar{S}^\delta_t, \frac{1}{\bar{S}^{\delta_*}_t}\right]$$

For the term involving $d\bar{S}^\delta_t$ note that, from (2.5) and taking into account (2.1) as well as the definition of the investment ratios $\pi_{\delta,t}$, one obtains

$$\begin{aligned} d\bar{S}^\delta_t &= \bar{S}^\delta_{t-}\left\{\sum_{k=1}^{m}\left(\sum_{j=1}^{d}\delta^j_t \frac{S^j_t}{S^\delta_t} b^{j,k}_t\right)\left(\theta^k_t dt + dW^k_t\right)\right. \\ &\quad \left. + \sum_{k=m+1}^{d}\left(\sum_{j=1}^{d}\delta^j_t \frac{S^j_{t-}}{S^\delta_{t-}} b^{j,k}_t\right)\left(\theta^k_t dt + dM^{k-m}_t\right)\right\} \\ &= \sum_{k=1}^{m}\left(\sum_{j=1}^{d}\delta^j_t \bar{S}^j_t b^{j,k}_t\right)\left(\theta^k_t dt + dW^k_t\right) \\ &\quad + \sum_{k=m+1}^{d}\left(\sum_{j=1}^{d}\delta^j_t \bar{S}^j_{t-} b^{j,k}_t\right)\left(\theta^k_t dt + dM^{k-m}_t\right) \end{aligned} \tag{3.7}$$

Next, using Itô's formula, from (3.6) we obtain for the inverse of the discounted GOP

$$d\left(\frac{1}{\bar{S}_t^{\delta_*}}\right) = \left[-\frac{1}{\bar{S}_t^{\delta_*}}\left(\sum_{k=1}^{m}(\theta_t^k)^2 + \sum_{k=m+1}^{d}\frac{\theta_t^k}{1-\frac{\theta_t^k}{\sqrt{\lambda_t^{k-m}}}}\left(\theta_t^k - \sqrt{\lambda_t^{k-m}}\right)\right)\right.$$
$$\left. + \frac{1}{\bar{S}_t^{\delta_*}}\sum_{k=1}^{m}\left(\theta_t^k\right)^2\right]dt - \frac{1}{\bar{S}_t^{\delta_*}}\sum_{k=1}^{m}\theta_t^k dW_t^k$$
$$+ \frac{1}{\bar{S}_{t-}^{\delta_*}}\sum_{k=m+1}^{d}\left[\left(1+\frac{\theta_t^k}{1-\frac{\theta_t^k}{\sqrt{\lambda_t^{k-m}}}}\frac{1}{\sqrt{\lambda_t^{k-m}}}\right)^{-1} - 1\right] dN_t^{k-m}$$

The terms in the last sum can be simplified in the following way

$$\frac{1}{S_{t-}^{\delta_*}}\left(\left(\frac{\sqrt{\lambda_t^{k-m}}}{\sqrt{\lambda_t^{k-m}} - \theta_t^k}\right)^{-1} - 1\right) = -\frac{\theta_t^k}{\bar{S}_{t-}^{\delta_*}\sqrt{\lambda_t^{k-m}}}$$

so that, by adding and subtracting $-\sum_{k=m+1}^{d}\theta_t^k\sqrt{\lambda_t^{k-m}}/\bar{S}_t^{\delta_*}dt$ and rearranging all the terms, we obtain

$$d\left(\frac{1}{\bar{S}_t^{\delta_*}}\right) = -\frac{1}{\bar{S}_t^{\delta_*}}\sum_{k=1}^{m}\theta_t^k dW_t^k - \frac{1}{\bar{S}_{t-}^{\delta_*}}\sum_{k=m+1}^{d}\theta_t^k dM_t^{k-m} \tag{3.8}$$

Finally, the last term is just

$$d\left[\bar{S}_t^{\delta}, \frac{1}{\bar{S}_t^{\delta_*}}\right]$$
$$= -\sum_{k=1}^{m}\left(\sum_{j=1}^{d}\delta_t^j\hat{S}_t^j b_t^{j,k}\right)\theta_t^k dt - \sum_{k=m+1}^{d}\left(\sum_{j=1}^{d}\delta_t^j\hat{S}_{t-}^j b_t^{j,k}\right)\frac{\theta_t^k}{\lambda_t^{k-m}}dN_t^{k-m}$$

Summing up all the components we obtain

$$d\left(\frac{\bar{S}_t^{\delta}}{\bar{S}_t^{\delta_*}}\right) = \sum_{k=1}^{m}\left(\sum_{j=1}^{d}\delta_t^j\hat{S}_t^j b_t^{j,k} - \hat{S}_t^{\delta}\theta_t^k\right)dW_t^k$$
$$+ \sum_{k=m+1}^{d}\left(\sum_{j=1}^{d}\delta_t^j\hat{S}_{t-}^j b_t^{j,k} - \hat{S}_{t-}^{\delta}\theta_t^k\right)dM_t^{k-m}$$
$$+ \sum_{k=m+1}^{d}\left(\sum_{j=1}^{d}\delta_t^j\hat{S}_t^j b_t^{j,k}\right)\theta_t^k dt - \sum_{k=m+1}^{d}\left(\sum_{j=1}^{d}\delta_t^j\hat{S}_{t-}^j b_t^{j,k}\right)\frac{\theta_t^k}{\lambda_t^{k-m}}dN_t^{k-m}$$

from which, observing that

$$\sum_{k=m+1}^{d}\left(\sum_{j=1}^{d}\delta_t^j\hat{S}_t^j b_t^{j,k}\right)\theta_t^k dt-\sum_{k=m+1}^{d}\left(\sum_{j=1}^{d}\delta_t^j\hat{S}_{t-}^j b_t^{j,k}\right)\frac{\theta_t^k}{\lambda_t^{k-m}}dN_t^{k-m}$$
$$=-\sum_{k=m+1}^{d}\left(\sum_{j=1}^{d}\delta_t^j\hat{S}_{t-}^j b_t^{j,k}\right)\frac{\theta_t^k}{\sqrt{\lambda_t^{k-m}}}\left(\frac{dN_t^{k-m}-\lambda_t^{k-m}dt}{\sqrt{\lambda_t^{k-m}}}\right),$$

we obtain

$$d\left(\frac{\bar{S}_t^\delta}{\bar{S}_t^{\delta_*}}\right)=\sum_{k=1}^{m}\left(\sum_{j=1}^{d}\delta_t^j\hat{S}_t^j b_t^{j,k}-\hat{S}_t^\delta\theta_t^k\right)dW_t^k$$
$$+\sum_{k=m+1}^{d}\left(\left(\sum_{j=1}^{d}\delta_t^j\hat{S}_{t-}^j b_t^{j,k}\right)\left(1-\frac{\theta_t^k}{\sqrt{\lambda_t^{k-m}}}\right)-\hat{S}_{t-}^\delta\theta_t^k\right)dM_t^{k-m}$$

which, indeed, is a nonnegative local martingale and thus a supermartingale. □

From this proof it actually follows that $\hat{Z}_t$ is an equivalent local martingale deflator (ELMD) in the sense that $\hat{Z}_t\,\bar{S}^\delta$ are local martingales.

Generalizing the notion of Benchmarked portfolio process (see Definition 3.9) we recall the following

Definition 3.12. An admissible portfolio process $S^{\tilde{\delta}}=\left(S_t^{\tilde{\delta}}\right)_{0\le t\le T}$ has *the numeraire property* if all admissible portfolio processes $S^\delta=\left(S_t^\delta\right)_{0\le t\le T}$, when denominated in terms of $S^{\tilde{\delta}}$, are supermartingales, i.e. if the process $S^\delta/S^{\tilde{\delta}}=\left(S_t^\delta/S_t^{\tilde{\delta}}\right)_{0\le t\le T}$ is a supermartingale for every admissible trading strategy δ.

Remark 3.13. As a corollary of Proposition 3.11 we have that the GOP has the numeraire property.

In continuous financial markets it can be shown that the numeraire portfolio is unique (see e.g. [10]). The proof in [10] can be carried over rather straightforwardly to the jump-diffusion case (see [19]) so that we have

Proposition 3.14. *The numeraire portfolio process* $S^{\tilde{\delta}}=\left(S_t^{\tilde{\delta}}\right)_{0\le t\le T}$ *is unique (in the sense of indistinguishability). Furthermore, there exists an*

unique admissible trading strategy $\tilde{\delta}$ such that $S^{\tilde{\delta}}$ is the numeraire portfolio, up to a null subset of $\Omega \times [0, T]$.

Remark 3.15. Since, as we have seen, the GOP has the numeraire property and the numeraire portfolio is unique, the GOP is the unique numeraire portfolio.

The numeraire property of the GOP plays a crucial role in the concept of *real world pricing* allowing one, also in the present jump-diffusion setting, to perform pricing of contingent claims in financial markets for which no ELMM may exist (see [10, 20]).

4. Supermartingale Deflators/Densities

Due to Propositions 3.11 and 3.14 (see Remarks 3.13 and 3.15), the GOP coincides with the unique numeraire portfolio and its discounted value is also the inverse of the supermartingale deflator $\hat{Z}_t$. This means that, if we express all price processes in terms of the GOP, the original probability measure P becomes an ESMM. We now investigate particular cases in which $\hat{Z}_t$ is not a martingale and so, since it will be shown to be also the only candidate for the density process of an ELMM, for these cases NFLVR fails to hold. Furthermore, when expressing the price processes in terms of the GOP, the physical measure P is not an ELMM.

We start by showing that for our financial market model the inverse of the discounted GOP is the only possible local martingale deflator; thus it is also the only candidate to be the Radon-Nikodym derivative (density process) of an ELMM. These concepts are in fact strictly linked to one another, since a supermartingale deflator D defines an ELMM if and only if D_T integrates to 1. On the other hand, the Radon-Nikodym derivative of an ELMM is of course a supermartingale deflator. We prove the claim of the uniqueness by studying changes of measure via the general Radon-Nikodym derivative when dealing with a jump-diffusion process, namely (see [2, 23])

$$\begin{aligned} L_t = \exp\left\{-\frac{1}{2}\sum_{k=1}^{m}\int_0^t \left(\varphi_s^k\right)^2 ds + \sum_{k=1}^{m}\int_0^t \varphi_s^k dW_s^k\right\} \\ \prod_{k=m+1}^{d}\left\{\exp\left[\int_0^t \left(1-\psi_s^{k-m}\right)\lambda_s^{k-m} ds\right]\prod_{n=1}^{N_t^{k-m}} \psi_{T_n}^{k-m}\right\} \end{aligned} \tag{4.1}$$

where φ_t is a square integrable predictable process and ψ_t is a positive predictable process, integrable with respect to λ_t. We will show that these coefficients have to satisfy a linear system whose only solution leads to the same dynamics as for the inverse of the discounted GOP.

Proposition 4.1. *Under Assumptions 2.1, 2.2 and 3.3, the inverse of the discounted GOP, namely $\hat{Z}_t$, is the only candidate to be the Radon-Nikodym derivative of an ELMM. Equivalently, it is the only ELMD in the sense of what was specified after the statement of Proposition 3.11.*

Proof. We start by defining the Wiener and the Poisson martingales W_t^Q and M_t^Q under the new measure Q defined by L_t in (4.1):

$$\begin{cases} dW_t^{Q,k} = dW_t^k - \varphi_t^k dt, & \text{for } k \in \{1,2,\dots,m\} \\ dM_t^{Q,k-m} = dN_t^{k-m} - \psi_t^{k-m}\lambda_t^{k-m} dt, & \text{for } k \in \{m+1,\dots,d\} \end{cases}$$

thereby obtaining the SDEs satisfied by the primary security accounts

$$dS_t^j = S_{t-}^j \left\{ \left(r_t + \sum_{k=1}^{d} b_t^{j,k}\theta_t^k + \sum_{k=1}^{m} b_t^{j,k}\varphi_t^k + \sum_{k=m+1}^{d} b_t^{j,k}\psi_t^{k-m}\sqrt{\lambda_t^{k-m}} - \sum_{k=m+1}^{d} b_t^{j,k}\sqrt{\lambda_t^{k-m}} \right) dt + \sum_{k=1}^{m} b_t^{j,k} dW_t^{Q,k} + \sum_{k=m+1}^{d} \frac{1}{\sqrt{\lambda_t^{k-m}}} b_t^{j,k} dM_t^{Q,k-m} \right\} \tag{4.2}$$

for every $j \in \{1,\dots,d\}$ and $t \in [0,T]$. If L_t is the Radon-Nikodym derivative of an ELMM, the drift term in (4.2) must be equal to r_t for every $t \in [0,T]$, so that the coefficients φ_t and ψ_t must be the solution to the linear system defined by the following equations

$$\sum_{k=1}^{m} b_t^{j,k}\varphi_t^k + \sum_{k=m+1}^{d} b_t^{j,k}\psi_t^{k-m}\sqrt{\lambda_t^{k-m}} = -\sum_{k=1}^{m} b_t^{j,k}\theta_t^k - \sum_{k=m+1}^{d} b_t^{j,k}\theta_t^k + \sum_{k=m+1}^{d} b_t^{j,k}\sqrt{\lambda_t^{k-m}}$$

for every $j \in \{1,\dots,d\}$ and $t \in [0,T]$. Since, by virtue of the standing Assumption 2.2 the generalized volatility matrix b_t has full rank, the linear system admits a unique solution which is given by

$$\begin{cases} \varphi_t^k = -\theta_t^k, & \text{for } k \in \{1,2,\dots,m\} \\ \psi_t^{k-m} = 1 - \frac{\theta_t^k}{\sqrt{\lambda_t^{k-m}}} & \text{for } k \in \{m+1,\dots,d\} \end{cases}$$

Plugging these terms into (4.1) we see that the Radon-Nikodym derivative has thus the following expression

$$
\begin{aligned}
L_t = \exp&\left\{ -\frac{1}{2}\sum_{k=1}^{m}\int_0^t \left(\theta_s^k\right)^2 ds - \sum_{k=1}^{m}\int_0^t \theta_s^k dW_s^k \right\} \\
&\prod_{k=m+1}^{d}\left\{ \exp\left[\int_0^t \theta_s^k \sqrt{\lambda_s^{k-m}} ds\right] \prod_{n=1}^{N_t^{k-m}} \left(1 - \frac{\theta_{T_n}^k}{\sqrt{\lambda_{T_n}^{k-m}}}\right)\right\}
\end{aligned}
\tag{4.3}
$$

which is precisely the inverse of the discounted GOP, since (see the equation (3.8) satisfied by $(\bar{S}_t^{\delta_*})^{-1}$)

$$
dL_t = -L_{t-}\left(\sum_{k=1}^{m} \theta_t^k dW_t^k + \sum_{k=m+1}^{d} \theta_t^k dM_t^{k-m}\right) \qquad \square
$$

Remark 4.2. Since the vector ψ_t in (4.1) has to be positive for every $t \in [0,T]$, we note that the condition $1 - \left(\theta_t^k / \sqrt{\lambda_t^{k-m}}\right) \geq 0$ has to hold true for every $k \in \{m+1, \ldots, d\}$. This means that if Assumption 3.3 is violated, and there is at least one $k \in \{m+1, \ldots, d\}$ for which $\sqrt{\lambda_t^{k-m}} < \theta_t^k$, there cannot exist an ELMM.

Remark 4.3. Notice that, by analogy to the 2nd FTAP, for our complete market here we have uniqueness of the ELMD.

We want to emphasize that, thanks to the proposition just proved, the financial market does not admit an ELMM as soon as the inverse of the discounted GOP is a strict supermartingale that is not a martingale. We therefore try to find some particular cases where the process $\hat{Z}_t$ is a supermartingale that is not a martingale (not even a local martingale). In the continuous case the most notorious example of a martingale deflator that does not yield an ELMM is the three-dimensional Bessel process β_t (see e.g. [7, 11, 20]): the GOP is simply obtained by placing all the wealth in the stock, but it has infinite expected growth rate and $\frac{1}{\beta_t}$ fails to integrate to 1. In the jump diffusion market that we are considering we note that, as $\theta_t^k \to \sqrt{\lambda_t^{k-m}}$ for any $k \in \{m+1, \ldots, d\}$, the GOP would explode and therefore $\hat{Z}_t$ would be close to zero as soon as a jump occurs (see (4.3)). In this case, $\hat{Z}_T$ will not integrate to 1, thus preventing the existence of an ELMM, and the GOP cannot be used as a numeraire.

4.1. *Supermartingale Densities that are not (local) Martingale Densities*

Let us see what happens if we violate Assumption 3.3, namely if we let $\sqrt{\lambda_t^{k-m}}$ be less than θ_t^k. For simplicity, in this entire section we will consider the particular case in which $d = 2$ and $m = 1$.

In order to be able to maximize the second sum in (3.2) we then have to impose a restriction on the possible trading strategies in the form of the following assumption.

Assumption 4.4. There exists a positive real number ψ such that for any admissible trading strategy $\pi = \{\pi_t = (\pi_t^0, \pi_t^1, \pi_t^2)^\top, \ t \in [0,T]\}$ we have

$$\pi_t^1 b_t^{1,2} + \pi_t^2 b_t^{2,2} \leq \psi \tag{4.4}$$

for all $t \in [0,T]$.

Remark 4.5. Assumption 4.4 can be seen as a *convex constraint* limiting the portfolio volatility belonging to the jump martingale M. This condition has a clearer financial interpretation when $b_t^2 := b_t^{1,2} = b_t^{2,2}$ for each $t \in [0,T]$. In this case we get that (4.4) is equivalent to a constraint on the minimum amount of wealth invested in the risk free asset $\pi_t^0 \geq 1 - \psi/b_t^2$. We emphasize that this applies also to the case in which $b_t^2 = -\sqrt{\lambda_t}$ for each $t \in [0,T]$, in which jumps are used to describe the default of the primary accounts.

In this case the optimal generalized portfolio volatilities are described by the following predictable process

$$\tilde{c}_t^k = \begin{cases} \theta_t^1, & \text{for} \quad k = 1 \\ \psi, & \text{for} \quad k = 2 \end{cases} \tag{4.5}$$

The first component $\tilde{c}_k^1$ follows from the first order conditions identifying the maximum growth rate, while $\tilde{c}_k^2$ is the maximum value obtainable in the constrained setting. From (2.5) we have that, for the case when c_t^k are given by (4.5), the discounted GOP must satisfy the following SDE

$$d\bar{S}_t^{\delta_*} = \bar{S}_{t-}^{\delta_*} \left\{\theta_t^1 \left(\theta_t^1 dt + dW_t\right) + \psi \left(\theta_t^2 dt + dM_t\right)\right\} \tag{4.6}$$

The convex constraints just introduced are the framework that enables us to provide a simple example of a market in which GOP denominated prices are strict supermartingales, as we show in the next theorem.

Theorem 4.6. *Under Assumptions 2.1, 2.2 and 4.4, the process $\hat{Z}_t$ and any benchmarked portfolio process are supermartingales which are not local martingales.*

Proof. By analogy to the proof of Proposition 3.11, we start by calculating the SDE for $\bar{S}_t^{\delta}/\bar{S}_t^{\delta_*}$, where δ_t is an arbitrary admissible trading strategy. According to the product formula we proceed by calculating separately the components $d\bar{S}_t^{\delta}$, $d\left(1/\bar{S}_t^{\delta_*}\right)$ and $d\left[\bar{S}_t^{\delta}, 1/\bar{S}_t^{\delta_*}\right]$. The first component can be obtained by adjusting the general formula in (3.7) to the case in which $d = 2$ so that

$$\begin{aligned} d\bar{S}_t^{\delta} &= \left(\delta_t^1 \bar{S}_t^1 b_t^{1,1} + \delta_t^2 \bar{S}_t^2 b_t^{2,1}\right)\theta_t^1 dt + \left(\delta_t^1 \bar{S}_t^1 b_t^{1,2} + \delta_t^2 \bar{S}_t^2 b_t^{2,2}\right)\theta_t^2 dt \\ &\quad + \left(\delta_t^1 \bar{S}_t^1 b_t^{1,1} + \delta_t^2 \bar{S}_t^2 b_t^{2,1}\right) dW_t + \left(\delta_t^1 \bar{S}_{t-}^1 b_t^{1,2} + \delta_t^2 \bar{S}_{t-}^2 b_t^{2,2}\right) dM_t \end{aligned}$$

The second term comes from applying the Itô formula to (4.6), namely

$$\begin{aligned} d\left(\frac{1}{\bar{S}_t^{\delta_*}}\right) &= -\frac{1}{\bar{S}_t^{\delta_*}}\left(\left(\theta_t^1\right)^2 + \psi\left(\theta_t^2 - \sqrt{\lambda_t}\right)\right) dt + \frac{1}{\bar{S}_t^{\delta_*}}\left(\theta_t^1\right)^2 dt \\ &\quad - \frac{1}{\bar{S}_t^{\delta_*}}\theta_t^1 dW_t + \frac{1}{\bar{S}_{t-}^{\delta_*}}\left[\left(1 + \psi\frac{1}{\sqrt{\lambda_t}}\right)^{-1} - 1\right] dN_t \\ &= -\frac{1}{\bar{S}_t^{\delta_*}}\psi\left(\theta_t^2 - \sqrt{\lambda_t}\right) dt - \frac{1}{\bar{S}_t^{\delta_*}}\theta_t^1 dW_t - \frac{1}{\bar{S}_{t-}^{\delta_*}}\frac{\psi}{\sqrt{\lambda_t} + \psi} dN_t \\ &= -\frac{1}{\bar{S}_t^{\delta_*}}\left(\psi\left(\theta_t^2 - \sqrt{\lambda_t}\right) + \frac{\psi\lambda_t}{\sqrt{\lambda_t} + \psi}\right) dt - \frac{1}{\bar{S}_{t-}^{\delta_*}}\left(\theta_t^1 dW_t + \frac{\psi\sqrt{\lambda_t}}{\sqrt{\lambda_t} + \psi} dM_t\right) \end{aligned} \tag{4.7}$$

Note that the drift term in the SDE satisfied by $1/\bar{S}_t^{\delta_*}$ is strictly negative as Assumption 3.3 does not hold and both ψ and λ_t are positive. This shows that the inverse of the discounted GOP, namely $\hat{Z}_t$, is a strict supermartingale. Finally,

$$\begin{aligned} &d\left[\bar{S}_t^{\delta}, \frac{1}{\bar{S}_t^{\delta_*}}\right] \\ &= -\left(\delta_t^1 \hat{S}_t^1 b_t^{1,1} + \delta_t^2 \hat{S}_t^2 b_t^{2,1}\right)\theta_t^1 dt - \left(\delta_t^1 \hat{S}_{t-}^1 b_t^{1,2} + \delta_t^2 \hat{S}_{t-}^2 b_t^{2,2}\right)\frac{\psi\sqrt{\lambda_t}}{\sqrt{\lambda_t} + \psi}\frac{1}{\lambda_t} dN_t \end{aligned}$$

On the other hand, for the benchmarked portfolios we have

$$
\begin{aligned}
d\left(\frac{\bar{S}_t^{\delta}}{\bar{S}_t^{\delta_*}}\right) &= \left(\delta_t^1 \hat{S}_t^1 b_t^{1,2} + \delta_t^2 \hat{S}_t^2 b_t^{2,2}\right) \theta_t^2 dt + \left(\delta_t^1 \hat{S}_t^1 b_t^{1,1} + \delta_t^2 \hat{S}_t^2 b_t^{2,1}\right) dW_t \\
&+ \left(\delta_t^1 \hat{S}_{t-}^1 b_t^{1,2} + \delta_t^2 \hat{S}_{t-}^2 b_t^{2,2}\right) dM_t - \left(\delta_t^1 \hat{S}_{t-}^1 b_t^{1,2} + \delta_t^2 \hat{S}_{t-}^2 b_t^{2,2}\right) \frac{\psi}{\sqrt{\lambda_t} + \psi} \frac{1}{\sqrt{\lambda_t}} dN_t \\
&- \hat{S}_t^{\delta} \left(\psi\left(\theta_t^2 - \sqrt{\lambda_t}\right) + \frac{\psi \lambda_t}{\sqrt{\lambda_t} + \psi}\right) dt - \hat{S}_{t-}^{\delta} \left(\theta_t^1 dW_t + \frac{\psi \sqrt{\lambda_t}}{\sqrt{\lambda_t} + \psi} dM_t\right) \\
&= \Bigg[\left(\delta_t^1 \hat{S}_t^1 b_t^{1,2} + \delta_t^2 \hat{S}_t^2 b_t^{2,2}\right) \left(\theta_t^2 - \frac{\psi \sqrt{\lambda_t}}{\sqrt{\lambda_t} + \psi}\right) \\
&\qquad - \hat{S}_t^{\delta} \left(\psi\left(\theta_t^2 - \sqrt{\lambda_t}\right) + \frac{\psi \lambda_t}{\sqrt{\lambda_t} + \psi}\right)\Bigg] dt \\
&+ \left(\delta_t^1 \hat{S}_t^1 b_t^{1,1} + \delta_t^2 \hat{S}_t^2 b_t^{2,1}\right) dW_t + \left(\delta_t^1 \hat{S}_{t-}^1 b_t^{1,2} + \delta_t^2 \hat{S}_{t-}^2 b_t^{2,2}\right) dM_t \\
&- \left(\delta_t^1 \hat{S}_{t-}^1 b_t^{1,2} + \delta_t^2 \hat{S}_{t-}^2 b_t^{2,2}\right) \frac{\psi}{\sqrt{\lambda_t} + \psi} \frac{dN_t - \lambda_t dt}{\sqrt{\lambda_t}} \\
&\qquad - \hat{S}_{t-}^{\delta} \left(\theta_t^1 dW_t + \frac{\psi \sqrt{\lambda_t}}{\sqrt{\lambda_t} + \psi} dM_t\right)
\end{aligned}
$$

from which, being condition (4.4) equivalent to $\delta_t^1 \hat{S}_t^1 b_t^{1,2} + \delta_t^2 \hat{S}_t^2 b_t^{2,2} \leq \psi \hat{S}_t^{\delta}$, we note that the drift term is negative, in fact

$$
\begin{aligned}
&\left(\delta_t^1 \hat{S}_t^1 b_t^{1,2} + \delta_t^2 \hat{S}_t^2 b_t^{2,2}\right) \left(\theta_t^2 - \frac{\psi \sqrt{\lambda_t}}{\sqrt{\lambda_t} + \psi}\right) - \hat{S}_t^{\delta} \left(\psi\left(\theta_t^2 - \sqrt{\lambda_t}\right) + \frac{\psi \lambda_t}{\sqrt{\lambda_t} + \psi}\right) \\
&\qquad \leq \hat{S}_t^{\delta} \left(\psi\left(\theta_t^2 - \frac{\psi \sqrt{\lambda_t}}{\sqrt{\lambda_t} + \psi}\right) - \psi\left(\theta_t^2 - \sqrt{\lambda_t}\right) - \frac{\psi \lambda_t}{\sqrt{\lambda_t} + \psi}\right) = 0
\end{aligned}
$$

This is because the factor $\theta_t^2 - \psi\sqrt{\lambda_t}/\left(\sqrt{\lambda_t} + \psi\right)$ is positive: the function $x \mapsto \psi x/(x + \psi)$ is always increasing, so that, since in this section we are violating Assumption 3.3, we have

$$
\theta_t^2 - \frac{\psi \sqrt{\lambda_t}}{\sqrt{\lambda_t} + \psi} \geq \theta_t^2 - \frac{\psi \theta_t^2}{\theta_t^2 + \psi} = \frac{\left(\theta_t^2\right)^2}{\theta_t^2 + \psi} > 0 \qquad \square
$$

The above theorem shows that, if trading is restricted, the investor maximizing the expected logarithmic utility function (the GOP maximizes also the expected log-utility) goes as far as the admissibility constraints, expressed in Assumption 4.4, allow. Notice that, since $\hat{Z}_t$ is a supermartingale without being a local martingale, it cannot be the density process of an ELMM.

For the sake of completeness we now study the property of $\hat{Z}_t$ when Assumption 3.3 holds true but we enforce Assumption 4.4 as well. We shall see that $\hat{Z}_t$ is again a supermartingale that is not a local martingale and that $S_t^{\delta_*}$ is the numeraire portfolio.

Proposition 4.7. *Under Assumptions 2.1, 2.2, 3.3 and 4.4, if there exists $B \subset [0,T]$ with positive Lebesgue measure such that for $t \in B$ one has $\psi < \theta_t^2 / \left(1 - \theta_t^2/\sqrt{\lambda_t}\right)$, then any benchmarked portfolio process is a supermartingale which is not a local martingale.*

Proof. We start by noting that the condition $\psi < \theta_t^2 / \left(1 - \theta_t^2/\sqrt{\lambda_t}\right)$ simply requires that Assumption 4.4 imposes a real constraint on the investor, who would otherwise construct the optimal portfolio as if the condition expressed in Assumption 4.4 was not present. Therefore, as soon as this condition holds, the discounted GOP dynamics is the same as (4.6), and the SDEs satisfied by the inverse of the discounted GOP, namely $\hat{Z}_t$, and by any benchmarked portfolio are the same as those obtained in the proof of Theorem 4.6. The only thing left to do is to check the negativity of the drift term in each of the dynamics. For the former, the drift term of $\hat{Z}_t$, we get from (4.7)

$$
\begin{aligned}
-\hat{Z}_t\left(\psi\left(\theta_t^2 - \sqrt{\lambda_t}\right) + \frac{\psi\lambda_t}{\sqrt{\lambda_t}+\psi}\right) &= -\hat{Z}_t\left(\frac{\psi\left(\theta_t^2 - \sqrt{\lambda_t}\right)\left(\sqrt{\lambda_t}+\psi\right) + \psi\lambda_t}{\sqrt{\lambda_t}+\psi}\right) \\
&= -\hat{Z}_t\left(\frac{\psi\left(\psi\left(\theta_t^2 - \sqrt{\lambda_t}\right) + \theta_t^2\sqrt{\lambda_t}\right)}{\sqrt{\lambda_t}+\psi}\right)
\end{aligned}
$$

which is negative since $\psi > 0$ and $\psi < \theta_t^2 / \left(1 - \theta_t^2/\sqrt{\lambda_t}\right)$. The latter follows in the same way as in the proof of Theorem 4.6, noting that $\theta_t^2 - \psi\sqrt{\lambda_t}/\left(\psi + \sqrt{\lambda_t}\right) > 0 \iff \psi < \theta_t^2/\left(1 - \theta_t^2/\sqrt{\lambda_t}\right)$. □

Acknowledgements

We are grateful to Claudio Fontana, Johannes Ruf and Ngoc Huy Chau for valuable comments.

References

[1] D. Becherer, The numeraire portfolio for unbounded semimartingales, *Finance and Stochastics* **5**, 327–341 (2001).

[2] P. Bremaud, *Point processes and queues: martingale dynamics.* Springer-Verlag (1981).

[3] N. H. Chau, PhD Thesis, University of Padova and Paris Diderot (in preparation).
[4] M. M. Christensen and K. Larsen, No arbitrage and the growth optimal portfolio, *Stochastic Analysis and Applications* **25**, 255–280 (2007).
[5] M M. Christensen and E. Platen, A general benchmark model for stochastic jump sizes, *Stochastic Analysis and Applications* **23**, 1017–1044 (2005).
[6] F. Delbaen, W.Schachermayer, A general version of the fundamental theorem of asset pricing, *Mathematische Annalen* **300**, 463–520 (1994).
[7] F. Delbaen and W. Schachermayer, Arbitrage possibilities in Bessel processes and their relations to local martingales, *Prob. Theory Rel. Fields* **102**, 357–366 (1995).
[8] F. Delbaen and W. Schachermayer, The fundamental theorem of asset pricing for unbounded stochastic processes, *Mathematische Annalen* **312**, 215–250 (1998).
[9] R. Fernholz and I. Karatzas, Stochastic portfolio theory: an overview. In: *Mathematical Modeling and Numerical Methods in Finance*, A. Bensoussan and Q. Zhang eds., Handbook of Numerical Analysis XV, 89–167. North Holland (2009).
[10] C. Fontana and W. J. Runggaldier, Diffusion-based models for financial markets without martingale measures. In: *Risk Measures and Attitudes* F. Biagini, A. Richter and H. Schlesinger, eds., EAA Series, 45–81 Springer Verlag London (2013).
[11] H. Hulley, The economic plausibility of strict local martingales in financial modelling. In: *Contemporary Quantitative Finance: Essays in Honour of Eckhard Platen*, C. Chiarella and A. Novikov eds., 53–75. Springer-Verlag (2010).
[12] H. Hulley and M. Schweizer, M6-on minimal market models and minimal martingale measures. In: *Contemporary Quantitative Finance: Essays in Honour of Eckhard Platen*, C. Chiarella and A. Novikov eds., 35–51. Springer-Verlag (2010).
[13] I. Karatzas and C. Kardaras, The numeraire portfolio in semimartingale financial models, *Finance and Stochastics*, **11**, 447–493 (2007).
[14] C. Kardaras, No-free-lunch equivalences for exponential Lévy models under convex constraints on Investment. *Mathematical Finance*, **19**, 161–187 (2009).
[15] C. Kardaras, Market viability via absence of arbitrage of the first kind, *Finance and Stochastics*, **16**, 651–667 (2012).
[16] C. Kardaras, D. Kreher and A. Nikeghbali, Strict local martingales and bubbles. Preprint 2011.
[17] D. Kramkov and W. Schachermayer, The asymptotic elasticity of utility functions and optimal investment in incomplete markets. *Annals of Applied Probability*, 904–950 (1999).
[18] M. Loewenstein and G. A. Willard, Local Martingales, arbitrage, and viability: Free snacks and cheap thrills, *Economic Theory* **16**, 135–161 (2000).
[19] J. Mancin, Master's Thesis, University of Padova 2012.

[20] E. Platen and D. Heath, *A Benchmark Approach to Quantitative Finance*, Springer-Verlag (2006).

[21] P. E. Protter, *Stochastic Integration and Differential Equations.* Springer-Verlag, Berlin, 2nd ed. (2004).

[22] J. Ruf, Hedging under arbitrage, *Mathematical Finance* **23**, 297–317 (2013).

[23] W. J. Runggaldier, Jump Diffusion Models. In: *Handbook of Heavy Tailed Distributions in Finance*, S.T.Rachev ed., 170–209, Elsevier-Amsterdam (2003).

[24] K. Takaoka, On the condition of no unbounded profit with bounded risk. *Finance and Stochastics*, forthcoming.

Arbitrages in a Progressive Enlargement Setting

Anna Aksamit*, Tahir Choulli†, Jun Deng†, Monique Jeanblanc*

Abstract: This paper completes the analysis of Choulli et al. [6] and contains two principal contributions. The first contribution consists in providing and analysing many practical examples of market models that admit classical arbitrages while they preserve the No Unbounded Profit with Bounded Risk under random horizon and when an honest time is incorporated for particular cases of models. For these markets, we calculate explicitly the arbitrage opportunities. The second contribution lies in providing simple proofs for the stability of the No Unbounded Profit with Bounded Risk under random horizon and after honest time satisfying additional important condition for particular cases of models.

1. Introduction

This paper studies a financial market in which some assets, with prices adapted with respect to a reference filtration $\mathbb{F}$, are traded. One then assumes that an agent has some extra information, and may use strategies that are predictable with respect to a larger filtration $\mathbb{G}$. This extra information is modeled by the knowledge of some random time τ, when this time occurs. We restrict our study to progressive enlargement of filtration setting, and we pay a particular attention to honest times. Our goal is to detect if the knowledge of τ allows for some arbitrage, i.e., if using $\mathbb{G}$-predictable strategies, the agent can make profit.

In this paper we consider two main notions of no-arbitrage, namely no classical arbitrage and No Unbounded Profit with Bounded Risk (NUPBR hereafter). To the best of our knowledge, there are no references for the case of classical arbitrages in a general setting. The goal of the present paper is firstly to introduce the problem, to solve it in some specific cases and to

*Laboratoire Analyse et Probabilités, Université d'Evry Val d'Essonne, Evry, France
†Mathematical and Statistical Sciences Depart., University of Alberta, Edmonton, Canada

give some explicit examples of classical arbitrages (with a proof different from the one in Fontana et al. [8]), and secondly to give, in some specific models, an easy proof of NUPBR condition.

In the case of honest times avoiding stopping times in a continuous filtration, the same problem was studied in Fontana et al. [8] where the authors have investigated several kinds of arbitrages. We refer the reader to that paper for an extensive list of related results in the literature.

The paper is organized as follows: Section 2 presents the problem and recalls some definitions and results on arbitrages and progressive enlargement of filtration. In Section 3 we study two classical situations in enlargement of filtration theory, namely immersion and positive density hypothesis cases. Section 4 concerns honest times, and we show that, in case of a complete market, there exist classical arbitrages before and after the honest time, and we give a way to construct these arbitrages. This fact is illustrated by many examples, where we exhibit these arbitrages in a closed form. In Section 5, we study some examples of non-honest times. In Section 6, we study NUPBR condition before a random time and after an honest time, in some specific examples.

2. General Framework

We consider a filtered probability space $(\Omega, \mathcal{A}, \mathbb{F}, \mathbb{P})$ where the filtration $\mathbb{F}$ satisfies the usual hypotheses and $\mathcal{F}_\infty \subset \mathcal{A}$, and a random time τ (i.e., a positive $\mathcal{A}$-measurable random variable). We assume that the financial market where a risky asset with price S (an $\mathbb{F}$-adapted positive process) and a riskless asset S^0 (assumed, for simplicity, to have a constant price so that the risk-free interest rate is null) are traded is arbitrage free. More precisely, without loss of generality we assume that S is a $(\mathbb{P}, \mathbb{F})$-(local) martingale. In this paper, the horizon is equal to ∞.

We denote by $\mathbb{G}$ the progressively enlarged filtration of $\mathbb{F}$ by τ, i.e., the smallest right-continuous filtration that contains $\mathbb{F}$ and makes τ a stopping time defined as

$$\mathcal{G}_t = \cap_{\epsilon>0} \mathcal{F}_{t+\epsilon} \vee \sigma(\tau \wedge (t+\epsilon)).$$

We recall that $(\mathcal{H}')$ hypothesis is said to hold between two filtrations $\mathbb{F}$ and $\mathbb{G}$ where $\mathbb{F} \subset \mathbb{G}$ if any $\mathbb{F}$-martingale is a $\mathbb{G}$-semimartingale. For a semimartingale X and a predictable process H, we use the notation $H \centerdot X$ for the stochastic integral $\int_0^\cdot H_s dX_s$ when it exists.

We start by an elementary remark: assume that there are no arbitrages using $\mathbb{G}$-predictable strategies and that $\mathbb{P}$ is the unique probability measure making S an $\mathbb{F}$-martingale. So, in particular, the $(S, \mathbb{F})$ market is complete (i.e., the market where (S, S^0) are traded). Then, roughly speaking, S would be a $(\mathbb{Q}, \mathbb{G})$-martingale for some equivalent martingale measure $\mathbb{Q}$, hence would be also a $(\mathbb{Q}, \mathbb{F})$-martingale[a] and $\mathbb{Q}$ will coincide with $\mathbb{P}$ on $\mathbb{F}$. This implies that any $(\mathbb{F}, \mathbb{Q})$-martingale is a $(\mathbb{G}, \mathbb{Q})$-martingale.

Another trivial remark is that, in the particular case where τ is an $\mathbb{F}$-stopping time, the enlarged filtration and the reference filtration are the same. Therefore, no-arbitrage conditions hold before and after τ.

2.1. *Illustrative examples*

We study here two basic examples, in order to show in a first step how arbitrages can occur in a Brownian filtration, and in a second step that discontinuous models present some difficulties.

2.1.1. *Brownian case*

Let $dS_t = S_t \sigma dW_t$, where W is a Brownian motion and σ a constant, be the price of the risky asset. This martingale S goes to 0 a.s. when t goes to infinity, hence the random time $\tau = \sup\{t : S_t = S^*\}$ where $S^* = \sup_{s \geq 0} S_s$ is a finite honest time, and obviously leads to an arbitrage before τ: at time 0, buy one share of S (at price S_0), borrow S_0, then, at time τ, reimburse the loan S_0 and sell the share of the asset at price S_τ. The gain is $S_\tau - S_0 > 0$ with an initial wealth null. There are also arbitrages after τ: at time τ, take a short position on S, i.e., hold a self financing portfolio with value V such that $dV_t = -dS_t, V_\tau = 0$. Usually shortselling positions are not admissible, since $V_t = -S_t + S_\tau$ is not bounded below. Here $-S_t + S_\tau$ is positive, hence shortselling is an arbitrage opportunity.

2.1.2. *Poisson case*

Let N be a Poisson process with intensity λ and M be its compensated martingale. We define the price process S as $dS_t = S_{t-}\psi dM_t, S_0 = 1$ with ψ is a constant satisfying $\psi > -1$ and $\psi \neq 0$, so that

$$S_t = \exp(-\lambda \psi t + \ln(1 + \psi) N_t) .$$

[a]Note that if S is a $(\mathbb{Q}, \mathbb{G})$-strict local martingale for some equivalent martingale measure $\mathbb{Q}$, one can not deduce that it is also a $(\mathbb{Q}, \mathbb{F})$-local martingale

Since $\frac{N_t}{t}$ goes to λ a.s. when t goes to infinity and $\ln(1+\psi)-\psi<0$, S_t goes to 0 a.s. when t goes to infinity. The random time

$$\tau = \sup\{t\,:\, S_t = S^*\}$$

with $S^* = \sup_{s\geq 0} S_s$ is a finite honest time.

If $\psi > 0$, then $S_\tau \geq S_0$ and an arbitrage opportunity is realized at time τ, with a long position in the stock. If $\psi < 0$, then the arbitrage is not so obvious. We shall discuss that with more details in Section 4.2.

There are arbitrages after τ, selling at time τ a contingent claim with payoff 1, paid at the first time ϑ after τ when $S_t > \sup_{s\leq\tau} S_s$. For $\psi > 0$, it reduces to $S_\tau = \sup_{s\leq\tau} S_s$, and, for $\psi < 0$, one has $S_{\tau-} = \sup_{s\leq\tau} S_s$. At time $t_0 = \tau$, the non informed buyer will agree to pay a positive price, the informed seller knows that the exercise will be never done.

2.2. *Admissible portfolio and arbitrages opportunities*

In this section, we recall the basic definitions on arbitrages, and we give sufficient conditions for no arbitrages in a market with zero interest rate. We refer to Fontana et al. [8] for details.

Let $\mathbb{K}$ be one of the filtrations $\{\mathbb{F},\mathbb{G}\}$. Note that, in order that the integral $\theta\centerdot S$ has a meaning for a $\mathbb{G}$ predictable process θ, one needs that S is a $\mathbb{G}$-semimartingale. This requires (on $\{t>\tau\}$) some hypotheses on τ.

For $a \in \mathbb{R}_+$, an element $\theta \in L^{\mathbb{K}}(S)$ is said to be an a-admissible $\mathbb{K}$-strategy if $(\theta\centerdot S)_\infty := \lim_{t\to\infty}(\theta\centerdot S)_t$ exists and $V_t(0,\theta) := (\theta\centerdot S)_t \geq -a$ $\mathbb{P}$-a.s. for all $t\geq 0$. We denote by $\mathcal{A}_a^{\mathbb{K}}$ the set of all a-admissible $\mathbb{K}$-strategies. A process $\theta\in L^{\mathbb{K}}(S)$ is called an *admissible* $\mathbb{K}$*-strategy* if $\theta \in \mathcal{A}^{\mathbb{K}} := \bigcup_{a\in\mathbb{R}_+}\mathcal{A}_a^{\mathbb{K}}$.

An admissible strategy yields an Arbitrage Opportunity if $V(0,\theta)_\infty \geq 0$ $\mathbb{P}$-a.s. and $\mathbb{P}\big(V(0,\theta)_\infty > 0\big) > 0$. In order to avoid confusions, we shall call these arbitrages *classical arbitrages*. If there exists no such $\theta\in\mathcal{A}^{\mathbb{K}}$ we say that the financial market $\mathcal{M}(\mathbb{K}) := (\Omega,\mathbb{K},\mathbb{P};S)$ satisfies the No Arbitrage (NA) condition.

No Free Lunch with Vanishing Risk (NFLVR) holds in the financial market $\mathcal{M}(\mathbb{K})$ if and only if there exists an Equivalent Martingale Measure in $\mathbb{K}$, i.e., a probability measure $\mathbb{Q}$, such that $\mathbb{Q}\sim\mathbb{P}$ and the process S is a $(\mathbb{Q},\mathbb{K})$-local martingale. If NFLVR holds, there are no classical arbitrages.

A non-negative $\mathcal{K}_\infty$-measurable random variable ξ with $\mathbb{P}(\xi>0)>0$ yields an Unbounded Profit with Bounded Risk if for all $x>0$ there exists an element $\theta^x \in \mathcal{A}_x^{\mathbb{K}}$ such that $V(x,\theta^x)_\infty := x + (\theta^x\centerdot S)_\infty \geq \xi$ $\mathbb{P}$-a.s.

If there exists no such random variable, we say that the financial market $\mathcal{M}(\mathbb{K})$ satisfies the No Unbounded Profit with Bounded Risk (NUPBR) condition.

We recall that NFLVR holds if and only if both NA and NUPBR hold (see Delbaen and Schachermayer [7] Corollary 3.4 and Kabanov [16] Proposition 3.6).

A strictly positive $\mathbb{K}$-local martingale $L = (L_t)_{t\geq 0}$ with $L_0 = 1$ and $L_\infty > 0$ $\mathbb{P}$-a.s. is said to be a local martingale deflator in $(S, \mathbb{K})$ on the time horizon $[0, \varrho]$ if the process LS^ϱ is a $\mathbb{K}$-local martingale; here ϱ is a $\mathbb{K}$-stopping time. The important result giving the characterisation of NUPBR condition for strictly positive price process is stated in Theorem 4.12 in Kabanov [16] and then generalized in Takaoka [19] Theorem 5. We recall it here.

Theorem 2.1. *Let S be a strictly positive $\mathbb{K}$-semimartingale. Then, the NUPBR condition holds in $\mathbb{K}$ if and only if there exists a local martingale deflator in $\mathbb{K}$.*

2.3. *Enlargement of filtration results*

We now recall some basic results on progressive enlargement of filtrations. The reader can refer to Jeulin [12] and Jeulin and Yor [13] for more information.

Let τ be a random time, i.e., a positive random variable. We define the right-continuous with left limits $\mathbb{F}$-supermartingale

$$Z_t := \mathbb{P}\left(\tau > t \mid \mathcal{F}_t\right).$$

Note that $Z_0 = 1$ if $\mathbb{P}(\tau > 0) = 1$. The optional decomposition of Z leads to an important $\mathbb{F}$-martingale that we denote by m, given by

$$m := Z + A^o, \tag{2.1}$$

where A^o is the $\mathbb{F}$-dual optional projection[b] of $A := 1\!\!1_{[\![\tau,\infty[\![}$ (so A^o is a non-decreasing process). Note that m is non-negative: indeed $m_t = \mathbb{E}(A^o_\infty + Z_\infty|\mathcal{F}_t)$.

A second important $\mathbb{F}$-supermartingale, defined through

$$\widetilde{Z}_t := \mathbb{P}\left(\tau \geq t \mid \mathcal{F}_t\right),$$

[b]See Appendix for the definition if needed

will play a particular rôle in the following. One has $\widetilde{Z} = Z + \Delta A^o$; hence the supermartingale $\widetilde{Z}$ admits a decomposition as

$$\widetilde{Z} = m - A^o_- . \tag{2.2}$$

We start with the following obvious (but useful) result

Lemma 2.2. *Assume that the financial market $(S, \mathbb{F})$ is complete and let φ be the $\mathbb{F}$-predictable process satisfying $m = 1 + \varphi \centerdot S$. If $m_\tau \geq 1$ and $\mathbb{P}(m_\tau > 1) > 0$, then the $\mathbb{G}$-predictable process $\varphi \mathbb{1}_{[\![0,\tau]\!]}$ is a classical arbitrage strategy in the market "before τ", i.e., in $(S^\tau, \mathbb{G})$.*

PROOF: The $\mathbb{F}$-predictable process φ exists due to the market completeness. Hence $\mathbb{1}_{[\![0,\tau]\!]}\varphi$ is a $\mathbb{G}$-predictable admissible self-financing strategy with initial value 1 and final value $m_\tau - 1$ satisfying $m_\tau - 1 \geq 0$ a.s. and $\mathbb{P}(m_\tau - 1 > 0) > 0$, so it is a classical arbitrage strategy in $(S^\tau, \mathbb{G})$. □

2.3.1. *Decomposition formula before τ*

In a first step, we restrict our attention to what happens before τ. Therefore, we do not require any extra hypothesis on τ, since, for any random time τ, any $\mathbb{F}$-martingale stopped at τ is a $\mathbb{G}$-semimartingale, as established by Jeulin [12] Prop. (4,16) : to any $\mathbb{F}$-local martingale X, we associate the $\mathbb{G}$-local martingale $\widehat{X}$ (stopped at time τ)

$$\widehat{X}_t := X^\tau_t - \int_0^{t\wedge\tau} \frac{d\langle X, m\rangle^{\mathbb{F}}_s}{Z_{s-}}, \tag{2.3}$$

where, as usual, X^τ is the stopped process defined as $X^\tau_t = X_{t\wedge\tau}$.

An interesting case is the one of pseudo-stopping times. We recall that a random time τ is a pseudo-stopping time if any $\mathbb{F}$-martingale stopped at τ is a $\mathbb{G}$-martingale (see Nikeghbali and Yor [18]). This is equivalent to the fact that the $\mathbb{F}$-martingale m is constantly equal to 1.

2.3.2. *Honest times and decomposition formula after τ*

We need to impose conditions on τ such that the ($\mathbb{F}$-martingale) price process S is a $\mathbb{G}$-semimartingale, so that one can define stochastic integrals of $\mathbb{G}$ predictable processes with respect to S. In this paper, we are not interested by necessary and sufficient conditions, these ones being far from tractable (see Jeulin [12] [III, 2,c]). Instead we focus here on honest times.

Theorem 2.3. *Let τ be a random time. Then the following conditions are equivalent:*

(a) *The random time* τ *is honest, i.e., for each* $t \geq 0$, *there exists an* $\mathcal{F}_t$-*measurable random variable* τ_t *such that* $\tau = \tau_t$ *on* $\{\tau < t\}$.
(b) $\widetilde{Z}_\tau = 1$ *on* $\{\tau < \infty\}$.
(c) *There exists an optional set* Λ *such that* $\tau(\omega) = \sup\{t : (\omega, t) \in \Lambda\}$ *on* $\{\tau < \infty\}$.
(d) $A^o_t = A^o_{t\wedge\tau}$.

PROOF: The equivalence among conditions (a), (b) and (c) is stated in Theorem (5,1) from Jeulin [12]. Implication (a)$\Rightarrow$ (d) comes from analogous arguments as in Azéma [5]. To finish the proof, we show implication (d)$\Rightarrow$(c). Let Λ be the support of the measure dA^o, i.e.,

$$\Lambda = \{(\omega, t)|\ \forall \varepsilon > 0\ A^o_t(\omega) > A^o_{t-\varepsilon}(\omega)\}.$$

The set Λ is optional since A^o is an optional process. Then, $[\![\tau]\!] \subset \Lambda$ and $A^o_t = A^o_{t\wedge\tau}$ imply that indeed τ is the end of Λ on $\{\tau < \infty\}$. □

In the case of honest time, any $\mathbb{F}$-martingale X is a $\mathbb{G}$-semimartingale with (predictable) decomposition (see Jeulin [12][Prop. (5,10)])

$$X_t = \widehat{X}_t + \int_0^{t\wedge\tau} \frac{d\langle X, m\rangle^{\mathbb{F}}_s}{Z_{s-}} - \int_{t\wedge\tau}^t \frac{d\langle X, m\rangle^{\mathbb{F}}_s}{1 - Z_{s-}}, \tag{2.4}$$

where $\widehat{X}$ is a $\mathbb{G}$-local martingale.

We would like to emphasize the role of $\widetilde{Z}$. As we shall see, this process will be important to prove the existence of arbitrage opportunities. We give also a simple characterisation of honest times avoiding $\mathbb{F}$- stopping times.

Lemma 2.4. *A random time* τ *is an honest time and avoids* $\mathbb{F}$-*stopping times if and only if* $Z_\tau = 1$ *a.s. on* $(\tau < \infty)$.

PROOF: Assume that τ is an honest time avoiding $\mathbb{F}$-stopping times. The honesty, by Theorem 2.3, implies that $\widetilde{Z}_\tau = 1$ and the avoiding property implies the continuity of A^o since for each $\mathbb{F}$-stopping time T, $\mathbb{E}(\Delta A^o_T) = \mathbb{P}(\tau = T < \infty) = 0$. Then, the relation $\widetilde{Z} = Z + \Delta A^o$ leads to the result. Assume now that $Z_\tau = 1$ on the set $\{\tau < \infty\}$. Then, on $\{\tau < \infty\}$ we have $1 = Z_\tau \leq \widetilde{Z}_\tau \leq 1$, so $\widetilde{Z}_\tau = 1$ and τ is an honest time. Furthermore, as $\Delta A^o_\tau = \widetilde{Z}_\tau - Z_\tau = 0$, for each $\mathbb{F}$ stopping time T we have

$$\begin{aligned}\mathbb{P}(\tau = T < \infty) &= \mathbb{E}(\mathbb{1}_{\{\tau=T\}}\mathbb{1}_{\{\Delta A^o_\tau=0\}}\mathbb{1}_{(T<\infty)}) \\ &= \mathbb{E}(\int_0^\infty \mathbb{1}_{\{u=T\}}\mathbb{1}_{\{\Delta A^o_u=0\}}dA^o_u) = 0.\end{aligned}$$

So τ avoids $\mathbb{F}$ stopping times. □

3. Some Particular Cases

3.1. *Immersion assumption, density hypothesis*

We recall that the filtration $\mathbb{F}$ is immersed in $\mathbb{G}$ under $\mathbb{Q}$ if any $(\mathbb{Q},\mathbb{F})$-local martingale is a $(\mathbb{Q},\mathbb{G})$-local martingale.

Lemma 3.1. *If the immersion property is satisfied under a probability $\mathbb{Q}$ on $\mathbb{G}$, such that S is a $(\mathbb{Q},\mathbb{F})$-martingale, all the three concepts of NFLVR, NA and NUPBR hold.*

PROOF: Let S be a $(\mathbb{Q},\mathbb{F})$-local martingale, then it is a $(\mathbb{Q},\mathbb{G})$-local martingale as well. □

One says that the random time τ satisfies the positive density hypothesis if there exists a positive $\mathcal{F}_t \otimes \mathcal{B}(\mathbb{R}^+)$-measurable function $(\omega,u) \to \alpha_t(\omega,u)$ which satisfies: for any Borel bounded function φ,

$$\mathbb{E}(\varphi(\tau)|\mathcal{F}_t) = \int_{\mathbb{R}_+} \varphi(u)\alpha_t(u)f(u)du, \quad \mathbb{P}-a.s.$$

where f is the density function of τ. In other terms, the conditional distribution of τ is characterized by the survival probability defined by

$$G_t(\theta) := \mathbb{P}(\tau > \theta|\mathcal{F}_t) = \int_\theta^\infty \alpha_t(u)f(u)du\,.$$

In that case Hypothesis $(\mathcal{H}')$ is satisfied (see Amendinger [3] or Grorud and Pontier [9]).

Lemma 3.2. *If S is a $(\mathbb{P},\mathbb{F})$-martingale and if the conditional law of τ with respect to $\mathbb{F}$ satisfies the positive density hypothesis then NFLVR, NA and NUPBR hold for $\mathbb{G}$*

PROOF: Indeed, under the positive density hypothesis, it can be proved (see Amendinger's thesis [3] and Grorud and Pontier [9]), that the probability $\mathbb{P}^*$, defined on $\mathbb{F}\vee\sigma(\tau)$ as

$$d\mathbb{P}^*|_{\mathcal{F}_t\vee\sigma(\tau)} = \frac{1}{\alpha_t(\tau)}d\mathbb{P}|_{\mathcal{F}_t\vee\sigma(\tau)}$$

satisfies the following assertions

(i) Under $\mathbb{P}^*$, τ is independent from $\mathcal{F}_t$ for any t

(ii) $\mathbb{P}^*|_{\mathcal{F}_t} = \mathbb{P}|_{\mathcal{F}_t}$

(iii) $\mathbb{P}^*|_{\sigma(\tau)} = \mathbb{P}|_{\sigma(\tau)}$

Note that immersion is satisfied under $\mathbb{P}^*$. It is now obvious that, if S is a $(\mathbb{P},\mathbb{F})$-martingale, NFLVR holds in the enlarged filtration $\mathbb{F}\vee\sigma(\tau)$, hence in $\mathbb{G}$. Indeed, the $(\mathbb{P},\mathbb{F})$-martingale S is - using the independence property - an $(\mathbb{P}^*,\mathbb{F}\vee\sigma(\tau))$-martingale, so that S, being $\mathbb{G}$-adapted, is a $(\mathbb{P}^*,\mathbb{G})$-martingale and $\mathbb{P}^*$ is an equivalent martingale measure. If S is only a $(\mathbb{P},\mathbb{F})$-local martingale, then one proceeds as follows. Let $\{T_n\}_{n\in\mathbb{N}}$ be an $\mathbb{F}$-localizing sequence for S, meaning that S^{T_n} is a $(\mathbb{P},\mathbb{F})$-martingale, for every $n\in\mathbb{N}$. Then, repeating previous resoning, it holds that S^{T_n} is a $(\mathbb{P}^*,\mathbb{G})$-martingale. Thus S is a $(\mathbb{P}^*,\mathbb{G})$-local martingale and $\mathbb{P}^*$ is an equivalent martingale measure.[c] □

4. Classical Arbitrages for a Class of Honest Times

Herein, we generalize the results obtained in Fontana et al. [8] – which are established for honest times avoiding $\mathbb{F}$-stopping times in a complete market with continuous filtration – to any complete market and to a much more broader class of honest times that will be defined below. Throughout this section, we denote by $\mathcal{T}_s$ the set of all $\mathbb{F}$-stopping times, $\mathcal{T}_h$ the set of all $\mathbb{F}$-honest times, and $\mathcal{R}$ the set of random times (r.t.) given by

$$\mathcal{R} := \Big\{\tau \ \text{ r.t. } \ \big|\ \exists\ \Gamma\in\mathcal{A} \text{ and } T\in\mathcal{T}_s \text{ such that } \tau = T1\!\!1_\Gamma + \infty 1\!\!1_{\Gamma^c}\Big\}, \quad (4.1)$$

Proposition 4.1. *The following inclusions hold*

$$\mathcal{T}_s\subset\mathcal{R}\subset\mathcal{T}_h. \quad (4.2)$$

PROOF: The first inclusion is clear. For the inclusion $\mathcal{R}\subset\mathcal{T}_h$, we give, for ease of the reader two different proofs. Let us take $\tau\in\mathcal{R}$.
1) On $(\tau<t)=(T<t)\cap\Gamma$, we have $\tau=T\wedge t$ and $T\wedge t$ is $\mathcal{F}_t$-measurable. Thus, τ is an honest time.
2) We want to show that on $(\tau<\infty)$, $\widetilde{Z}_\tau = 1$. Indeed, $\widetilde{Z}_t = 1\!\!1_{(T\geq t)}\mathbb{P}(\Gamma|\mathcal{F}_t)+\mathbb{P}(\Gamma^c|\mathcal{F}_t)$, so that

$$\begin{aligned}1\!\!1_{(\tau<\infty)}\widetilde{Z}_\tau &= 1\!\!1_\Gamma 1\!\!1_{(T<\infty)}\widetilde{Z}_T = 1\!\!1_\Gamma 1\!\!1_{(T<\infty)}(1\!\!1_{(T\geq T)}\mathbb{P}(\Gamma|\mathcal{F}_T)+\mathbb{P}(\Gamma^c|\mathcal{F}_T))\\ &= 1\!\!1_\Gamma 1\!\!1_{(T<\infty)} = 1\!\!1_{(\tau<\infty)}.\end{aligned}$$

This proves that τ is an honest time. □

The following theorem represents our principal result in the general framework.

Theorem 4.2. *Assume that $(S,\mathbb{F})$ is a complete market and let φ be an $\mathbb{F}$-predictable process satisfying $m=1+\varphi\centerdot S$. Then the following assertions*

[c]We thank C. Fontana for the comment on this proof.

hold.
(a) *If τ is an honest time, and $\tau \notin \mathcal{R}$, then the $\mathbb{G}$-predictable process $\varphi^b = \varphi 1\!\!1_{[\![0,\tau]\!]}$ is a classical arbitrage strategy in the market "before τ", i.e., in $(S^\tau, \mathbb{G})$.*
(b) *If τ is an honest time, which is not an $\mathbb{F}$-stopping time, and if $\{\tau = \infty\} \in \mathcal{F}_\infty$, then the $\mathbb{G}$-predictable process $\varphi^a = -\varphi 1\!\!1_{]\!]\tau,\nu]\!]}$, with $\mathbb{G}$-stopping time defined as*

$$\nu := \inf\{t > \tau : \widetilde{Z}_t \leq \frac{1 - \Delta A^o_\tau}{2}\}, \tag{4.3}$$

is a classical arbitrage strategy in the market "after τ", i.e., in $(S - S^\tau, \mathbb{G})$.

PROOF: (a) From $m = \widetilde{Z} + A^o_-$ and $\widetilde{Z}_\tau = 1$, we deduce that $m_\tau \geq 1$. Since $\tau \notin \mathcal{R}$, one has $\mathbb{P}(m_\tau > 1) = \mathbb{P}(A^o_{\tau-} > 0) > 0$. Then, by Lemma 2.2, process $\varphi^b = \varphi 1\!\!1_{[\![0,\tau]\!]}$ is an arbitrage strategy in $(S^\tau, \mathbb{G})$.
(b) From $m = Z + A^o$ and Theorem 2.3 (iv), one obtains that, for $t > \tau$, $m_t - m_\tau = Z_t - Z_\tau \geq -1$. On the other hand, using $m = \widetilde{Z} + A^o_-$, one obtains that, for $t > \tau$, $m_t - m_\tau = \widetilde{Z}_t - 1 + \Delta A^o_\tau$. Assumption $\{\tau = \infty\} \in \mathcal{F}_\infty$ ensures that $\widetilde{Z}_\infty = 1\!\!1_{\{\tau=\infty\}}$ and in particular $\{\tau < \infty\} \subset \{\widetilde{Z}_\infty = 0\}$. So, $\mathbb{G}$-stopping time ν defined in (4.3) satisfies $\{\nu < \infty\} = \{\tau < \infty\}$. Then,

$$m_\nu - m_\tau = \widetilde{Z}_\nu - 1 + \Delta A^o_\tau \leq \frac{\Delta A^o_\tau - 1}{2} \leq 0,$$

and, as τ is not an $\mathbb{F}$-stopping time,

$$\mathbb{P}(m_\nu - m_\tau < 0) = \mathbb{P}(\Delta A^o_\tau < 1) > 0.$$

Hence $-\int_\tau^{t\wedge\nu} \varphi_s dS_s = m_{\tau\wedge t} - m_{t\wedge\nu}$ is the value of an admissible self-financing strategy $\varphi^a = -\varphi 1\!\!1_{]\!]\tau,\nu]\!]}$ with initial value 0 and terminal value $m_\tau - m_\nu \geq 0$ satisfying $\mathbb{P}(m_\tau - m_\nu > 0) > 0$. This ends the proof of the theorem. □

Remark 4.3. We recall that if τ is a finite honest time (hence $\mathcal{F}_\infty$-measurable) and is not an $\mathbb{F}$-stopping time, then the density hypothesis is not satisfied and immersion does not hold. Indeed:

(i) Density hypothesis would hold if, under some equivalent probability measure, τ would be independent from $\mathcal{F}_\infty$.
(ii) The immersion property is equivalent to $\mathbb{P}(\tau > t|\mathcal{F}_t) = \mathbb{P}(\tau > t|\mathcal{F}_\infty)$ which, for a finite honest time is $1\!\!1_{\tau>t}$. Then, one should have $\mathbb{P}(\tau > t|\mathcal{F}_t) = 1\!\!1_{\tau>t}$ and τ would be a stopping time.

Remark 4.4. The completeness of the market is a necessary condition to conclude. See Fontana et al. [8] for a counter example.

4.1. *Classical arbitrage opportunities in a Brownian filtration*

In this subsection, we develop practical market models S and honest times τ within the Brownian filtration for which one can compute explicitly the arbitrage opportunities for both before and after τ. For other examples of honest times, and associated classical arbitrages we refer the reader to Fontana et al. [8] (Note that the arbitrages constructed in that paper are different from our arbitrages). Throughout this subsection, we assume given a one-dimensional Brownian motion W and $\mathbb{F}$ is its augmented natural filtration. The market model is represented by the bank account whose process is the constant one and one stock whose price process is given by

$$S_t = \exp(\sigma W_t - \frac{1}{2}\sigma^2 t), \qquad \sigma > 0 \text{ given.}$$

It is worth mentioning that in this context of Brownian filtration, for any process V with locally integrable variation, its $\mathbb{F}$-dual optional projection is equal to its $\mathbb{F}$-dual predictable projection, i.e., $V^{o,\mathbb{F}} = V^{p,\mathbb{F}}$. All the examples given below correspond to random times that are the end of optional sets, hence are honest times.

4.1.1. *Last passage time at a given level*

Proposition 4.5. *Consider the following random times*

$$\tau := \sup\{t \,:\, S_t = a\} \qquad \text{and} \qquad \nu := \inf\{t > \tau \mid S_t \leq \frac{a}{2}\},$$

where $0 < a < 1$. Then, the following assertions hold.
(a) *The model "before τ" $(S^\tau, \mathbb{G})$ admits a classical arbitrage opportunity given by the $\mathbb{G}$-predictable process*

$$\varphi^b = \frac{1}{a} 1\!\!1_{\{S<a\}} I_{]\!]0,\tau]\!]}.$$

(b) *The model "after τ" $(S - S^\tau, \mathbb{G})$ admits a classical arbitrage opportunity given by $\mathbb{G}$-predictable process*

$$\varphi^a = -\frac{1}{a} 1\!\!1_{\{S<a\}} I_{]\!]\tau,\nu]\!]}.$$

Proof: It is clear that τ is a finite honest time [Theorem 2.3 (c)], and does not belong to the set $\mathcal{R}$ defined in (4.1). Thus τ fulfills the assumptions of Theorem 4.2. We now compute the predictable process φ such that

$m = 1 + \varphi \centerdot S$. To this end, we calculate Z as follows. Using Jeanblanc et al. [11][exercise 1.2.3.10], we derive

$$1 - Z_t = \mathbb{P}\left(\sup_{t<u} S_u \leq a|\mathcal{F}_t\right) = \mathbb{P}\left(\sup_u \widetilde{S}_u \leq \frac{a}{S_t}|\mathcal{F}_t\right) = \Phi\left(\frac{a}{S_t}\right)$$

where $\widetilde{S}_u = \exp(\sigma \widetilde{W}_u - \frac{1}{2}\sigma^2 u)$, $\widetilde{W}$ independent of $\mathcal{F}_t$ and $\Phi(x) = \mathbb{P}\left(\sup_u \widetilde{S}_u \leq x\right) = \mathbb{P}(\frac{1}{U} \leq x) = \mathbb{P}(\frac{1}{x} \leq U) = (1 - \frac{1}{x})^+$, where U is a random variable with uniform law. Thus we get $Z_t = 1 - (1 - \frac{S_t}{a})^+$ (in particular $Z_\tau = \widetilde{Z}_\tau = 1$), and

$$dZ_t = 1\!\!1_{\{S_t \leq a\}} \frac{1}{a} dS_t - \frac{1}{2a} d\ell_t^a$$

where ℓ^a is the local time of the S at the level a (see page 252 of He et al. [10] for the definition of the local time). Therefore, we deduce that

$$m = 1 + \varphi \centerdot S.$$

Note that $\nu := \inf\{t > \tau \mid S_t \leq \frac{a}{2}\} = \inf\{t > \tau \mid 1 - (1 - \frac{S_t}{a})^+ \leq \frac{1}{2}\}$, so ν coincides with (4.3). Theorem 4.2 ends the proof of the proposition. □

4.1.2. *Last passage time at a level before maturity*

Our second example of random time, in this subsection, takes into account finite horizon. In this example, we introduce the following notation

$$H(z, y, s) := e^{-zy} \mathcal{N}\left(\frac{zs - y}{\sqrt{s}}\right) + e^{zy} \mathcal{N}\left(\frac{-zs - y}{\sqrt{s}}\right), \tag{4.4}$$

where $\mathcal{N}(x)$ is the cumulative distribution function of the standard normal distribution.

Proposition 4.6. *Consider the following random time (an honest time)*

$$\tau_1 := \sup\{t \leq 1 \, : \, S_t = b\}$$

where b is a positive real number, $0 < b < 1$. Let V and β be given by

$$V_t := \alpha - \gamma t - W_t \quad \textit{with } \alpha = \frac{\ln b}{\sigma} \textit{ and } \gamma = -\frac{\sigma}{2}$$

$$\beta_t := e^{\gamma V_t} \left(\gamma H(\gamma, |V_t|, 1 - t) + sgn(V_t)\, H'_x(\gamma, |V_t|, 1 - t)\right),$$

with H defined in (4.4), and let ν be as in (4.3). Then, the following assertions hold.

(a) *The model "before τ_1" $(S^{\tau_1}, \mathbb{G})$ admits a classical arbitrage opportunity given by the $\mathbb{G}$-predictable process*

$$\varphi^b := \frac{1}{\sigma S_t} \beta_t I_{[\![0,\tau_1]\!]}.$$

(b) *The model "after τ_1" $(S - S^{\tau_1}, \mathbb{G})$ admits a classical arbitrage opportunity given by $\mathbb{G}$-predictable process*

$$\varphi^a := -\frac{1}{\sigma S_t} \beta_t I_{]\!]\tau_1,\nu]\!]}.$$

PROOF: The proof of this proposition follows from Theorem 4.2 as long as we can write the martingale m as an integral stochastic with respect to S. This is the main focus of the remaining part of this proof. By Theorem 2.3 (c), the time τ_1 is honest and finite. Honest time τ_1 can be seen as

$$\tau_1 = \sup\{t \leq 1 \,:\, \gamma t + W_t = \alpha\} = \sup\{t \leq 1 \,:\, V_t = 0\}.$$

Setting $T_0(V) = \inf\{t \,:\, V_t = 0\}$, we obtain, using standard computations (see Jeanblanc et al. [11] p. 145-148)

$$1 - Z_t = (1 - e^{\gamma V_t} H(\gamma, |V_t|, 1-t)) 1\!\!1_{\{T_0(V) \leq t \leq 1\}} + 1\!\!1_{\{t>1\}},$$

where H is given in (4.4). In particular $Z_\tau = \widetilde{Z}_\tau = 1$. Using Itô's lemma, we obtain the decomposition of $1 - e^{\gamma V_t} H(\gamma, |V_t|, 1-t)$ as a semimartingale. The martingale part of Z is given by $dm_t = \beta_t dW_t = \frac{1}{\sigma S_t} \beta_t dS_t$, which ends the proof. □

4.2. Arbitrage opportunities in a Poisson filtration

Throughout this subsection, we suppose given a Poisson process N, with intensity rate $\lambda > 0$, and natural filtration $\mathbb{F}$. The stock price process is given by

$$dS_t = S_{t-} \psi dM_t, \quad S_0 = 1, \qquad M_t := N_t - \lambda t,$$

or equivalently $S_t = \exp(-\lambda \psi t + \ln(1+\psi) N_t)$, where $\psi > -1$. In what follows, we introduce the notation

$$\alpha := \ln(1+\psi), \qquad \mu := \frac{\lambda \psi}{\ln(1+\psi)} \quad \text{and} \quad Y_t := \mu t - N_t,$$

so that $S_t = \exp(-\ln(1+\psi) Y_t)$. We associate to the process Y its ruin probability, denoted by $\Psi(x)$ given by, for $x \geq 0$,

$$\Psi(x) = \mathbb{P}(T^x < \infty), \qquad \text{with} \quad T^x = \inf\{t : x + Y_t < 0\}. \tag{4.5}$$

Below, we describe our first example of honest time and the associated arbitrage opportunity.

4.2.1. *Last passage time at a given level*

Proposition 4.7. *Suppose that $\psi > 0$ and let $a := -\frac{1}{\alpha}\ln b$ and*

$$\varphi := \frac{\Psi(Y_- - a - 1)1\!\!1_{\{Y_- \geq a+1\}} - \Psi(Y_- - a)1\!\!1_{\{Y_- \geq a\}} + 1\!\!1_{\{Y_- < a+1\}} - 1\!\!1_{\{Y_- < a\}}}{\psi S_-}.$$

For $0 < b < 1$, consider the following random time

$$\tau := \sup\{t : S_t \geq b\} = \sup\{t : Y_t \leq a\}, \tag{4.6}$$

Then the following assertions hold.
(a) *The random time τ is an honest time.*
(b) *The model "before τ" $(S^\tau, \mathbb{G})$ admits a classical arbitrage opportunity given by the $\mathbb{G}$-predictable process $\varphi^b := \varphi I_{[\![0,\tau_1]\!]}$.*
(c) *The model "after τ" $(S - S^\tau, \mathbb{G})$ admits a classical arbitrage opportunity given by the $\mathbb{G}$-predictable process $\varphi^a := -\varphi I_{]\!]\tau_1,\nu]\!]}$, with ν as in (4.3).*

Proof: Since $\psi > 0$, one has $\mu > \lambda$ so that Y goes to $+\infty$ as t goes to infinity, and τ is finite. The supermartingale Z associated with the time τ is

$$Z_t = \Psi(Y_t - a)1\!\!1_{\{Y_t \geq a\}} + 1\!\!1_{\{Y_t < a\}} = 1 + 1\!\!1_{\{Y_t \geq a\}}\left(\Psi(Y_t - a) - 1\right),$$

where Ψ is defined in (4.5) (see Aksamit et al. [2] for more details on this example).

We set $\theta = \dfrac{\mu}{\lambda} - 1$, and deduce that $\Psi(0) = (1+\theta)^{-1}$ (see Asmussen [4]). Define $\vartheta_1 = \inf\{t > 0 : Y_t = a\}$ and then, for each $n > 1$, $\vartheta_n = \inf\{t > \vartheta_{n-1} : Y_t = a\}$. It can be proved that the times $(\vartheta_n)_n$ are $\mathbb{F}$-predictable stopping times (Aksamit et al. [2]). The $\mathbb{F}$-dual optional projection A^o of the process $1\!\!1_{[\![\tau,\infty]\!]}$ equals

$$A^o = \frac{\theta}{1+\theta}\sum_n 1\!\!1_{[\![\vartheta_n,\infty]\!]}.$$

Indeed, for any $\mathbb{F}$-optional process U we have

$$\mathbb{E}(U_\tau) = \mathbb{E}(\sum 1\!\!1_{\{\tau=\vartheta_n\}}U_{\vartheta_n}) = \mathbb{E}(\sum \mathbb{E}(1\!\!1_{\{\tau=\vartheta_n\}}|\mathcal{F}_{\vartheta_n})U_{\vartheta_n})$$

and $\mathbb{E}(1\!\!1_{\{\tau=\vartheta_n\}}|\mathcal{F}_{\vartheta_n}) = \mathbb{P}(T^0 = \infty) = 1 - \Psi(0) = 1 - \frac{1}{1+\theta}$.

As a result the process A^o is predictable, and hence $Z = m - A^o$ is the Doob-Meyer decomposition of Z. Thus we can get

$$\Delta m = Z - {}^pZ$$

where ${}^pZ$ is the $\mathbb{F}$-predictable projection[d] of Z . To calculate ${}^pZ$, we write the process Z in a more adequate form. To this end, we first remark that

$$\mathbb{1}_{\{Y \geq a\}} = \mathbb{1}_{\{Y_- \geq a+1\}}\Delta N + (1 - \Delta N)\mathbb{1}_{\{Y_- \geq a\}}$$
$$\mathbb{1}_{\{Y < a\}} = \mathbb{1}_{\{Y_- < a+1\}}\Delta N + (1 - \Delta N)\mathbb{1}_{\{Y_- < a\}}.$$

Then, we obtain

$$\Delta m = \big(\Psi(Y_- - a - 1)\mathbb{1}_{\{Y_- \geq a+1\}} - \Psi(Y_- - a)\mathbb{1}_{\{Y_- \geq a\}} + \mathbb{1}_{\{Y_- < a+1\}} - \mathbb{1}_{\{Y_- < a\}}\big)\,\Delta N = \psi S_- \varphi \Delta M = \varphi \Delta S.$$

Since the two martingales m and S are purely discontinuous, we deduce that $m = 1 + \varphi \bullet S$. Therefore, the proposition follows from Theorem 4.2. □

4.2.2. *Time of supremum on fixed time horizon*

The second example requires the following notations $S_t^* := \sup_{s \leq t} S_s$, and

$$\Psi(x,t) := \mathbb{P}(S_t^* > x), \quad \widehat{\Phi}(t) := \mathbb{P}(S_t^* \leq 1), \quad \widetilde{\Phi}(x,t) := \mathbb{P}(S_t^* < x) \quad (4.7)$$

Proposition 4.8. *Consider the random time τ defined by*

$$\tau = \sup\{t \leq 1 : S_t = S_t^*\}, \quad (4.8)$$

where $S_t^* = \sup_{s \leq t} S_s$. Then, the following assertions hold.
(a) *The random time τ is an honest time.*
(b) *For $\psi > 0$, define the $\mathbb{G}$-predictable process φ as*

$$\begin{aligned}\varphi_t := &\mathbb{1}_{\{t<1\}}\left[\Psi\left(\max(\tfrac{S_{t-}^*}{S_{t-}(1+\psi)},1), 1-t\right) - \Psi\left(\tfrac{S_{t-}^*}{S_{t-}}, 1-t\right)\right] \\ &+ \mathbb{1}_{\{S_{t-}^* < S_{t-}(1+\psi)\}}\,\widehat{\Phi}(1-t) \\ &+ \left[\mathbb{1}_{\{\max(S_{1-}^*, S_{1-}(1+\psi)) = S_0\}} - \mathbb{1}_{\{\max(S_{1-}^*, S_{1-}) = S_0\}}\right]\mathbb{1}_{\{t=1\}}\,.\end{aligned}$$

Then, $\varphi^b := \varphi\mathbb{1}_{[\![0,\tau]\!]}$ is an arbitrage opportunity for the model $(S^\tau, \mathbb{G})$, and $\varphi^a := -\varphi I_{]\!]\tau,\nu]\!]}$ is an arbitrage opportunity for the model $(S - S^\tau, \mathbb{G})$. Here Ψ and $\widehat{\Phi}$ are defined in (4.7), and ν is defined similarly as in (4.3).
(c) *For $-1 < \psi < 0$, define the $\mathbb{G}$-predictable process*

$$\varphi_t := \frac{\psi I_{\{S_t^* = S_{t-}\}}\Phi(\frac{1}{1+\psi}, 1-t) + \Psi(\widehat{\frac{S_t^*}{S_{t-}(1+\psi)}}, 1-t) - \Psi(\frac{S_t^*}{S_{t-}}, 1-t)}{\psi S_{t-}}.$$

[d]Note that here we talk about predictable projection and not about dual predictable projection.

Then, $\varphi^b := \varphi 1\!\!1_{[\![0,\tau]\!]}$ *is an arbitrage opportunity for the model* $(S^\tau, \mathbb{G})$*, and* $\varphi^a := -\varphi I_{]\!]\tau,\nu]\!]}$ *is an arbitrage opportunity for the model* $(S - S^\tau, \mathbb{G})$.

PROOF: Note that, if $-1 < \psi < 0$ the process S^* is continuous, $S_\tau < S^*_\tau = \sup_{t\in[0,1]} S_t$ on the set $(\tau < 1)$ and $S_{\tau-} = S^*_{\tau-} = \sup_{t\in[0,1]} S_t$. If $\psi > 0$, $S_{\tau-} < S^*_{\tau-} < \sup_{t\in[0,1]} S_t$ on the set $(\tau < 1)$. Define the sets $(E_n)_{n=0}^\infty$ such that $E_0 = \{\tau = 1\}$ and $E_n = \{\tau = T_n\}$ with $n \geq 1$. The sequence $(E_n)_{n=0}^\infty$ forms a partition of Ω. Then, $\tau = 1\!\!1_{E_0} + \sum_{n=1}^\infty T_n 1\!\!1_{E_n}$. Note that τ is not an $\mathbb{F}$ stopping time since $E_n \notin \mathcal{F}_{T_n}$ for any $n \geq 1$.

The supermartingale Z associated with the honest time τ is

$$\begin{aligned} Z_t &= \mathbb{P}(\sup_{s\in(t,1]} S_s > \sup_{s\in[0,t]} S_s|\mathcal{F}_t) = \mathbb{P}(\sup_{s\in[0,1-t]} \widehat{S}_s > \frac{S^*_t}{S_t}|\mathcal{F}_t) \\ &= 1\!\!1_{(t<1)}\Psi(\frac{S^*_t}{S_t}, 1-t), \end{aligned}$$

with $\widehat{S}$ an independent copy of S and $\Psi(x,t)$ is given by (4.7).

As $\{\tau = T_n\} \subset \{\tau \leq T_n\} \subset \{Z_{T_n} < 1\}$, we have

$$Z_\tau = 1\!\!1_{\{\tau=1\}} Z_1 + \sum_{n=1}^\infty 1\!\!1_{\{\tau=T_n\}} Z_{T_n} < 1, \quad \text{and} \;\; \{\widetilde{Z} = 0 < Z_-\} = \emptyset.$$

In the following we will prove assertion (b). Thus, we suppose that $\psi > 0$, and we calculate

$$\begin{aligned} A^o_t &= \mathbb{P}(\tau = 1|\mathcal{F}_1)1\!\!1_{\{t\geq 1\}} + \sum_n \mathbb{P}(\tau = T_n|\mathcal{F}_{T_n})1\!\!1_{\{t\geq T_n\}} \\ &= 1\!\!1_{\{S^*_1 = S_0\, t\geq 1\}} + \sum_n 1\!\!1_{\{T_n<1\, S^*_{T_n-} < S_{T_n}\}}\mathbb{P}(\sup_{s\in[T_n,1[} S_s \leq S_{T_n}|\mathcal{F}_{T_n})1\!\!1_{\{t\geq T_n\}} \\ &= 1\!\!1_{\{S^*_1 = S_0\, t\geq 1\}} + \sum_n 1\!\!1_{\{T_n<1,\, S^*_{T_n-} < S_{T_n-}(1+\psi)\}}\, \widehat{\Phi}(1-T_n)1\!\!1_{\{t\geq T_n\}}, \end{aligned}$$

with $\widehat{\Phi}$ is given by (4.7). As before, we write

$$\begin{aligned} A^o_t &= 1\!\!1_{\{S^*_1=S_0\}}1\!\!1_{\{t\geq 1\}} + \sum_{s\leq t} 1\!\!1_{\{s<1\}}1\!\!1_{\{S^*_{s-}<S_{s-}(1+\psi)\}}\, \widehat{\Phi}(1-s)\Delta N_s \\ &= 1\!\!1_{\{S^*_1=S_0\}}1\!\!1_{\{t\geq 1\}} + \int_0^{t\wedge 1} 1\!\!1_{\{S^*_{s-}<S_{s-}(1+\psi)\}}\, \widehat{\Phi}(1-s)\, dM_s \\ &\quad +\lambda \int_0^{t\wedge 1} 1\!\!1_{\{S^*_{s-}<S_{s-}(1+\psi)\}}\, \hat{\Phi}(1-s) ds. \end{aligned}$$

Remark that we have

$$\begin{aligned} 1\!\!1_{\{S^*_1=S_0\}} = &\left[1\!\!1_{\{\max(S^*_{1-}, S_{1-}(1+\psi))=S_0\}} - 1\!\!1_{\{\max(S^*_{1-}, S_{1-})=S_0\}}\right]\Delta M_1 \\ &+1\!\!1_{\{\max(S^*_{1-}, S_{1-})=S_0\}}. \end{aligned}$$

and

$$\Delta m = \Delta Z + \Delta A^o = Z - {}^p(Z) + \Delta A^o - {}^p(\Delta A^o).$$

Then we re-write the process Z as follows

$$Z = \mathbb{1}_{[\![0,1[\![}\Psi\left(\max(\frac{S^*_-}{S_-(1+\psi)},1),1-t\right)\Delta M + (1-\Delta M)I_{[\![0,1[\![}\Psi\left(\frac{S^*_-}{S_-},1-t\right).$$

This implies that

$$Z - {}^p(Z) = \mathbb{1}_{[\![0,1[\![}\left[\Psi\left(\max(\frac{S^*_-}{S_-(1+\psi)},1),1-t\right) - \Psi\left(\frac{S^*_-}{S_-},1-t\right)\right]\Delta M.$$

Thus by combining all these remarks, we deduce that

$$\Delta m = Z - {}^p(Z) + \Delta A^o - {}^p(\Delta A^o) = \varphi\Delta S,$$

where we used the fact that ${}^p(\widetilde{Z}) = Z_-$ and $\widetilde{Z} = Z + \Delta A^o$. Then, the assertion (b) follows immediately from Theorem 4.2.
Next, we will prove assertion (c). Suppose that $-1 < \psi < 0$, and we calculate

$$\begin{aligned}
A^o_t &= \mathbb{P}(\tau = 1|\mathcal{F}_1)\mathbb{1}_{\{t\geq 1\}} + \sum_n \mathbb{P}(\tau = T_n|\mathcal{F}_{T_n})\mathbb{1}_{\{t\geq T_n\}} \\
&= \mathbb{1}_{\{S^*_1=S_1,\, t\geq 1\}} + \sum_n \mathbb{1}_{\{T_n<1, S^*_{T_n}=S_{T_n-}\}}\mathbb{P}(\sup_{s\in[T_n,1[} S_s < S_{T_n-}|\mathcal{F}_{T_n})\mathbb{1}_{\{t\geq T_n\}} \\
&= \mathbb{1}_{\{S^*_1=S_1, t\geq 1\}} + \sum_n \mathbb{1}_{\{T_n<1, S^*_{T_n}=S_{T_n-}\}}\widetilde{\Phi}(\frac{S_{T_n-}}{S_{T_n}}, 1-T_n)\mathbb{1}_{\{t\geq T_n\}},
\end{aligned}$$

with $\widetilde{\Phi}(x,t)$ is given by (4.7). In order to find the compensator of A^o, we write

$$\begin{aligned}
A^o_t &= \mathbb{1}_{\{S^*_1=S_1\}}\mathbb{1}_{\{t\geq 1\}} + \sum_{s\leq t}\mathbb{1}_{\{s<1\}}\mathbb{1}_{\{S^*_s=S_{s-}\}}\widetilde{\Phi}(\frac{1}{1+\psi}, 1-s)\,\Delta N_s \\
&= \mathbb{1}_{\{S^*_1=S_1\}}\mathbb{1}_{\{t\geq 1\}} + \int_0^{t\wedge 1}\mathbb{1}_{\{S^*_s=S_{s-}\}}\widetilde{\Phi}(\frac{1}{1+\psi}, 1-s)\,dM_s \\
&+ \lambda\int_0^{t\wedge 1}\mathbb{1}_{\{S^*_s=S_{s-}\}}\widetilde{\Phi}(\frac{1}{1+\psi}, 1-s)\,ds.
\end{aligned}$$

As a result, due to the continuity of the process S^*, we get

$$\begin{aligned}
A^o_t - {}^p(A^o)_t &= I_{\{S^*_t=S_{t-}\}}\widetilde{\Phi}(\frac{1}{1+\psi}, 1-t)\Delta M_t, \\
Z_t - {}^pZ_t &= \left[\Psi(\frac{S^*_t}{S_{t-}(1+\psi)}, 1-t) - \Psi(\frac{S^*_t}{S_{t-}}, 1-t)\right]\Delta N_t.
\end{aligned}$$

This implies that

$$\begin{aligned}\Delta m_t &= Z_t - {}^p Z_t + A^o_t - {}^p(A^o)_t \\ &= \left\{\psi I_{\{S^*_t = S_{t-}\}} \widetilde{\Phi}\left(\frac{1}{1+\psi}, 1-t\right) + \Psi\left(\frac{S^*_t}{S_{t-}(1+\psi)}, 1-t\right)\right\} \Delta N_t \\ &\quad - \left\{\Psi\left(\frac{S^*_t}{S_{t-}}, 1-t\right)\right\} \Delta N_t.\end{aligned}$$

Since m and S are purely discontinuous $\mathbb{F}$-local martingales, we conclude that m can be written in the form of

$$m = m_0 + \varphi \cdot S,$$

and the proof of the assertion (c) follows immediately from Theorem 4.2. This ends the proof of the proposition. □

4.2.3. *Time of overall supremum*

Below, we will present our last example of this subsection. The analysis of this example is based on the following three functions. Here $S^* = \sup_s S_s$.

$$\Psi(x) = \mathbb{P}(S^* > x), \quad \widehat{\Phi} = \mathbb{P}(S^* \leq 1), \text{ and } \quad \widetilde{\Phi}(x) = \mathbb{P}(S^* < x). \tag{4.9}$$

Proposition 4.9. *Consider the random time τ given by*

$$\tau = \sup\{t : S_t = S^*_t\}. \tag{4.10}$$

Then, the following assertions hold.
(a) *The random time τ is an honest time.*
(b) *For $\psi > 0$, define the $\mathbb{G}$-predictable process φ as*

$$\varphi_t := \frac{\mathbb{1}_{\{S^*_{t-} < S_{t-}(1+\psi)\}} \widehat{\Phi} + \Psi\left(\max(\frac{S^*_{t-}}{S_{t-}(1+\psi)}, 1\right) - \Psi(\frac{S^*_{t-}}{S_{t-}})}{S_{t-}\psi}.$$

Then, $\varphi^b := \varphi \mathbb{1}_{[\![0,\tau]\!]}$ is an arbitrage opportunity for the model $(S^\tau, \mathbb{G})$, and $\varphi^a := -\varphi I_{]\!]\tau,\nu]\!]}$ is an arbitrage opportunity for the model $(S - S^\tau, \mathbb{G})$. Here Ψ and $\widehat{\Phi}$ are defined in (4.9), and ν is defined in similar way as in (4.3).
(c) *For $-1 < \psi < 0$, define the $\mathbb{G}$-predictable process φ as*

$$\varphi := \frac{\Psi(\frac{S^*}{S_-(1+\psi)}) - \Psi(\frac{S^*}{S_-}) + \mathbb{1}_{\{S^* = S_-\}} \widetilde{\Phi}(\frac{1}{1+\psi})\psi}{\psi S_-}.$$

Then, $\varphi^b := \varphi \mathbb{1}_{[\![0,\tau]\!]}$ is an arbitrage opportunity for the model $(S^\tau, \mathbb{G})$, and $\varphi^a := -\varphi I_{]\!]\tau,\nu]\!]}$ is an arbitrage opportunity for the model $(S - S^\tau, \mathbb{G})$. Here again ν is defined as in (4.3).

PROOF: It is clear that τ is an $\mathbb{F}$ honest time.
Let us note that τ is finite and, as before, if $-1 < \psi < 0$, $S_\tau < S^*_\tau = \sup_t S_t$ and S^* is continuous and if $\psi > 0$, $S_\tau = S^*_\tau = \sup_t S_t$.
The supermartingale Z associated with the honest time τ is

$$Z_t = \mathbb{P}(\sup_{s\in(t,\infty]} S_s > \sup_{s\in[0,t]} S_s|\mathcal{F}_t) = \mathbb{P}(\sup_{s\in[0,\infty]} \widehat{S}_s > \frac{S^*_t}{S_t}|\mathcal{F}_t) = \Psi(\frac{S^*_t}{S_t}),$$

with $\widehat{S}$ an independent copy of S and Ψ is given by (4.9). As a result, we deduce that $Z_\tau < 1$. In the following, we will prove assertion (b). We suppose that $\psi > 0$, denoting by $(T_n)_n$ the sequence of jumps of the Poisson process N, we derive

$$\begin{aligned}
A^o_t &= \sum_n \mathbb{P}(\tau = T_n|\mathcal{F}_{T_n})1\!\!1_{\{t\geq T_n\}} \\
&= \sum_n 1\!\!1_{\{S^*_{T_n-}<S_{T_n}\}}\mathbb{P}(\sup_{s\geq T_n} S_s \leq S_{T_n}|\mathcal{F}_{T_n})1\!\!1_{\{t\geq T_n\}} \\
&= \sum_n 1\!\!1_{\{S^*_{T_n-}<S_{T_n-}(1+\psi)\}}\widehat{\Phi}1\!\!1_{\{t\geq T_n\}},
\end{aligned}$$

with $\widehat{\Phi} = \mathbb{P}(\sup_s S_s \leq 1)$ given by (4.9).

We continue to find compensator of A^o

$$\begin{aligned}
A^o_t &= \sum_{s\leq t} 1\!\!1_{\{S^*_{s-}<S_{s-}(1+\psi)\}}\widehat{\Phi}\Delta N_s \\
&= \int_0^t 1\!\!1_{\{S^*_{s-}<S_{s-}(1+\psi)\}}\widehat{\Phi}dM_s + \lambda\int_0^t 1\!\!1_{\{S^*_{s-}<S_{s-}(1+\psi)\}}\widehat{\Phi}ds.
\end{aligned}$$

Now as we did for the previous propositions, we calculate the jumps of m. To this end, we re-write Z as follows

$$Z = \left[\Psi\left(\max(\frac{S^*_-}{S_-(1+\psi)},1)\right) - \Psi(\frac{S^*_-}{S_-})\right]\Delta M + \Psi(\frac{S^*_-}{S_-}).$$

This implies that

$$Z - {}^pZ = \left[\Psi\left(\max(\frac{S^*_-}{S_-(1+\psi)},1)\right) - \Psi(\frac{S^*_-}{S_-})\right]\Delta M.$$

Hence, we derive

$$\Delta m = \left[1\!\!1_{\{S^*_{s-}<S_{s-}(1+\psi)\}}\widehat{\Phi} + \Psi\left(\max(\frac{S^*_-}{S_-(1+\psi)},1)\right) - \Psi(\frac{S^*_-}{S_-})\right]\Delta M.$$

Since both martingales m and M are purely discontinuous, we deduce that $m = m_0 + \varphi\centerdot S$. Then, the proposition follows immediately from Theorem 4.2.

In the following, we will prove assertion (c). To this end, we suppose that $\psi < 0$, and we calculate

$$\begin{aligned} A_t^o &= \sum_n \mathbb{P}(\tau = T_n | \mathcal{F}_{T_n}) 1\!\!1_{\{t \geq T_n\}} \\ &= \sum_n 1\!\!1_{\{S^*_{T_n} = S_{T_n-}\}} \mathbb{P}(\sup_{s \geq T_n} S_s < S_{T_n-} | \mathcal{F}_{T_n}) 1\!\!1_{\{t \geq T_n\}} \\ &= \sum_n 1\!\!1_{\{S^*_{T_n} = S_{T_n-}\}} \widetilde{\Phi}(\frac{S_{T_n-}}{S_{T_n}}) 1\!\!1_{\{t \geq T_n\}}, \end{aligned}$$

with $\widetilde{\Phi}(x) = \mathbb{P}(\sup_s S_s < x)$. Therefore,

$$\begin{aligned} A_t^o &= \sum_{s \leq t} 1\!\!1_{\{S^*_s = S_{s-}\}} \widetilde{\Phi}(\frac{1}{1+\psi}) \Delta N_s \\ &= \int_0^t 1\!\!1_{\{S^*_s = S_{s-}\}} \widetilde{\Phi}(\frac{1}{1+\psi}) dM_s + \lambda \int_0^t 1\!\!1_{\{S^*_s = S_{s-}\}} \widetilde{\Phi}(\frac{1}{1+\psi}) ds. \end{aligned}$$

Since in the case of $\psi < 0$, the process S^* is continuous, we obtain

$$\begin{aligned} Z - {}^pZ &= \left[\Psi(\frac{S^*}{S_-(1+\psi)}) - \Psi(\frac{S^*}{S_-}) \right] \Delta N, \\ A^o - {}^p(A^o) &= 1\!\!1_{\{S^* = S_-\}} \widetilde{\Phi}(\frac{1}{1+\psi}) \Delta M. \end{aligned}$$

Therefore, we conclude that

$$\begin{aligned} \Delta m &= Z - {}^pZ + A^o - {}^p(A^o) \\ &= \left\{ \Psi(\frac{S^*}{S_-(1+\psi)}) - \Psi(\frac{S^*}{S_-}) + 1\!\!1_{\{S^* = S_-\}} \widetilde{\Phi}(\frac{1}{1+\psi}) \psi \right\} \Delta N. \end{aligned}$$

This implies that the martingale m has the form of $m = 1 + \varphi \cdot S$, and assertion (c) follows immediately from Theorem 4.2, and the proof of the proposition is completed. □

5. Arbitrage Opportunities for Non-honest Random Times

This section is our second main part of the paper. Herein, we develop a number of practical examples of market models and examples of random times that are not honest times and we study the existence of classical arbitrages. This section contains two subsections that treat two different situations.

5.1. *In a Brownian filtration: Emery's example*

We present here an example where τ is a pseudo stopping-time [18]. Let S be defined through $dS_t = \sigma S_t dW_t$, where W is a Brownian motion and σ a constant. Let

$$\tau = \sup\{t \leq 1 \,:\, S_1 - 2S_t = 0\}$$

with $\sup\{\emptyset\} = \infty$, that is the last time before 1 at which the price is equal to half of its terminal value at time 1.

Lemma 5.1. *(a) The random time τ is an $\mathbb{F}$-pseudo-stopping time and the process S^τ is a $\mathbb{G}$-martingale.*
(b) The process S is a $\mathbb{G}$-semimartingale.

PROOF: (a) Note that

$$\{\tau \leq t\} = \left\{\inf_{t\leq s\leq 1} 2S_s \geq S_1\right\} = \left\{\inf_{t\leq s\leq 1} 2\frac{S_s}{S_t} \geq \frac{S_1}{S_t}\right\}$$

Since $\frac{S_s}{S_t}, s \geq t$ and $\frac{S_1}{S_t}$ are independent from $\mathcal{F}_t$,

$$\mathbb{P}\left(\inf_{t\leq s\leq 1} 2\frac{S_s}{S_t} \geq \frac{S_1}{S_t}|\mathcal{F}_t\right) = \mathbb{P}\left(\inf_{t\leq s\leq 1} 2S_{s-t} \geq S_{1-t}\right) = \Phi(1-t)$$

where $\Phi(u) = \mathbb{P}(\inf_{s\leq u} 2S_s \geq S_u)$. It follows that the supermartingale Z is a deterministic decreasing function, hence, τ is a pseudo-stopping time and S is a $\mathbb{G}$-martingale up to time τ.
(b) Let us consider initial enlargement of $\mathbb{F}$ with W_1, namely the filtration $\mathbb{F} \vee \sigma(W_1)$. By Theorem 3 in Jeulin and Yor [14] S is a $\mathbb{G}$-semimartingale as $\int_0^1 \exp\left\{\sigma W_s - \frac{\sigma^2}{2}s\right\}\frac{1}{\sqrt{1-s}}ds < \infty$. The random time τ in an $\mathbb{F}\vee\sigma(W_1)$-honest time as

$$\tau = \sup\{t \leq 1 : W_t + \frac{1}{\sigma}\ln 2 + \frac{1}{2}\sigma(1-t) = W_1\}.$$

Thus, denoting by $\mathcal{G}'_t = \bigcap_{s>t} \mathcal{F}_s \vee \sigma(W_1) \vee \sigma(\tau \wedge s)$, S is $\mathbb{G}'$-semimartingale (this follows from (2.4)). Finally, by Stricker's theorem, as $\mathbb{G} \subset \mathbb{G}'$ and S is $\mathbb{G}$-adapted, we conclude that S is a $\mathbb{G}$-semimartingale. See Aksamit [1] for the details on the decomposition for $\mathbb{F}$-martingales in the case of progressive enlargement with Emery's times. □

Proposition 5.2. *(a) NFLVR property holds in the model "before τ" $(S^\tau, \mathbb{G})$.*
(b) NA and NUPBR properties fail to hold in the model "after τ" $(S - S^\tau, \mathbb{G})$.

PROOF: (a) NFLVR property holds up to τ as S remains a $\mathbb{G}$-martingale up to time τ by Lemma 5.1 (a).

(b) There are obviously classical arbitrages after τ, since, at time τ, one knows the value of S_1 and $S_1 > S_\tau$. In fact, for $t > \tau$, one has $S_t > S_\tau$, and the arbitrage occurs at any time before 1. The arbitrage strategy is given by $\varphi_t = 1\!\!1_{]\!]\tau,1]\!]}(t)$.
The NUPBR condition is not satisfied after τ. Indeed, NUPBR condition is equivalent to the following statement: there does not exist an arbitrage of the first kind, i.e., the random variable $\xi \geq 0$ with $\mathbb{P}(\xi > 0) > 0$ such that for every $x > 0$ there exists a strategy $\theta^x \in \mathcal{A}^x(\mathbb{G})$ satisfying $V(x, \theta^x))_\infty > \xi$. Here it is enough to take the random variable $\xi := \frac{1}{2}S_1$ as an arbitrage of the first kind, since, for $t > \tau$ and $x > 0$, one has, for $\theta = 1\!\!1_{]\!]\tau,1]\!]}$ $x + \int_\tau^1 \theta_s dS_s = x + S_1 - S_\tau = x + \xi > \xi$. □

5.2. *In a Poisson filtration*

This subsection develops similar examples of random times – as in the Brownian filtration of the previous subsection – and shows that the effects of these random times on the market's economic structure differ tremendously from the one of the previous subsection.

In this section, we will work on a Poisson process N with intensity λ and the compensated martingale $M_t = N_t - \lambda t$. Denote

$$T_n = \inf\{t \geq 0 : N_t \geq n\}, \text{ and } H_t^n = 1\!\!1_{\{T_n \leq t\}}, \quad n = 1, 2.$$

The stock price S is described by

$$dS_t = S_{t-}\psi dM_t, \quad \text{where,} \quad \psi > -1, \text{ and } \psi \neq 0,$$

or equivalently, $S_t = S_0 \exp(-\lambda\psi t + \ln(1 + \psi)N_t)$. Then,

$$M_t^1 := H_t^1 - \lambda(t \wedge T_1) := H_t^1 - A_t^1,$$
$$M_t^2 := H_t^2 - (\lambda(t \wedge T_2) - \lambda(t \wedge T_1))) := H_t^2 - A_t^2$$

are two $\mathbb{F}$-martingales. Remark that if $\psi \in (-1, 0)$, between T_1 and T_2, the stock price increases; if $\psi > 0$, between T_1 and T_2, the stock process decreases. This would be the starting point of the existence of arbitrages.

5.2.1. *Convex combination of two jump times*

Below, we present an example of random time that avoids stopping times and the non-arbitrage property fails.

Proposition 5.3. *Consider the random time $\tau = k_1T_1 + k_2T_2$ that avoids $\mathbb{F}$ stopping times, where $k_1 + k_2 = 1$ and $k_1, k_2 > 0$. Then the following properties hold:*
(a) *The random time τ is not an honest time.*
(b) $\widetilde{Z}_\tau = Z_\tau = e^{-\lambda k_1(T_2-T_1)} < 1$, *and* $\{\widetilde{Z} = 0 < Z_-\} = [\![T_2]\!]$.
(c) *There is a classical arbitrage before τ, given by*

$$\varphi_t := -e^{-\lambda \frac{k_2}{k_1}(t-T_1)} \left(\mathbb{1}_{\{N_{t-}\geq 1\}} - \mathbb{1}_{\{N_{t-}\geq 2\}}\right) \frac{1}{\psi S_{t-}} \mathbb{1}_{\{t\leq\tau\}}.$$

(d) *There exist arbitrages after τ: if $\psi \in (-1, 0)$, buy at τ and sell before T_2; if $\psi > 0$, short sell at τ and buy back before T_2.*

PROOF: First, we compute the supermartingale Z:

$$\mathbb{P}(\tau > t|\mathcal{F}_t) = \mathbb{1}_{T_1>t} + \mathbb{1}_{\{T_1\leq t\}}\mathbb{1}_{\{T_2>t\}}\mathbb{P}(k_1T_1 + k_2T_2 > t|\mathcal{F}_t)$$

On the set $E = (T_1 \leq t) \cap (T_2 > t)$, the quantity $\mathbb{P}(k_1T_1 + k_2T_2 > t|\mathcal{F}_t)$ is $\mathcal{F}_{T_1}$-measurable. It follows that, on E,

$$\mathbb{P}(k_1T_1 + k_2T_2 > t|\mathcal{F}_t) = \frac{\mathbb{P}(k_1T_1 + k_2T_2 > t, T_2 > t|\mathcal{F}_{T_1})}{\mathbb{P}(T_2 > t|\mathcal{F}_{T_1})} = e^{-\lambda\frac{k_1}{k_2}(t-T_1)},$$

where we used the independence property of T_1 and $T_2 - T_1$. Therefore, we deduce that,

$$Z_t := \mathbb{P}(\tau > t|\mathcal{F}_t) = \mathbb{1}_{\{T_1>t\}} + \mathbb{1}_{\{T_1\leq t\}}\mathbb{1}_{\{T_2>t\}}e^{-\lambda\frac{k_1}{k_2}(t-T_1)}.$$

Since $Z_t = (1 - H_t^1) + H_t^1(1 - H_t^2)e^{-\lambda\frac{k_1}{k_2}(t-T_1)}$, we deduce, using the fact that $e^{-\lambda(t-T_1)}dH_t^1 = dH_t^1$,

$$\begin{aligned}
dZ_t &= e^{-\lambda\frac{k_1}{k_2}(t-T_1)}(-H^2dH_t^1 - H_t^1dH_t^2) - \lambda\frac{k_1}{k_2}H_t^1(1 - H_t^2)e^{-\lambda\frac{k_1}{k_2}(t-T_1)}dt \\
&= -e^{-\lambda\frac{k_1}{k_2}(t-T_1)}dH_t^2 - \lambda\frac{k_1}{k_2}H_t^1(1 - H_t^2)e^{-\lambda\frac{k_1}{k_2}(t-T_1)}dt \\
&= dm_t - e^{-\lambda\frac{k_1}{k_2}(t-T_1)}dA_t^2 - \lambda\frac{k_1}{k_2}H_t^1(1 - H_t^2)e^{-\lambda\frac{k_1}{k_2}(t-T_1)}dt,
\end{aligned}$$

where

$$dm_t = -e^{-\lambda\frac{k_1}{k_2}(t-T_1)}dM_t^2.$$

Hence

$$m_\tau = 1 - \int_0^\tau e^{-\lambda\frac{k_1}{k_2}(t-T_1)}dM_t^2 = 1 + \int_{T_1}^\tau e^{-\lambda\frac{k_1}{k_2}(t-T_1)}\lambda dt > 1.$$

Now we will start proving the proposition.
i) Since τ avoids stopping times, $Z = \widetilde{Z}$. Note that $\widetilde{Z}_\tau = Z_\tau = e^{-\lambda k_1(T_2-T_1)} < 1$. Hence, τ is not an honest time. Thus, we deduce that both assertions (a) and (b) hold.
ii) Now, we will prove assertion (c). We will describe explicitly the arbitrage strategy. Note that $\{T_2 \leq t\} = \{N_t \geq 2\}$. We deduce that

$$M_t^2 = 1\!\!1_{\{T_2 \leq t\}} - A_t^2 = 1\!\!1_{\{N_t \geq 2\}} - A_t^2 = 1\!\!1_{\{N_{t-} \geq 1\}} \Delta N_t + 1\!\!1_{\{N_{t-} \geq 2\}}(1 - \Delta N_t) - A_t^2.$$

Hence,

$$\begin{aligned} \Delta M_t^2 = M_t^2 - {}^p(M^2)_t &= \left(1\!\!1_{\{N_{t-} \geq 1\}} - 1\!\!1_{\{N_{t-} \geq 2\}}\right) \Delta N_t \\ &= \left(1\!\!1_{\{N_{t-} \geq 1\}} - 1\!\!1_{\{N_{t-} \geq 2\}}\right) \Delta M_t. \end{aligned}$$

Since M^2 and M are both purely discontinuous, we have $m_t = 1 + (\phi \centerdot M)_t = 1 + (\varphi \centerdot S)_t$, where

$$\phi_t = -e^{-\lambda \frac{k_1}{k_2}(t-T_1)} \left(I_{\{N_{t-} \geq 1\}} - I_{\{N_{t-} \geq 2\}}\right), \text{ and } \varphi_t = \phi_t \frac{1}{\psi S_{t-}}.$$

iii) Arbitrages after τ: At time τ, the value of T_2 is known for the one who has $\mathbb{G}$ information. If $\psi > 0$, then the price process decreases before time T_2, however, waiting up time T_2 does not lead to an arbitrage. Setting $\Delta = T_2 - \tau$ (which is known at time τ), there is an arbitrage selling short S at time τ for a delivery at time $\tau + \frac{1}{2}\Delta$. The strategy is admissible, since between T_1 and T_2, the quantity S_t is bounded by $S_0(1 + \varphi)$. This ends the proof of the proposition. □

5.2.2. *Minimum of two scaled jump times*

We give now an example of a non honest random time, which does not avoid $\mathbb{F}$ stopping time and induces classical arbitrage opportunities.

Proposition 5.4. *Consider the same market as before, and define* $\tau = T_1 \wedge aT_2$*, where* $0 < a < 1$ *and* $\beta = \lambda(1/a - 1)$*. Then, the following properties hold:*
(a) τ *is not an honest time and does not avoid* $\mathbb{F}$*-stopping times,*
(b) $Z_\tau = 1\!\!1_{\{T_1 > aT_2\}} e^{-\beta a T_2}(\beta a T_2 + 1) < 1$ *and* $\widetilde{Z}_\tau = e^{-\beta a T_2}(\beta a T_2 + 1) < 1$*, and* $\{\widetilde{Z} = 0 < Z_-\} = \emptyset$.
(c) *There exists a classical arbitrage before* τ *given by*

$$\varphi_t = -e^{-\beta t}(\beta t + 1)\left(1\!\!1_{\{N_{t-} \geq 0\}} - 1\!\!1_{\{N_{t-} \geq 1\}}\right) \frac{1}{\psi S_{t-}} 1\!\!1_{\{t \leq \tau\}}.$$

(d) *There exist arbitrages after τ: if $\psi \in (-1,0)$, buy at τ and sell before τ/a; if $\psi > 0$, short sell at τ and buy back before τ/a.*

PROOF: First, let us compute the supermartingale Z,

$$\begin{aligned} Z_t &= \mathbb{1}_{\{T_1>t\}}\mathbb{P}(aT_2 > t|\mathcal{F}_t) = \mathbb{1}_{\{T_1>t\}}\frac{\mathbb{P}(aT_2 > t, T_1 > t)}{\mathbb{P}(T_1 > t)} \\ &= \mathbb{1}_{\{T_1>t\}}e^{\lambda t}\,\mathbb{E}(\mathbb{1}_{T_1>t}e^{-\lambda(\frac{t}{a}-T_1)^+}) \\ &= \mathbb{1}_{\{T_1>t\}}e^{\lambda t}\int_t^{t/a} e^{-\lambda(\frac{t}{a}-x)}\lambda e^{-\lambda x}dx + \mathbb{1}_{\{T_1>t\}}e^{\lambda t}\int_{t/a}^{\infty}\lambda e^{-\lambda y}dy \\ &= \mathbb{1}_{\{T_1>t\}}e^{-\beta t}(\beta t + 1), \end{aligned}$$

where $\beta = \lambda(1/a - 1)$. In particular $Z_\tau = \mathbb{1}_{\{T_1>aT_2\}}e^{-\beta aT_2}(\beta aT_2 + 1) < 1$. Similar computation as above leads to $\widetilde{Z}_t = Z_{t-} = \mathbb{1}_{\{T_1\geq t\}}e^{-\beta t}(\beta t + 1)$. This proves assertions (a) and (b).
i) Here, we will prove assertion (c). Thanks to Itô's formula, we have

$$\begin{aligned} dZ_t &= -e^{-\beta t}(\beta t + 1)dH_t^1 - \mathbb{1}_{t\leq T_1}\beta^2 e^{-\beta t}t dt \\ &= -e^{-\beta t}(\beta t + 1)dM_t^1 - e^{-\beta t}(\beta t + 1)dA_t^1 - \mathbb{1}_{t\leq T_1}\beta^2 e^{-\beta t}t dt. \end{aligned}$$

Therefore,

$$dm_t = -e^{-\beta t}(\beta t + 1)dM_t^1.$$

Hence

$$\begin{aligned} m_\tau = 1 &+ \mathbb{1}_{\{aT_2<T_1\}}\lambda\left(\frac{2(1-e^{-\beta aT_2})}{\beta} - aT_2e^{-\beta aT_2}\right) \\ &+ \mathbb{1}_{\{T_1<aT_2\}}\left(2\lambda\frac{1-e^{-\beta T_1}}{\beta} - \lambda T_1e^{-\beta T_1} - T_1\beta e^{-\beta T_1} - e^{-\beta T_1} + 1\right) \end{aligned}$$

and, using the fact that when $x > 0$, $1 - e^{-x} - xe^{-x} > 0$ and $2\lambda(1 - e^{-x}) - \lambda xe^{-x} - x\beta e^{-x} - \beta e^{-x} + 1 > 0$, one obtains $m_\tau > 1$; hence the existence of classical arbitrages.

Now, we describe explicitly the arbitrage strategy. Notice that $\{T_1 \leq t\} = \{N_t \geq 1\}$. We deduce that

$$M_t^1 = \mathbb{1}_{\{T_1\leq t\}} - A_t^1 = \mathbb{1}_{\{N_t\geq 1\}} - A_t^1 = \mathbb{1}_{\{N_{t-}\geq 0\}}\Delta N_t + \mathbb{1}_{\{N_{t-}\geq 1\}}(1-\Delta N_t) - A_t^1.$$

Hence,

$$\Delta M_t^1 = M_t^1 - {}^p(M^1)_t = \left(I_{\{N_{t-}\geq 0\}} - I_{\{N_{t-}\geq 1\}}\right)\Delta M_t.$$

Since M^1 and M are both purely discontinuous, we have $m = 1 + \varphi \bullet S$, where

$$\varphi_t = -e^{-\beta t}(\beta t + 1)\left(\mathbb{1}_{\{N_{t-}\geq 0\}} - \mathbb{1}_{\{N_{t-}\geq 1\}}\right)\frac{1}{\psi S_{t-}}.$$

ii) The proof of assertion (d) follows the same proof of assertion (d) of Proposition 5.6. This ends the proof of the proposition. □

6. NUPBR for Particular Models

In this section, we address some interesting practical models, for which we prove that the NUPBR remains valid up to τ. The originality of this part – as we mentioned in the introduction and the abstract – lies in the simplicity of the proof. A general and complete analysis about the NUPBR is addressed in full generality in Choulli et al. [6]. Throughout this section, we will assume that $Z > 0$.

6.1. *Before τ*

Let $\widehat{m}$ be the $\mathbb{G}$-martingale stopped at time τ associated with m by (2.3), on $\{t \leq \tau\}$

$$\widehat{m}_t := m_t^\tau - \int_0^t \frac{d\langle m, m\rangle_s^{\mathbb{F}}}{Z_s}.$$

6.1.1. *Case of continuous filtration*

We start with the particular case of continuous martingales and prove that, for any random time τ, NUPBR holds before τ.

We note that the continuity assumption implies that the martingale part of Z is continuous and that the optional and Doob-Meyer decompositions of Z are the same.

Proposition 6.1. *Assume that all $\mathbb{F}$-martingales are continuous. Then, for any random time τ, NUPBR holds before τ. A $\mathbb{G}$-local martingale deflator for S^τ is given by $dL_t = -\frac{L_t}{Z_t}d\widehat{m}_t$.*

PROOF: We make a use of Theorem 2.1 and we provide a $\mathbb{G}$-local martingale deflator for S^τ. Define the positive $\mathbb{G}$-local martingale L as $dL_t = -\frac{L_t}{Z_t}d\widehat{m}_t$. Then, if SL is a $\mathbb{G}$-local martingale, NUPBR holds. Recall that, using (2.3) again,

$$\widehat{S}_t := S_t^\tau - \int_0^{t\wedge\tau} \frac{d\langle S, m\rangle_s^{\mathbb{F}}}{Z_s}$$

is a $\mathbb{G}$-local martingale. From integration by parts, we obtain (using that the bracket of continuous martingales does not depend on the filtration)

$$\begin{aligned} d(LS^\tau)_t &= L_t dS_t^\tau + S_t dL_t + d\langle L, S^\tau\rangle_t^{\mathbb{G}} \\ &\overset{\mathbb{G}-\text{mart}}{=} L_t \frac{1}{Z_t} d\langle S, m\rangle_t^{\mathbb{F}} + \frac{1}{Z_t} L_t d\langle S, \widehat{m}\rangle_t^{\mathbb{G}} \\ &\overset{\mathbb{G}-\text{mart}}{=} L_t \frac{1}{Z_t} \left(d\langle S, m\rangle_t - d\langle S, m\rangle_t\right) = 0 \end{aligned}$$

where $X \overset{\mathbb{G}-\text{mart}}{=} Y$ is a notation for $X - Y$ is a $\mathbb{G}$-local martingale. □

Remark 6.2. If τ is an honest time and Predictable Representation Property holds with respect to S then, as a consequence of Theorem 4.2, the NA condition does not hold, hence NFLVR condition does not hold neither. That in turn implies that all the $\mathbb{G}$-local martingale deflators for S^τ are strict $\mathbb{G}$-local martingales.

6.1.2. *Case of a Poisson filtration*

We assume that S is an $\mathbb{F}$-martingale of the form $dS_t = S_{t-}\psi_t dM_t$, where ψ is a predictable process, satisfying $\psi > -1$ and $\psi \neq 0$, where M is the compensated martingale of a standard Poisson process.

In a Poisson setting, from Predictable Representation Property, $dm_t = \nu_t dM_t$ for some $\mathbb{F}$-predictable process ν, so that, on $t \leq \tau$,

$$d\widehat{m}_t = dm_t - \frac{1}{Z_{t-}} d\langle m, m\rangle_t = dm_t - \frac{1}{Z_{t-}} \lambda \nu_t^2 dt$$

Proposition 6.3. *In a Poisson setting, for any random time τ, NUPBR holds before τ since*

$$L = \mathcal{E}\left(-\frac{1}{Z_- + \nu} \centerdot \widehat{m}\right) = \mathcal{E}\left(-\frac{\nu}{Z_- + \nu} \centerdot \widehat{M}\right),$$

is a $\mathbb{G}$-local martingale deflator for S^τ.

PROOF: We make a use of Theorem 2.1 and we are looking for a $\mathbb{G}$-local martingale deflator of the form $dL_t = L_{t-}\kappa_t d\widehat{m}_t$ (and $\psi_t \kappa_t > -1$) so that

L is positive and $S^\tau L$ is a $\mathbb{G}$-local martingale. Integration by parts formula leads to (on $t \le \tau$)

$$\begin{aligned} d(LS)_t &= L_{t-}dS_t + S_{t-}dL_t + d[L,S]_t \\ &\overset{\mathbb{G}-\text{mart}}{=} L_{t-}S_{t-}\psi_t \frac{1}{Z_{t-}} d\langle M, m\rangle_t + L_{t-}S_{t-}\kappa_t \psi_t \nu_t dN_t \\ &\overset{\mathbb{G}-\text{mart}}{=} L_{t-}S_{t-}\psi_t \frac{1}{Z_{t-}} \nu_t \lambda dt + L_{t-}S_{t-}\kappa_t\psi_t\nu_t\lambda(1+\frac{1}{Z_{t-}}\nu_t)dt \\ &= L_{t-}S_{t-}\psi_t\nu_t\lambda\left(\frac{1}{Z_{t-}} + \kappa_t(1+\frac{1}{Z_{t-}}\nu_t)\right)dt. \end{aligned}$$

Therefore, for $\kappa_t = -\frac{1}{Z_{t-}+\nu_t}$, one obtains a deflator. Note that

$$dL_t = L_{t-}\kappa_t d\widehat{m}_t = -L_{t-}\frac{1}{Z_{t-}+\nu_t}\nu_t d\widehat{M}_t$$

is indeed a positive $\mathbb{G}$-local martingale, since $\frac{1}{Z_{t-}+\nu_t}\nu_t < 1$. □

Remark 6.4. If τ is an honest time and Predictable Representation Property holds with respect to S then all the $\mathbb{G}$-local martingale deflators for S^τ are strict $\mathbb{G}$-local martingales.

6.1.3. *Lévy processes*

Assume that $S = \psi \star (\mu - \nu)$ where μ is the jump measure of a Lévy process and ν its compensator. Here, $\psi \star (\mu - \nu)$ is the process $\int_0^\cdot \int \psi(x,s)(\mu(dx,ds) - \nu(dx,ds)$. The martingale m admits a representation as $m = \psi^m \star (\mu - \nu)$. Then, using (2.3), the $\mathbb{G}$-compensator of μ is $\nu^{\mathbb{G}}$ where

$$\nu^{\mathbb{G}}(dt,dx) = \frac{1}{Z_{t-}}\left(Z_{t-} + \psi^m(t,x)\right)\nu(dt,dx)$$

i.e., S admits a $\mathbb{G}$-semimartingale decomposition of the form

$$S = \psi \star (\mu - \nu^{\mathbb{G}}) - \psi \star (\nu - \nu^{\mathbb{G}})$$

Proposition 6.5. *Consider the positive $\mathbb{G}$-local martingale*

$$L := \mathcal{E}\left(-\frac{\psi^m}{Z_- + \psi^m} I_{]\!]0,\tau]\!]} \star (\mu - \nu^{\mathbb{G}})\right).$$

Then L is a $\mathbb{G}$-local martingale deflator for S^τ, and hence S^τ satisfies NUPBR.

PROOF: We make a use of Theorem 2.1 and our goal is to find a positive $\mathbb{G}$-local martingale L of the form

$$dL_t = L_{t-}\kappa_t d\widehat{m}_t$$

so that LS^τ is a $\mathbb{G}$-local martingale.

From integration by parts formula

$$\begin{aligned} d(SL) &\overset{\mathbb{G}-\text{mart}}{=} -L_-\psi \star (\nu - \nu^{\mathbb{G}}) + d[S,L] = -L_-\psi \star (\nu - \nu^{\mathbb{G}}) + L_-\psi\psi^m\kappa \star \mu \\ &\overset{\mathbb{G}-\text{mart}}{=} -L_-\psi \star (\nu - \nu^{\mathbb{G}}) + L_-\psi\psi^m\kappa \star \nu^{\mathbb{G}} \\ &= \quad -L_-\psi\left(1 - (1+\psi^m\kappa)\frac{1}{Z_-}(Z_- + \psi^m)\right) \star \nu \end{aligned}$$

Hence the possible choice $\kappa = -\frac{1}{Z_- + \psi^m}$. It can be checked that indeed, L is a positive $\mathbb{G}$-local martingale (see Choulli et al. [6]). □

6.2. *After τ*

We now assume that τ is an honest time, which satisfies $Z_\tau < 1$ (for integrability reasons). This condition and Lemma 2.4 imply in particular that τ does not avoids $\mathbb{F}$-stopping times. For the further discussion on the condition $Z_\tau < 1$ we refer the reader to Aksamit et al. [2]. Note also that, in the case of continuous filtration, and $Z_\tau = 1$, NUPBR fails to hold after τ (see Fontana et al. [8]).

In view of (2.4), for any $\mathbb{F}$-martingale X (in particular for m and S)

$$\widehat{X}_t := X_t - \int_0^{t\wedge\tau} \frac{d\langle X, m\rangle_s^{\mathbb{F}}}{Z_s} + \int_{t\wedge\tau}^t \frac{d\langle X, m\rangle_s^{\mathbb{F}}}{1 - Z_s}$$

is a $\mathbb{G}$-local martingale.

6.2.1. *Case of continuous filtration*

We start with the particular case of continuous martingales and prove that, for any honest time τ such that $Z_\tau < 1$, NUPBR holds after τ.

Proposition 6.6. *Assume that τ is an honest time, which satisfies $Z_\tau < 1$ and that all $\mathbb{F}$-martingales are continuous. Then, for any honest time τ, NUPBR holds after τ. A $\mathbb{G}$-local martingale deflator for $S - S^\tau$ is given by $dL_t = -\frac{L_t}{1-Z_t} d\widehat{m}_t$.*

PROOF: We use Theorem 2.1 as usual. The proof is based on Itô's calculus. Looking for a $\mathbb{G}$-local martingale deflator of the form $dL_t = L_t\kappa_t d\widehat{m}_t$, and using the integration by parts formula, we obtain that, for $\kappa = -(1-Z)^{-1}$, the process $L(S - S^\tau)$ is a $\mathbb{G}$-local martingale. □

Remark 6.7. If Predictable Representation Property holds with respect to S then, as a consequence of Theorem 4.2, the NA condition does not hold, hence NFLVR condition does not hold neither. That in turn implies that all the $\mathbb{G}$-local martingale deflators for $S - S^\tau$ are strict $\mathbb{G}$-local martingales.

6.2.2. *Case of a Poisson filtration*

We assume that S is an $\mathbb{F}$-martingale of the form $dS_t = S_{t-}\psi_t dM_t$ with ψ is a predictable process, satisfying $\psi > -1$.

The decomposition formula (2.4) reads after τ as

$$\widehat{S}_t = (\mathbb{1}_{]\tau,\infty[} \cdot S)_t + \lambda \int_{t\vee\tau}^{t} \frac{1}{1-Z_{s-}} \nu_s \psi_s S_{s-} ds.$$

Proposition 6.8. *Let $\mathbb{F}$ be a Poisson filtration and τ be an honest time satisfying $Z_\tau < 1$. Then NUPBR holds after τ since*

$$L = \mathcal{E}\left(\frac{1}{1-Z_- - \nu} \bullet \widehat{m}\right) = \mathcal{E}\left(\frac{\nu}{1-Z_- - \nu} \mathbb{1}_{]\tau,\infty[} \bullet \widehat{M}\right),$$

is a $\mathbb{G}$-local martingale deflator for $S - S^\tau$.

PROOF: We make a use of Theorem 2.1 and we are looking for a $\mathbb{G}$-local martingale deflator of the form $dL_t = L_{t-}\kappa_t d\widehat{m}_t$ (and $\psi_t\kappa_t > -1$) so that L is positive $\mathbb{G}$-local martingale and $(S - S^\tau)L$ is a $\mathbb{G}$-local martingale. Integration by parts formula leads to

$$\begin{aligned} d(L(S-S^\tau))_t &= L_{t-}d(S-S^\tau)_t + (S_{t-} - S^\tau_{t-})dL_t + d[L, S - S^\tau]_t \\ &\overset{\mathbb{G}-\text{mart}}{=} -\lambda L_{t-}S_{t-}\nu_t\psi_t \frac{1}{1-Z_{t-}} \mathbb{1}_{\{t>\tau\}}dt + L_{t-}S_{t-}\kappa_t\psi_t\nu_t\mathbb{1}_{\{t>\tau\}}dN_t \\ &\overset{\mathbb{G}-\text{mart}}{=} \lambda L_{t-}S_{t-}\psi_t\nu_t\mathbb{1}_{\{t>\tau\}}\left(-\frac{1}{1-Z_{t-}} + \kappa_t(1 - \frac{1}{1-Z_{t-}}\nu_t)\right) dt. \end{aligned}$$

Therefore, for $\kappa_t = \frac{1}{1-Z_{t-}-\nu_t}$, one obtains a $\mathbb{G}$-local martingale deflator. Note that

$$dL_t = L_{t-}\kappa_t d\widehat{m}_t = L_{t-}\frac{1}{1-Z_{t-} - \nu_t}\nu_t\mathbb{1}_{\{t>\tau\}}d\widehat{M}_t$$

is indeed a positive $\mathbb{G}$-local martingale, since $\frac{1}{1-Z_{t-}-\nu_t}\nu_t\Delta N_t > -1$. □

Remark 6.9. If Predictable Representation Property holds with respect to S, then all the $\mathbb{G}$-local martingale deflators for $S - S^\tau$ are strict $\mathbb{G}$-local martingales.

6.2.3. *Lévy processes*

Assume that $S = \psi\star(\mu-\nu)$ where μ is the jump measure of a Lévy process and ν its $\mathbb{F}$-compensator. Then, by (2.4), the $\mathbb{G}$-compensator of μ is $\nu^{\mathbb{G}}$ where

$$\nu^{\mathbb{G}}(dt,dx) = \left(1 + 1\!\!1_{\{t\leq\tau\}}\frac{1}{Z_{t-}}\psi^m(t,x) - 1\!\!1_{\{t>\tau\}}\frac{1}{1-Z_{t-}}\psi^m(t,x)\right)\nu(dt,dx)$$

i.e., S admits a $\mathbb{G}$-semimartingale decomposition of the form

$$S = \psi\star(\mu-\nu^{\mathbb{G}}) - \psi\star(\nu-\nu^{\mathbb{G}})$$

Proposition 6.10. *Assume that τ be an honest time satisfying $Z_\tau < 1$ in a Lévy framework. Then, the positive $\mathbb{G}$-local martingale*

$$L := \mathcal{E}\left(\frac{\psi^m}{1-Z_- - \psi^m} I_{]\!]\tau,\infty[\![} \star (\mu - \nu^{\mathbb{G}})\right),$$

is a $\mathbb{G}$-local martingale deflator for $S - S^\tau$, and hence $S - S^\tau$ satisfies NUPBR.

Proof: We use Theorem 2.1 again. Our goal is to find a positive $\mathbb{G}$-local martingale L of the form

$$dL_t = L_{t-}\kappa_t 1\!\!1_{\{t>\tau\}}d\widehat{m}_t$$

so that $L(S - S^\tau)$ is a $\mathbb{G}$-local martingale.

From the integration by parts formula

$$\begin{aligned}
d(L(S-S^\tau)) &\overset{\mathbb{G}-\text{mart}}{=} -L_-d(S-S^\tau) + d[S,L] \\
&= -L_-\psi\frac{\psi^m}{1-Z_-}1\!\!1_{]\tau,\infty[}\star\nu + L_-\kappa\psi\psi^m 1\!\!1_{]\tau,\infty[}\star\mu \\
&\overset{\mathbb{G}-\text{mart}}{=} -L_-\psi\frac{\psi^m}{1-Z_-}1\!\!1_{]\tau,\infty[}\star\nu + L_-\kappa\psi\psi^m 1\!\!1_{]\tau,\infty[}\star\nu^{\mathbb{G}} \\
&= -L_-\psi\psi^m 1\!\!1_{]\tau,\infty[}\left(-\frac{1}{1-Z_-} + \kappa(1-\frac{\psi^m}{1-Z_-})\right)\star\nu
\end{aligned}$$

Hence the possible choice $\kappa = \frac{1}{1-Z_- - \psi^m}$. □

Conclusions

In this paper we have treated the question whether the no-arbitrage conditions are stable with respect to progressive enlargement of filtration. We focused on two components of No Free Lunch with Vanishing Risk concept, namely on No Arbitrage Opportunity and No Unbounded Profit with

Bounded Risk. The problem was divided into stability before and after random time containing extra information.

The question regarding No Arbitrage Opportunity condition was answered in the case of Brownian filtration and Poissonian filtration for special case of honest time, moreover particular examples of non-honest times were described. Both, Brownian and Poissonian filtrations possess the important, and crucial from our problem point of view, property of Predictable Representation Property. One may further investigate similar problem without assuming market completness. One may as well consider other examples/classes of non-honest random times.

Afterwards, we handled the stability of NUPBR concept in some very particular situations, namely in continuous martingale case, standard Poisson process case and Lévy process case. We provided results with simple proofs in those particular situations. We emphasize again that in full generality the problem is solved in Choulli et al. [6] revealing as well results within progressive enlargement of filtration theory.

Combining results on NA and NUPBR conditions, we concluded (in Remarks 6.2, 6.4, 6.7, 6.9) that some $\mathbb{G}$-local martingales are in fact $\mathbb{G}$-strict local martingales. That provides a way to construct strict local martingale in enlarged Brownian and Poissonian filtrations.

A.1. Appendix

Let $(A_t, t \geq 0)$ be an integrable increasing process (not necessarily $\mathbb{F}$-adapted). There exists a unique integrable $\mathbb{F}$-optional increasing process $(A^o_t, t \geq 0)$, called the dual optional projection of A such that

$$\mathbb{E}\left(\int_{[0,\infty[} U_s dA_s\right) = \mathbb{E}\left(\int_{[0,\infty[} U_s dA^o_s\right)$$

for any positive $\mathbb{F}$-optional process U.
There exists a unique integrable $\mathbb{F}$-predictable increasing process $(A^p_t, t \geq 0)$, called the dual predictable projection of A such that

$$\mathbb{E}\left(\int_{[0,\infty[} U_s dA_s\right) = \mathbb{E}\left(\int_{[0,\infty[} U_s dA^p_s\right)$$

for any positive $\mathbb{F}$-predictable process U.

Acknowledgements

This research benefited from the support of the "Chair Markets in Transition", under the aegis of Louis Bachelier laboratory, a joint initiative of Ecole Polytechnique, Université d'Evry Val d'Essonne and Fedération Bancaire Française.

References

[1] Aksamit, A., *Random times, enlargement of filtration and arbitrages*, PhD Thesis, Université d'Evry, 2014.
[2] Aksamit, A., Choulli, T., and Jeanblanc, M. *Thin random times and their applications to finance*, Working paper, 2013.
[3] Amendinger, J., *Initial enlargement of filtrations and additional information in financial markets*, PhD Thesis, Technischen Universität Berlin, 1999.
[4] Asmussen, S., *Ruin Probability*, World Scientific, 2000.
[5] Azéma, J., *Quelques applications de la théorie générale des processus. I*, Inventiones Mathematicae 18, pp. 293–336, Springer, 1972.
[6] Choulli T., Aksamit A., Deng J., and Jeanblanc M. *Non-arbitrage up to random horizon and after honest times for semimartingale models* Preprint, 2013. http://arxiv.org/abs/1310.1142
[7] Delbaen, F., Schachermayer, W., *A general version of the fundamental theorem of asset pricing*, Mathematische Annalen, 300: 463-520, 1994.
[8] Fontana, C., Jeanblanc, M. and Song, S. *On arbitrages arising with honest times.* To appear in Finance and Stochastics.
[9] Grorud, A., and Pontier, M., *Insider trading in a continuous time market model*, International Journal of Theoretical and Applied Finance, 1, pp. 331-347, 1998.
[10] He, S. W., Wang, C. K., Yan, J. A., *Semimartingale theory and stochastic calculus*, CRC Press, 1992.
[11] Jeanblanc, M., Yor, M., and Chesney, M., *Mathematical Methods for Financial Markets*, Springer, 2009.
[12] Jeulin, T., *Semi-Martingales et Grossissement d'une Filtration*, Lecture Notes in Mathematics, vol. 833, Springer, Berlin - Heidelberg - New York, 1980.
[13] Jeulin, T., and Yor, M., *Grossissement d'une filtration et semi-martingales : formules explicites.* In C. Dellacherie, P-A. Meyer, and M. Weil, editors, *Séminaire de Probabilités XII*, volume 649 of *Lecture Notes in Mathematics*, pp. 78–97, Springer-Verlag, 1978.
[14] Jeulin, T., and Yor, M., *Inégalité de Hardy, semimartingales, et faux-amis* Séminaire de Probabilités XIII, pp. 332-359, Springer Berlin Heidelberg, 1979.
[15] Kabanov, Y., *On the FTAP of Kreps-Delbaen-Schachermayer*, In: Kabanov, Y. et al. (eds.): Statistics and Control of Stochastic Processes: The Liptser Festschrift, pp. 191–203, World Scientific, Singapore, 1997.

[16] Karatzas, I., and Kardaras, C., *The numeraire portfolio in semimartingale financial models*, Finance and Stochastics 11, pp. 447–493, 2007.

[17] Kardaras, C., *Finitely additive probabilities and the fundamental theorem of asset pricing* In: Chiarella, C., Novikov, A. (eds.): Contemporary Quantitative Finance: Essays in Honour of Eckhard Platen, pp. 19–34, Springer, Berlin Heidelberg, 2010.

[18] Nikeghbali, A., and Yor, M., *A definition and some characteristic properties of pseudo-stopping times*, Annals of Probability, 33(5), pp. 1804-1824, 2005.

[19] Takaoka, K., *A note on the condition of no unbounded profit with bounded risk*, To appear in Finance and Stochastics, 2013.

PART 2
Credit Risk

Pricing Credit Derivatives with a Structural Default Model

Sébastien Hitier*, Ying Zhu†

Abstract. When considering credit risk in computational finance, it is customary to refer to either of two approaches: structural or reduced form models. While the latter is adopted by practitioners for its direct applicability to derivatives pricing, the former is used where a causal link between default and firm's assets is needed.

This entails that structural models are both more difficult and more rewarding to put in place. They are used in a broader range of application such as: optimal capital structure of firm's assets between equity and debt, estimation of default frequency without credit market data, or default correlation from other financial data.

We show in this paper how a structural model can be calibrated and used for risk neutral derivatives pricing, as is usually done for reduced form models, and show what the properties of this model are when pricing more advanced derivatives such as CDS options.

The main steps are to show what the variables of interest are for CDS and CDS option calibration and pricing: we show how the forward survival probability and progressive conditional probability can be derived from the SDE satisfied by the survival probability; and show applications to CDS pricing and the implied CDS option volatility surface.

1. Background on Credit Risk Models

1.1. *Reasons for success of the reduced form model*

The reduced form model in general uses a totally inaccessible stopping time to model default (see Bielecki and Rutkowski [1]). As the most robust model is often achieved by the most parsimonious hypothesis, a simple reduced form model where default is modeled as the first jump of an exogenous Poisson process provides the most natural calibration to a market price of credit risk charged as an insurance premium per year.

*Quantitative Research, BNP Paribas, email: sebastien.hitier@bnpparibas.com
†student of Ecole Polytechnique - X2010

In such case, the reduced form models do not consider the relation between default and firm value. Default is not related to basic market observables but comes from an exogenous variable that is independent of all the default free market information.

Reduced form models have from the start been devised with the intention of pricing derivatives. The focus of the approach presented in 1996 by Madan and Unal [9] is to show how credit curves calibrated on bonds and possibly other derivatives can be used to strip default probabilities and reprice other bonds. The publication of such tractable methods as reduced form model coincided with the creation of the CDS market, for which such models became market standard, because default intensity maps itself so well to the cost of market protection.

1.2. *Versatility of the structural models*

The structural model has a more colorful history in comparison. This approach models default as the passage of a firm's asset process below a certain barrier. Owing to 1973 Merton's [10] work, credit structural models are amongst the earliest models in modern mathematical finance. His approach considers the firm's equity as a call option on the firm's asset. While this work is an important stage in the academic tradition concerned with optimal corporation capital structure, as pioneered in 1958 by Modigliani and Miller [11] and pursued in 1994 by Leland [7], it is focused on the relative value of corporate debt and equity, and not on the replication of credit derivatives using such debt. Structural default models provide a link between the credit quality of a firm and the firm's economic and financial conditions. This makes this model more intuitive. Such intuition behind structural model has helped with the definition of multi-credit setup as in Vasicek's [15] model, or for cross-asset modeling such as debt-equity in Finkelstein et al. [5].

1.3. *Objectives of this paper*

An observer will discover the relative dearth of references dealing with replication of credit derivatives using a structural model. This is a niche for which our paper is intended. Looking back at Merton's [10] model, a firm defaults if, at the time of servicing the debt, its assets are below its outstanding debt. An extension more consistent with observed market CDS price was later given by Black and Cox [2]. In this approach, a default occurs as soon as firm's asset value falls below a certain threshold. In contrast to the Merton approach, a default may occur at any time.

In the context of our structural (firm value) model in the following paragraphs, default is caused by the value of the firm going below its debt level. Mathematically, the default time is modeled as the first hitting time of a sufficiently smooth deterministic barrier by a continuous stochastic process (a Brownian motion). The question of the implied density of hitting time, given a time varying deterministic barrier, has been studied in many prior works, such as: Novikov [12], Giorno et al. [6], and more recently Peskir [13], and Cheng et al. [3]. As we are concerned here with repricing correctly the CDS market, rather than deriving a closed-form solution, we focused on tractable numerical method based on Fokker-Plank and Kolmogorov equation. What we show here is a way to calibrate a structural model solely to the credit market and a numerical application for derivatives pricing.

The paper is organised as follows: we start with a model description, where the default time is modeled as the hitting time of a normalised asset process. In the next section, using the Feynman-Kac formula, we derive the Kolmogorov and Fokker-Plank equations, which give the forward survival probability and progressive conditional probability needed to price credit derivatives. From these two variables of interest, we show how to calculate the forward spread and risky annuity and further obtain the ATM CDS option volatility surfaces.

2. Structural Model Setup

2.1. *Model setup*

In a structural model, default time τ is defined as the hitting time of a normalised asset process X. The default time and asset (distance to default) process X are defined as:

$$\begin{aligned} \tau &:= \inf \{t \in \mathbb{R}_+, X_t \leq b(t)\}, \\ dX_t &:= \sigma_t \, dW_t, \end{aligned}$$

where σ_t is a deterministic function, so that the variable X_t is a Gaussian variable. The structural process X is calibrated to a barrier b, which is a continuous deterministic function for $0 \leq t \leq T$. The following graph gives us an example of trajectories of X and a calibrated barrier b. The deterministic barrier b can be calibrated by ensuring that expected survival probability is consistent with CDS prices for every time step as we propagate forward the value of t.

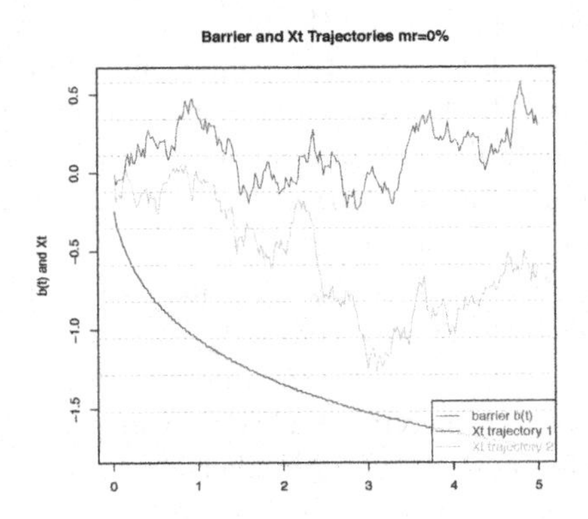

Fig. 1. Barrier and X trajectories

2.2. *Variables of interest*

While the process X is Markovian, the model, as far as it is concerned with forward probability of survival requires the information of $\{X_t, 1_{\{\tau>t\}}\}$ to avoid being path dependent. The information of X_t per se is not sufficient. The density of X_t is known, since this variable is Gaussian. The pricing variables of interest in a credit model given information at time t are: the progressive conditional probability, which is used to discount risky cashflow from t to 0, and the forward survival probability, which is used to discount from $T > t$ to t. They are defined as follows:

- the *progressive conditional probability* $Q(x,t)$

$$Q(x,t) := \mathbb{P}\left(\tau > t | X_t = x\right),$$

- the *forward survival probability* $P_T(x,t)$

$$P_T(x,t) := \mathbb{P}\left(\tau > T | \tau > t, X_t = x\right).$$

The progressive conditional probability formula is used during the barrier calibration phase to compute the expected survival probability to t and to ensure that CDS prices are correct for all t. The formula for forward survival probability is used for more advanced applications such as option pricing along with the progressive conditional probability.

3. PDEs for Progressive Conditional Probability and Forward Survival Probability

3.1. *Feynman-Kac applications*

Feynman-Kac 1: Backward Kolmogorov equation

The PDE for survival probabilities can be derived from the Feynman-Kac formula, from which the forward and backward Kolmogorov equations can be derived. The main idea of Feynman-Kac formula is presented below.

For a real-valued deterministic function u defined in $\mathbb{R} \times [0, T]$ and $u \in C^0(\mathbb{R} \times [0, T]) \cap C^{1,2}(\mathbb{R}, [0, T])$, and $\psi(\cdot)$ a deterministic function $\mathbb{R} \to \mathbb{R}$ which may be named as payoff function or value function, we can assert the following equivalence: we have for all $0 \le t \le T$,

$$\begin{cases} dX_t := \mu(X_t, t)dt + \sigma(X_t, t)dW_t, \\ u(x, t) := \mathbb{E}[\psi(X_T)|X_t = x], \\ u(x, T) := \psi(x), \end{cases}$$

if and only if u satisfies the following Feynman-Kac formula with terminal condition

$$\begin{cases} \frac{\partial u}{\partial t} = \mathcal{L}u, \\ u(x, T) = \psi(x), \end{cases}$$

where $\mathcal{L}$ is the infinitesimal generator of the Feynman-Kac equation:

$$\mathcal{L}u := -\mu(x, t)\frac{\partial u}{\partial x} - \frac{1}{2}\sigma^2(x, t)\frac{\partial^2 u}{\partial x^2}.$$

Feynman-Kac 2: Forward Kolmogorov equation

The forward Kolmogorov equation (also called the Fokker-Plank equation) is useful to compute the progressive conditional probability $Q(x, t)$. Our goal is to find the probability density of X_t. Define the unconditional probability density $p(x, t) := \frac{\partial}{\partial x}\mathbb{P}(X_t \le x)$. Then the PDE for $p(x, t)$ reads

$$-\frac{\partial p}{\partial t} = \mathcal{L}^\star p$$

where $\mathcal{L}^\star$ is the adjoint of the diffusion generator

$$\mathcal{L}^\star p := \frac{\partial \mu(x)p}{\partial x} - \frac{1}{2}\frac{\partial^2 \sigma^2(x, t)p}{\partial x^2}.$$

As $\delta(\cdot)$ refers to the Dirac delta function, the initial condition for the previous equation is:

$$p(x, 0) = \delta(x - X_0).$$

Notice that, in our model, $\mu \equiv 0$ and σ_t is a deterministic function depending only on time, and thus $\mathcal{L} = \mathcal{L}^\star$.

3.2. *Forward survival probability $P_T(x,t)$*

Our goal is to calculate the forward survival probability

$$P_T(x,t) := \mathbb{P}(\tau > T | X_t = x, \tau > t).$$

Using the Feynman-Kac formula, we obtain the following PDE with terminal and boundary conditions

$$\begin{cases} \frac{\partial P_T}{\partial t} = \mathcal{L} P_T, \\ P_T(x,T) = 1_{\{x > b(T)\}}, \\ P_T(x,t) = 0, \ x \leq b(t). \end{cases}$$

Given sufficiently regular $\mu(\cdot)$, $\sigma(\cdot,\cdot)$ and $b(\cdot)$, we have the existence and uniqueness of the solution of the parabolic PDE. Hence we can conclude that the PDE above gives as solution the forward survival probability $P_T(x,t)$.

3.3. *Progressive conditional probability $Q(x,t)$*

Let us now consider the joint survival probability density

$$q(x,t) := \frac{\partial}{\partial x} \mathbb{P}(\tau > t, X_t \leq x).$$

Since default is a knocked-out on barrier b, $q(x,t)$ becomes zero if $x \leq b(t)$. From the Kolmogorov forward equation, q satisfies the following PDE with initial and boundary conditions

$$\begin{cases} -\frac{\partial q}{\partial t} = \mathcal{L}^\star q, \\ q(x,0) = \delta(x - X_0), \\ q(x,t) = 0, \ x \leq b(t). \end{cases}$$

Given sufficiently regular $\mu(\cdot)$, $\sigma(\cdot,\cdot)$ and $b(\cdot)$, this system has a unique solution. We conclude similarly that the PDE above gives us the joint survival probability $q(x,t)$ as solution.

The calibration of the barrier b by the survival probability is an inverse first passage problem studied by Cheng et al. [3]. They proved that our inverse first passage problem here of calibrating $b(t)$ by the survival probability is well-posed and has a unique (weak) solution.

Besides, the unconditional probability density $p(x,t) := \frac{\partial}{\partial x}\mathbb{P}(X_t \leq x)$ can be obtained explicitly since X_t is a Gaussian variable. A good remark here is that, it would be a better choice to compute $p(x,t)$ numerically by the Fokker-Plank equation in the same way as the calculation of $q(x,t)$, for the reason of numerical stability in the computation of the conditional survival probability $Q(x,t)$. We see easily that p satisfies the following PDE with initial condition

$$-\frac{\partial p}{\partial t} = \mathcal{L}^{\star} p,$$
$$p(x,0) = \delta(x - X_0).$$

Hence using the definitions of q and p, we have the progressive conditional (survival) probability $Q(x,t)$

$$Q(x,t) := \mathbb{P}(\tau > t | X_t = x) = \frac{\frac{\partial}{\partial x}\mathbb{P}(\tau > t, X_t \leq x)}{\frac{\partial}{\partial x}\mathbb{P}(X_t \leq x)} = \frac{q(x,t)}{p(x,t)}.$$

To solve numerically the PDEs and to get simulation results for those probabilities and densities, we implement the finite differences methods with a *Crank Nicolson scheme*. Here are our numerical results for progressive conditional probability and forward survival probability.

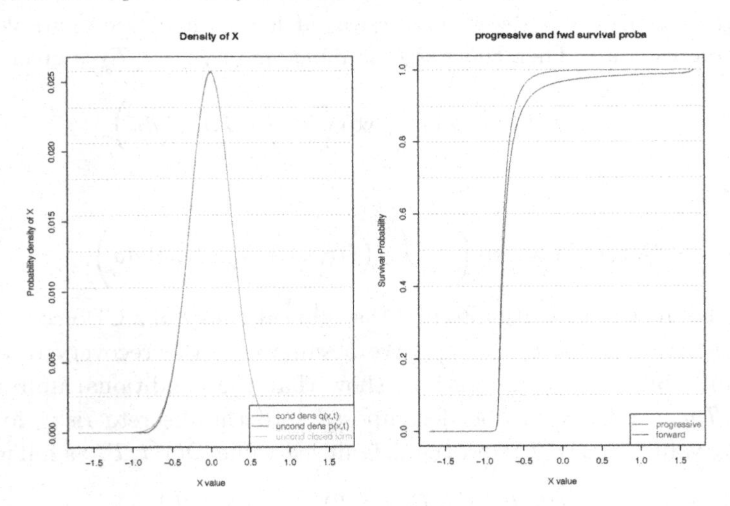

Fig. 2. Progressive conditional and forward survival probabilities

4. CDS Pricing

The CDS market is now using to standard "IMM" maturities, which roll every 3 months to avoid maturity mismatch for dealers, as well as standardised contract coupon c of either 100 or 500 basis points. In the CDS market, when a spread is quoted, the market convention for converting it to upfront is to use a flat model with a piecewise flat interest rate curve published by Markit (as the previous business day close) with no cross-currency or funding basis.

4.1. *Conditional pricing of a CDS*

In this subsection, we consider the valuation of a CDS conditional on the event $\{X_t = x, \tau > t\}$ assuming that $\mathbb{P}$ is the pricing probability measure. To this end, for given x and t, we define the *conditional forward hazard rate* $\lambda(x,t,u)$ for the date $u > t$ by postulating that, for all $T \geq t$,

$$P_T(x,t) = \exp\Big(- \int_t^T \lambda(x,t,u)\, du\Big)$$

or, equivalently,

$$\lambda(x,t,u) := -\frac{d \ln P_u(x,t)}{du}.$$

Let $f(t,u)$ represent a deterministic (or, at least, an $\mathcal{F}_t^X$-measurable) forward interest rate. Then the *risky discount factor* $\widehat{D}(x,t,T)$ is given by

$$\widehat{D}(x,t,T) = P_T(x,t) \exp\Big(- \int_t^T f(t,u)\, du\Big)$$

or, more explicitly,

$$\widehat{D}(x,t,T) = \exp\Big(- \int_t^T \big(f(t,u) + \lambda(x,t,u)\big)\, du\Big).$$

We are now in a position to find the value at time t of a CDS conditional on the event $\{X_t = x, \tau > t\}$. We assume that the recovery rate R is constant. Standard computations show that the conditional upfront fee $U(x,t,T)$ of a CDS can be decomposed into the discrete risky forward annuity value $A_d(x,t,T)$ and the default leg value $D(x,t,T)$ as follows

$$U(x,t,T) = D(x,t,T) - cA_d(x,t,T)$$

where c denotes the standardised contract coupon,

$$\begin{aligned} A_d(x,t,T) &= \sum_i 1_{\{t_{i+1}>t\}} \widehat{D}(x,t,t_{i+1}) \frac{t_{i+1} - \max(t,t_i)}{360} \\ &\quad + \sum_i 1_{\{t_{i+1}>t\}} \int_{t_i}^{t_{i+1}} \widehat{D}(x,t,u)\lambda(x,t,u) \frac{u - t_i}{360}\, du \end{aligned}$$

and

$$D(x,t,T) = (1-R)\int_t^T \lambda(x,t,u)\widehat{D}(x,t,u)\,du.$$

The first term of the risky annuity formula corresponds to ACT/360 basis cashflows PV conditional on survival, while the integral term corresponds to accrual payment made in case of default. Note the term $\max(t, t_i)$ in the formula, which signifies that the credit annuity generally has a full first coupon and is considered net of accruals. We note that when $t_{i+1} - t_i \to 0$, then the value $A_d(x,t,T)$ of the discrete annuity converges towards its much simpler continuous limit $A(x,t,T)$ given by

$$A(x,t,T) = \int_t^T \widehat{D}(x,t,u)\,du.$$

4.2. *Approximate representations for conditional values*

The CDS quotes convention is that a flat credit term structure is used for each credit tenor. Mimicking this convention, we fix T and we define the *conditional flat forward hazard rate* $\lambda_{t,T}(x)$ by setting

$$P_T(x,t) := e^{-\lambda_{t,T}(x)(T-t)}$$

or, equivalently,

$$\lambda_{t,T}(x) = -\frac{\ln P_T(x,t)}{T-t} = \frac{\int_t^T \lambda(x,t,u)\,du}{T-t}. \tag{4.1}$$

Assuming that the forward hazard curve $\lambda(x,t,u)$, $u \in [t,T]$ is flat enough, we may use the following approximation

$$\begin{aligned} D(x,t,T) &= (1-R)\int_t^T \lambda(x,t,u)\widehat{D}(x,t,u)\,du \\ &\approx (1-R)\int_t^T \lambda_{t,T}(x)\widehat{D}(x,t,u)\,du \\ &= (1-R)\lambda_{t,T}(x)A(x,t,T). \end{aligned}$$

Consequently, the continuous forward par spread $S_{t,T}(x)$ satisfies

$$S_{t,T}(x) := \frac{D(x,t,T)}{A(x,t,T)} \approx (1-R)\lambda_{t,T}(x), \tag{4.2}$$

so that the continuous par spread and the flat forward hazard rate are approximately proportional with a factor $(1-R)$.

Let us now focus on the term $A(x,t,T)$. If we postulate, in addition to the flatness of the forward hazard curve, that the forward term structure is flat, so that we may assume that $f(t,u) \approx r$ for a constant r, then we may approximate $A(x,t,T)$ by a closed-form expression in terms of r and $\lambda_{t,T}(x)$, specifically,

$$\begin{aligned} A(x,t,T) &= \int_t^T \widehat{D}(x,t,u)\,du \\ &= \int_t^T P_T(x,u)\exp\Big(-\int_t^u f(t,s)\,ds\Big)du \\ &\approx \int_t^T \exp\big((r+\lambda_{t,T}(x))(u-t)\big)du \\ &= \frac{1-\exp(-(r+\lambda_{t,T}(x))(T-t))}{r+\lambda_{t,T}(x)}, \end{aligned}$$

From the last equality, we may in turn deduce that under the assumption that the interest rates are low, the continuous risky forward annuity $A(x,t,T)$ may be approximated as follows

$$A(x,t,T) \approx \frac{1-P_T(x,t)}{\lambda_{t,T}(x)}. \tag{4.3}$$

Let us stress once again that the above computations were done conditional on the event $\{X_t = x,\ \tau > t\}$.

4.3. *Forward continuous annuity conditional on* $X_t = x$

The previous sections dealt with the simplified pricing of a forward CDS from t to T conditional on $\{X_t = x,\ \tau > t\}$. We now proceed to the computation of the value of a CDS conditional on $\{X_t = x\}$. Since

$$\mathbb{P}(t<\tau<T|X_t=x) = \mathbb{P}(t<\tau|X_t=x)\mathbb{P}(t<\tau<T|X_t=x,\tau>t),$$

we obtain the following expression for the value at time t of the continuous risky annuity conditional on the event $\{X_t = x\}$ (but not on $\{\tau > t\}$)

$$A_{t,T}(x) = Q(x,t)A(x,t,T). \tag{4.4}$$

4.4. *Forward spread and risky annuity at time 0*

We henceforth assume that the forward interest rates $f(t,u)$, $u \in [t,T]$ are low. Then the forward spread $\bar{S}(t,T)$ and the value of continuous annuity

$\bar{A}(t,T)$ at time 0 can be calculated in terms of $Q(x,t)$, $P_T(x,t)$ and X_t as follows

$$\bar{A}(t,T) := \mathbb{E}[A_{t,T}(X_t)] = \mathbb{E}[Q(X_t,t)A(X_t,t,T)]$$

and

$$\bar{S}(t,T) := \frac{\mathbb{E}[A_{t,T}(X_t)S_{t,T}(X_t)]}{\mathbb{E}[A_{t,T}(X_t)]}$$

where, from (4.3), we may approximate $A(X_t,t,T)$ as follows

$$A(X_t,t,T) \approx (T-t)\frac{P_T(X_t,t)-1}{\ln P_T(X_t,t)},$$

whereas (4.2) yield the following approximation for $S_{t,T}(X_t)$

$$S_{t,T}(X_t) \approx (R-1)\frac{1}{T-t}\ln P_T(X_t,t)$$

These approximations are used as equalities in our implementation of the structural default model. We present below the numerical simulation results of spread, risky annuity and forward default leg for the option with knock out on default.

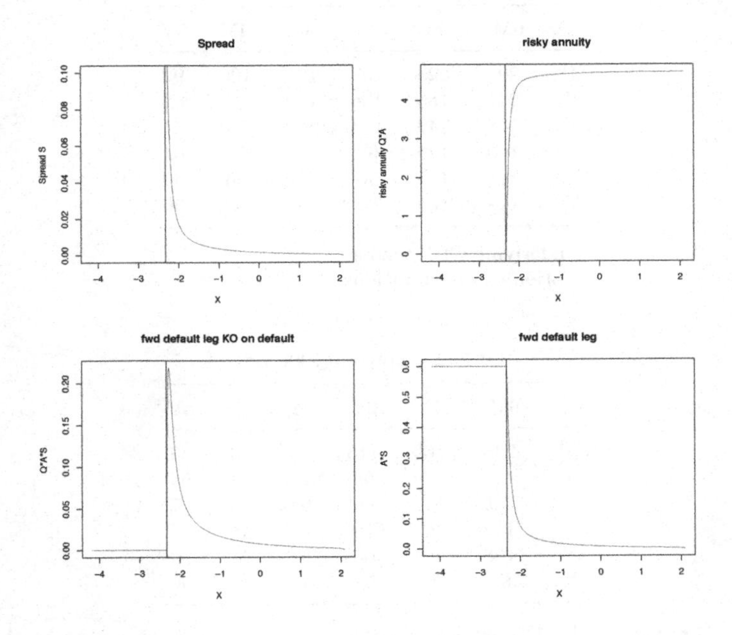

Fig. 3. Spread, risky annuity and default leg (with knock out on default or not)

5. Volatility Surface of CDS Option

The price V_0 of a CDS option maturing at t and with strike K is defined as

$$V_0 := \mathbb{E}[A_{t,T}(X_t)(S_{t,T}(X_t) - K)^+].$$

Given the forward spread $\bar{S}(t,T)$ and the value of continuous annuity $\bar{A}(t,T)$ at time 0, we also have the Black price of CDS option, which defines the implied volatility σ of the option

$$V_0 = \bar{A}(t,T)\,\mathrm{BS}(\bar{S}(t,T), K, \sigma, t).$$

Hence at-the-money CDS option implied volatility σ_{im} can be found by solving the following equation

$$\mathbb{E}[A_{t,T}(X_t)(S_{t,T}(X_t) - K)^+] = \bar{A}(t,T)\,\mathrm{BS}(\bar{S}(t,T), K, \sigma_{\mathrm{im}}, t).$$

Tables 1 and 2 give us the numerical results of ATM vol surfaces with different option expiry durations (1st column) and underlying CDS tenors (1st row):

Table 1. Spread=300 bp

	6M[a]	1Y	2Y	3Y	4Y	5Y
6M[b]	298	223	158	126	106	92
1Y	241	186	136	110	94	82
2Y	188	149	113	93	80	71
3Y	160	129	99	83	72	64
4Y	141	115	90	76	66	59
5Y	128	105	83	70	62	56

[a] Underlying CDS tenors
[b] Option expiry durations

Table 2. Spread=1000 bp

	6M[a]	1Y	2Y	3Y	4Y	5Y
6M[b]	235	169	115	88	73	63
1Y	188	139	98	77	64	55
2Y	145	111	80	64	54	47
3Y	122	95	70	57	48	42
4Y	108	85	63	52	44	39
5Y	98	77	58	48	41	36

[a] Underlying CDS tenors
[b] Option expiry durations

6. Conclusions

We now look at the lessons learned from this exercise in risk neutral pricing with the structural model. The relative dearth of references concerning themselves with credit market replication of credit derivatives using a structural model can be explained by the relative complexity of fitting a structural model to the market.

Reduced form models have from the start been setup with the intention of pricing derivatives. The publication of such tractable methods as reduced form model coincided with the creation of the CDS market, for which those models became market standard because default intensity maps itself so well to the cost of market protection. While the latter is adopted by practitioners for its direct applicability to derivatives pricing, the former is used where a causal link between default and firm's assets is needed.

This entails that structural models are both more difficult and more rewarding to put in place, as they help in a broader range of applications such as: optimal capital structure of a firm's asset between equity and debt, estimation of default frequency without credit market data, or default correlation from other financial data.

What we established first are variables of interest conditional on which CDS pricing is possible, and their forward and backward propagation equations. Authors who do not consider risk neutral pricing, and thus do not need to put such an emphasis on model calibration, usually eschew this technicality to focus on other features of the model. We computed the prices implied for CDS and the CDS option volatility surfaces implied by this model. It should be noted that these implied volatilities are far higher than what is quoted in the CDS market or what could be gathered from historical volatility studies. This makes this structural model unsuitable for callable credit product pricing and therefore begs the question of how to extend structural models to control its implied credit market volatility.

Acknowledgements. The authors wish to thank Adrien Godet, Xi Chen and Wei Ding for their prior work on the subject and fruitful discussion on the subject, and Jean-Paul Laurent and Trung Tien Do for their feedback on this article. The authors thank warmly an anonymous referee, who rewrite some part of this paper, making it more clear and correcting some missprints and mistakes. The views expressed here are the authors' own and do not necessarily represent those of their employer.

References

[1] Bielecki, T. R. and Rutkowski, M. (2002). *Credit Risk: Modeling, Valuation and Hedging*, Springer.

[2] Black, F. and Cox, J. C. (1976). Valuing corporate securities: some effects of bond indenture provisions, *Journal of Finance* **31(2)**, 351–367.

[3] Cheng, L., Chen, X., Chadam, J. and Saunders, D. (2006). Analysis of an inverse first passage problem from risk management, *SIAM Journal on Mathematical Analysis* **38(3)**, 845–873.

[4] Elizalde, A. (2003). *Credit Risk Models II: Structural Models*, Financial Mathematics, King's College London.

[5] Finkelstein, V. et al. (2001). *CreditGrades Technical Document*, RiskMetrics Group.

[6] Giorno, V., Nobile, A.G., Ricciardi, L. M. and Sato, S. (1989). On the evaluation of first-passage-time probability densities via non-singular integral equations, *Adv. Appl. Prob.* **21**, 20–36.

[7] Leland, H. E. (1994). Corporate debt value, bond covenants, and optimal capital structure, *Journal of Finance* **49(4)**, 1213–1252.

[8] Leland, H. E. and Toft, K. B. (1996). Optimal capital structure, endogenous bankruptcy, and the term structure of credit spreads, *Journal of Finance* **51(3)**, 987–1019.

[9] Madan, D. B. and Unal, H. (1996). Pricing the risks of default, presented at the Wharton Financial Institutions Center Conference on Risk Management in Banking.

[10] Merton, R. C. (1974). On the pricing of corporate debt: the risk structure of interest rates, *Journal of Finance* **29(2)**, 449–470.

[11] Modigliani, F. and Miller, M. H. (1958). The cost of capital, corporation finance and the theory of investment, *American Economic Review* **48(3)**, 261–297.

[12] Novikov, A. A. (1981), A martingale approach to first passage problems and a new condition for Wald's identity, in *Stochastic Differential Systems, Lecture Notes in Control and Information Sciences 36*, pp. 146–156.

[13] Peskir, G. (2002). On integral equations arising in the first-passage problem for Brownian motion, *Journal Integral Equations and Applications* **14**, 397–423.

[14] Schönbucher, P. J. (2003). A note on survival measures and the pricing of options on credit default swaps, working paper.

[15] Vasicek, O. (1987). Probability of loss on loan portfolio, working paper.

Reduced-Form Modeling of Counterparty Risk on Credit Derivatives

Stéphane Crépey*

In [8, 10], a basic reduced-form counterparty risk modeling approach has been proposed under a rather standard immersion hypothesis between the reference filtration and the filtration progressively enlarged by the default times of the two parties. This paper shows how a suitable extension of this approach, fully developed in [11] and [17], can be applied beyond such a basic immersion setup, in dynamic copula models of counterparty risk on credit derivative underlyings.

1. Introduction

For dealing with counterparty risk on credit derivatives (see e.g. the papers [4, 6] and the books [7] and [14]), one needs a credit portfolio model with the following features. First, as the CVA is an option on the future values of the clean price (i.e. mark-to-market) of the underlyings, it should be a dynamic Markovian model, in which clean prices of credit derivatives in the future can be assessed consistently and computed numerically. Second, the model should be calibrated to relevant data sets: individual CDS data, if the targeted application consists of counterparty risk computations on CDSs, and tranches data for CVA computation on CDOs. The model should therefore be a bottom-up model of portfolio credit risk so that individual names are represented in the model (see [5]). But, for tractability of the calibration, the model should also enjoy a copula-like separation property between the individual and the dependence model parameters, as well as efficient pricing schemes for vanillas (CDSs and/or CDOs). One possibility is to use "informationally dynamized copula models", resulting from the introduction of a suitable filtration on the top of a static copula model for the default times of the two parties and the reference entities. The simplest example is the dynamic Gaussian copula (DGC) model of [16], which can

*Université d'Évry Val d'Essonne, Laboratoire Analyse et Probabilités,
Email: stephane.crepey@univ-evry.fr

be sufficient to deal with CVA on CDSs. If CDOs are also present in the reference portfolio, then a Gaussian copula dependence structure is not rich enough. Instead, one can use the dynamic Marshall-Olkin (DMO) copula model of [3, 4].

However, in a bilateral counterparty risk perspective with the related funding issue, the pricing equation for counterparty risk and funding costs (we refer to an aggregate adjustment as a TVA, which stands for total valuation adjustment) is nonlinear (see [8, 10, 23, 24]). In the case of credit derivative underlyings, the problem is also very high-dimensional. From a practical point of view, for nonlinear and very high-dimensional problems, any numerical scheme based, even to some extent, on dynamic programming, such as "purely backward" deterministic PDE schemes, but also "hybrid forward/backward" simulation/regression BSDE schemes, are ruled out by the curse of dimensionality (see e.g. [12]). In this case the only feasible TVA schemes are "purely forward" simulation schemes, such as the expansions of [19] in vanilla cases with explicit clean pricing formulas, or the CVA branching particles of [20] in exotic cases.

But purely forward schemes cannot deal with the implicit terminal condition $\xi(P_{\tau-} - \Theta_{\tau-})$ of the full TVA equations of [8, 10] (recalled as equation (2.2) below) or [23, 24]. To tackle this issue, this paper shows how a suitable reduced-form developed in [11] and [17] can be applied, beyond the basic immersion setup of [8, 10], in the above-mentioned dynamic copula credit portfolio models.

Outline Sect. 2 briefly reviews the extended reduced-form counterparty risk modeling approach of [11]. Sect. 3 and 4 show how this approach can be applied to the modeling of counterparty risk on credit derivatives in the dynamic copula models of [16] (dynamic Gaussian copula) and [3, 4] (dynamic Marshall-Olkin copula or common shock model).

All the cash-flows that appear in the paper are assumed to be integrable under the prevailing pricing measure. "Martingale" is meant everywhere as local martingale, but true martingality is assumed whenever necessary.

2. Marked Default Time Reduced-Form Modeling

We denote by T the time horizon of a contract (or netted set of contracts) with promised dividends dD_t from the bank to its counterparty. The two parties are defaultable, with first-to-default time and related survival process denoted by τ and $J_t = \mathbb{1}_{t<\tau}$. Therefore, the effective horizon of the problem is

$$\bar{\tau} = \tau \wedge T$$

(as there are no actual cash-flows after that time). We denote by $(\Omega, \mathcal{G}_{\bar{\tau}}, \mathbb{G}, \mathbb{Q})$ a prevailing pricing stochastic basis over $[0, \bar{\tau}]$, by $\mathbb{E}_t$ the $(\mathcal{G}_t, \mathbb{Q})$-conditional expectation operator, by r_t an OIS rate (overnight indexed swap rate, the best market proxy for a risk-free rate) and by $\beta_t = e^{-\int_0^t r_s ds}$ the corresponding discount factor. Accordingly, the clean value process P_t of the contract (mark-to-market ignoring counterparty risk and funding features) is defined, for $t \in [0, T]$, by the "classical textbook formula"

$$\beta_t P_t = \mathbb{E}_t \left(\int_t^T \beta_s dD_s \right). \tag{2.1}$$

The counterparty exposure at default for the bank, i.e. loss for the bank resulting from the default of one of the two parties at time τ (actually in fact gain if the bank is the defaulting one, up to the DVA debate for which we refer the reader to the discussion Sect. 2 in [13]), is denoted by $\xi(\pi)$, a random variable parameterized by the wealth π of the bank right before time τ (see [9]). Moreover, the bank needs to fund its position. Using the OIS rate r_t as a reference funding rate, we denote by $g_t = g_t(\pi)$ a random function such that $(r_t\pi + g_t(\pi))dt$ represents the bank's funding cost over $(t, t+dt)$, depending on the bank's wealth π. Given the data ξ and g_t, it is shown in [10] how the TVA process Θ can consistently be modeled on $[0, \bar{\tau}]$ as a solution, provided it exists, to the following BSDE written in integral form over the random time interval $[0, \bar{\tau}]$:

$$\beta_t \Theta_t = \mathbb{E}_t \Big[\beta_{\bar{\tau}} \mathbb{1}_{\tau<T} \xi(P_{\tau-} - \Theta_{\tau-}) + \int_t^{\bar{\tau}} \beta_s g_s(P_s - \Theta_s) ds \Big], \ t \in [0, \bar{\tau}]. \tag{2.2}$$

The object of this paper is to show how a suitable reduced-form approach for z (2.2), developed in [11] and [17] to go beyond the standard immersion case of [10], applies in the dynamic copula models of [16] and [3].

2.1. *Pre-default Setup*

We assume that τ is endowed with a mark e in a finite set E, i.e. $\tau = \min_{e \in E} \tau_e$ with $\mathbb{Q}(\tau^e \neq \tau^{e'}) = 1$ (for $e' \neq e$), where τ_e has a pre-default intensity λ_t^e. We also postulate that for every mark e, there exists a predictable random function $\widetilde{\xi}_t^e(\pi)$ such that

$$\mathbb{1}_{\{\tau=\tau_e\}}\xi(\pi) = \mathbb{1}_{\{\tau=\tau_e\}}\widetilde{\xi}_\tau^e(\pi). \tag{2.3}$$

For every semimartingale Y, we then have the compensated martingale defined, for $t \in [0, \bar{\tau}]$, as:

$$\xi(Y_{t-})dJ_t + \lambda_t \cdot \widetilde{\xi}_t(Y_t)dt, \tag{2.4}$$

where we denote $\lambda_t \cdot \chi_t = \sum_E \lambda_t^e \chi_t^e$, for every random function χ_t^e. For every $t \in [0, \bar{\tau}]$ and real ϑ, we write

$$f_t(P_t - \vartheta) = g_t(P_t - \vartheta) + \lambda_t \cdot \widetilde{\xi}_t(P_t - \vartheta) - \widetilde{r}_t\vartheta, \tag{2.5}$$

where $\widetilde{r}_t = r_t + \gamma_t$, in which $\gamma_t = \sum_{e\in E} \lambda_t^e = \lambda_t \cdot \mathbb{1}_E$ is a pre-default intensity of τ. Let there be given, assumed to exist (a set of assumptions, relaxing the usual immersion hypothesis as discussed in [11], that we denote collectively by **(A)**):

(A.1) a filtration $\mathbb{F}$ over $[0, T]$ such that $\mathcal{F}_t \subseteq \mathcal{G}_t$ on $[0, \bar{\tau}]$ and $\mathbb{F}$-semimartingales stopped at τ are $\mathbb{G}$-semimartingales,

(A.2) a probability measure $\mathbb{P}$ equivalent to $\mathbb{Q}$ on $\mathcal{F}_{\bar{\tau}}$ such that an $(\mathbb{F}, \mathbb{P})$-martingale stopped at $\tau-$ is a $(\mathbb{G}, \mathbb{Q})$-martingale[a],

(A.3) an $\mathbb{F}$-progressive random function $\widetilde{f}_t(\vartheta)$ such that $f_t(P_t - \vartheta)dt = \widetilde{f}_t(\vartheta)dt$ on $[0, \bar{\tau}]$,

and let $\widetilde{\mathbb{E}}_t$ stand for the conditional expectation given $\mathcal{F}_t$ under $\mathbb{P}$. It is proven in [11] that if an $(\mathbb{F}, \mathbb{P})$-semimartingale $\widetilde{\Theta}$ satisfies the following equation over $[0, T]$:

$$\widetilde{\Theta}_t = \widetilde{\mathbb{E}}_t \int_t^T \widetilde{f}_s(\widetilde{\Theta}_s)ds, \quad t \in [0, T], \tag{2.6}$$

then the $(\mathbb{G}, \mathbb{Q})$-semimartingale Θ defined as $\Theta = \widetilde{\Theta}$ on $[0, \bar{\tau})$ and $\Theta_{\bar{\tau}} = \mathbb{1}_{\tau<T}\xi(P_{\tau-} - \widetilde{\Theta}_{\tau-})$ satisfies the full TVA equation (2.2) on $[0, \bar{\tau}]$. Note that the equation (2.6) defines a BSDE (in integral form) for the process $\widetilde{\Theta}_t$, referred to henceforth as the pre-default TVA BSDE. Moreover, as it is well known (see e.g. [12]), in the pre-default Markovian case where there exists an $\mathbb{R}^q$-valued $(\mathbb{F}, \mathbb{P})$-Markov factor process X such that, over $[0, T]$,

$$\widetilde{f}_t(\vartheta) = \widetilde{f}(t, X_t, \vartheta) \tag{2.7}$$

[a] By a process stopped at $\tau-$, we mean the process deprived of its jump at τ.

for some Borel function $\widetilde{f}(t,x,\vartheta)$, then a solution $\widetilde{\Theta}_t$ to (2.6) can be represented as $\widetilde{\Theta}(t,X_t)$, for some Borel function $\widetilde{\Theta}(t,x)$ satisfying the following semilinear P(I)DE:

$$\begin{cases} \widetilde{\Theta}(T,x) = 0\,,\ x \in \mathbb{R}^q \\ (\partial_t + \mathcal{A})\,\widetilde{\Theta}(t,x) + \widetilde{f}(t,x,\widetilde{\Theta}(t,x)) = 0 \text{ on } [0,T)\times\mathbb{R}^q, \end{cases} \tag{2.8}$$

where $\mathcal{A}$ is the $(\mathbb{F},\mathbb{P})$-generator of the Markov process X_t. Any BSDE/P(I)DE numerical scheme for (2.8) can then be used for computing the TVA $\widetilde{\Theta}$.

The practical conclusion of this section is that one can model a TVA process Θ as sketched above in terms of a solution $\widetilde{\Theta}_t = \widetilde{\Theta}(t,X_t)$, assumed to exist (for which we refer the reader to [10], [12] and the references there), to some Markovian pre-default TVA BSDE (2.6). In the coming two sections, we show how this approach can be applied to the modeling of the TVA on credit derivatives in the dynamic copula models of [16] and [3], [4]. Let $N = \{-1,0,1,\dots,n\}$, $N^\star = \{1,\dots,n\}$. In both cases, we dynamize a copula model for the $(\tau_i)_{i\in N}$ by introduction of a suitable filtration, where $\tau_{-1} = \tau_b$ and $\tau_0 = \tau_c$ are used to model the default times of the the bank and its counterparty, whereas the $(\tau_i)_{i\in N^\star}$ represent the default times of a pool of reference entities underlying a credit portfolio between the two parties.

For any Euclidean vector $\mathbf{k}$, we denote

$$\mathrm{supp}(\mathbf{k}) = \{i;\ k_i \neq 0\},\ \ \mathrm{supp}^c(\mathbf{k}) = \{i;\ k_i = 0\}. \tag{2.9}$$

In the financial interpretation, $\mathrm{supp}(\mathbf{k})$ denotes the obligors who have defaulted in the portfolio-state $\mathbf{k}$ and therefore $\mathrm{supp}^c(\mathbf{k})$ corresponds to the survivors.

3. Dynamic Gaussian Copula TVA Model

3.1. *Model of Default Times*

We consider a multivariate Brownian motion $\mathbf{B} = (B^i)_{i\in N}$ with pairwise correlation ϱ in its own completed filtration $\mathbb{F}^{\mathbf{B}}$, under a pricing probability measure $\mathbb{Q}$. For any $i \in N$, let h_i be an increasing function from $\mathbb{R}_+^\star$ to $\mathbb{R}$ with $\lim_{0+} h_i(s) = -\infty$ and $\lim_{+\infty} h_i(s) = +\infty$. For every $i \in N$, we define a random time

$$\tau_i = h_i^{-1}\Big(\int_0^{+\infty} \varsigma(u)dB_u^i\Big), \tag{3.1}$$

where $\varsigma(\cdot)$ is a square integrable function with unit L^2-norm. So the τ_i jointly follow the standard (static) Gaussian copula model of [21], with correlation parameter ϱ and with marginal distribution function $\Phi \circ h_i$ of τ_i, where Φ is the the standard normal distribution function. In order to make the model dynamic as required by counterparty risk applications, we introduce a model filtration $\mathbb{G} = \mathbb{G}^{gc}$ given as $\mathbb{F}^{\mathbf{B}}$ progressively enlarged by the τ_i, so, for every t,

$$\mathcal{G}_t^{gc} = \mathcal{F}_t^{\mathbf{B}} \vee \big(\bigvee_{i \in N} (\tau_i \wedge t) \big).$$

Let

$$\mathbf{m}_t = (m_t^i)_{i \in N} \text{ with } m_t^i = \int_0^t \varsigma(u) dB_u^i, \;\; \mathbf{k}_t = (k_t^i)_{i \in N} \text{ with } k_t^i = \tau_i \mathbb{1}_{\{\tau_i \le t\}}.$$

The above "informational dynamization" of the Gaussian copula (DGC, i.e. dynamic Gaussian copula) was introduced in [16] (see also [18] for a structural perspective), where one can find that for explicit processes γ_t^i and λ_t^i of the form

$$\gamma_t^i, \lambda_t^i = \gamma_i, \lambda_i \, (t, \mathbf{m}_t, \mathbf{k}_t), \tag{3.2}$$

the $\mathbb{G}$-Brownian motions $dW_t^i = dB_t^i - \gamma_t^i dt$ and the $\mathbb{G}$-compensated default indicator processes $dM_t^i = d\mathbb{1}_{\tau_i \le t} - \lambda_t^i dt$ have the $\mathbb{G}$-martingale representation property. Moreover, the process $(\mathbf{m}_t, \mathbf{k}_t)$ is $\mathbb{G}$-Markov, with a generator $\mathcal{L} = \mathcal{L}^{gc}$ acting on every function $u = u(t, \mathbf{m}, \mathbf{k})$ as

$$\mathcal{L}^{gc} u = \mathcal{L}_{\mathbf{m}} u + \sum_{i \in N} \lambda_i \delta_i u, \tag{3.3}$$

where

$$\mathcal{L}_{\mathbf{m}} u = \varsigma \sum_{i \in N} \gamma_i \partial_{m_i} u + \frac{\varsigma^2}{2} \left(\sum_{i \in N} \partial^2_{m_i^2} u + \varrho \sum_{i \neq j \in N} \partial^2_{m_i, m_j} u \right) \tag{3.4}$$

and

$$\delta_i u(t, m, \mathbf{k}) = u(t, m, \mathbf{k}^{i,t}) - u(t, m, \mathbf{k})$$

in which $\mathbf{k}^{i,t}$ stands for $\mathbf{k}$ with component i replaced by t.

Remark 3.1. For the Markov sake, time t has to be included in the state variables $(t, \mathbf{m}_t, \mathbf{k}_t)$ above (see [16]) and likewise in $(t, \mathbf{m}_t, \widetilde{\mathbf{k}}_t)$ below. In the statement preceding (3.3), a more rigorous statement from the point of view of operator theory would be that the right-hand side in this equality defines the time s-generator of the time-inhomogeneous Markov process $(t, \mathbf{m}_t, \mathbf{k}_t)$ (and one would typically write $\mathcal{A}_s u$ on the left-hand side). But technicalities of this kind are immaterial in this paper, where we only need the corresponding "Itô-Markov formulas", which are established by direct SDE computations without reference to operator theory.

3.2. *Pre-default TVA Model*

A DGC setup will now be used as a TVA credit derivatives model, for a mark space of τ given as

$$E = \{-1, 0\},$$

under the related assumption (2.3), discussed in [11], that for $i = -1, 0$, there exists a function $\widetilde{\xi}_i$ satisfying

$$\xi(\pi) = \widetilde{\xi}_i(\tau, \mathbf{m}_\tau, \mathbf{k}_{\tau-}; \pi) = \widetilde{\xi}^i_\tau(\pi) \tag{3.5}$$

on the event $\{\tau = \tau_i\}$. By reduced DGC model we mean $(t, X_t) = (t, \mathbf{m}_t, \widetilde{\mathbf{k}}_t)$ (cf. the remark 3.1), where $\widetilde{\mathbf{k}}_t = (k^i_t)_{i \in N^\star}$, considered relatively to the reference filtration $\mathbb{F} = \mathbb{F}^{gc}$ generated by the Brownian motion $\mathbf{B}$ and the default times of references names $i \in N^\star$, i.e. for every t

$$\mathcal{F}^{gc}_t = \mathcal{F}^{\mathbf{B}}_t \vee \big(\bigvee_{i \in N^\star} (\tau_i \wedge t) \big).$$

The reduced model has properties similar to the full DGC model, with fundamental $(\mathbb{F}, \mathbb{Q})$-martingales

$$d\overline{W}^i_t = dB^i_t - \bar{\gamma}^i_t dt,\ i \in N;\ d\overline{M}^i_t = d\mathbb{1}_{\tau_i \le t} - \bar{\lambda}^i_t dt,\ i \in N^\star, \tag{3.6}$$

for explicit processes of the form

$$\bar{\gamma}^i_t = \bar{\gamma}_i(t, \mathbf{m}_t, \widetilde{\mathbf{k}}_t),\ \ \bar{\lambda}^i_t = \bar{\lambda}_i(t, \mathbf{m}_t, \widetilde{\mathbf{k}}_t). \tag{3.7}$$

Let us change the measure to $\mathbb{P}$ such that the "$(\mathbb{F}, \mathbb{P})$-intensities" of the $B^i_t, i \in N$ and of the $\mathbb{1}_{\tau_i \le t}, i \in N^\star$ are respectively equal, letting $\mathbf{k} = (0, 0, \widetilde{\mathbf{k}})$ for every $\widetilde{\mathbf{k}} \in \mathbb{R}^n_+$, to

$$\begin{aligned} \widetilde{\gamma}^i_t &= \widetilde{\gamma}_i(t, \mathbf{m}_t, \widetilde{\mathbf{k}}_t) := \gamma_i(t, \mathbf{m}_t, \mathbf{k}_t) \\ \widetilde{\lambda}^i_t &= \widetilde{\lambda}_i(t, \mathbf{m}_t, \widetilde{\mathbf{k}}_t) := \lambda_i(t, \mathbf{m}_t, \mathbf{k}_t). \end{aligned} \tag{3.8}$$

Thus, the fundamental $(\mathbb{F}, \mathbb{P})$-martingales are written as

$$d\widetilde{W}^i_t = dB^i_t - \widetilde{\gamma}^i_t dt,\ i \in N;\ d\widetilde{M}^i_t = d\mathbb{1}_{\tau_i \le t} - \widetilde{\lambda}^i_t dt,\ i \in N^\star. \tag{3.9}$$

Note that before τ, we have:

$$\gamma_t^i = \widetilde{\gamma}_t^i, \ \ \lambda_t^i = \widetilde{\lambda}_t^i$$

and therefore, until τ, we have:

$$W_t^i = \widetilde{W}_t^i, \ \ M_t^i = \widetilde{M}_t^i.$$

The $(\mathbb{F}, \mathbb{P})$-generator of (t, X_t) acts on every function $u = u(t, \mathbf{m}, \widetilde{\mathbf{k}})$ as

$$\mathcal{A}u = \mathcal{A}^{gc}u = \mathcal{A}_{\mathbf{m}}u + \sum_{i \in N^\star} \widetilde{\lambda}_i \delta_i u, \tag{3.10}$$

where (compare (3.4)):

$$\mathcal{A}_{\mathbf{m}}u = \varsigma \sum_{i \in N} \widetilde{\gamma}_i \partial_{m_i} u + \frac{\varsigma^2}{2} \left(\sum_{i \in N} \partial^2_{m_i^2} u + \varrho \sum_{i \neq j \in N} \partial^2_{m_i, m_j} u \right). \tag{3.11}$$

Before τ, every dependence in $\mathbf{k}_t$ reduces to a dependence in $\widetilde{\mathbf{k}}_t$. Under the additional assumption that the processes $r_t, g_t(\pi)$ (for every real π) and P_t are given before τ as continuous functions of (t, X_t) (functions denoted below by the same letters as the related processes), then, in view of (2.5), we have:

$$f_t(P_t - \vartheta)dt = \widetilde{f}(t, X_t, \vartheta)dt, \ \ t \in [0, \bar{\tau}]$$

for the function $\widetilde{f} = \widetilde{f}^{mo}$ such that, writing $\mathbf{k} = (0, 0, \widetilde{\mathbf{k}})$ for every $\widetilde{\mathbf{k}} \in \mathbb{R}^n_+$:

$$\begin{aligned} \widetilde{f}^{gc}(t, \mathbf{m}, \widetilde{\mathbf{k}}, \vartheta) + r(t, \mathbf{m}, \mathbf{k})\vartheta &= g(t, \mathbf{m}, \mathbf{k}, P(t, \mathbf{m}, \mathbf{k}) - \vartheta) \\ &+ \lambda_{-1}(t, \mathbf{m}, \mathbf{k})\widetilde{\xi}_{-1}(t, \mathbf{m}, \mathbf{k}; P(t, \mathbf{m}, \mathbf{k}) - \vartheta) \\ &+ \lambda_0(t, \mathbf{m}, \mathbf{k})\widetilde{\xi}_0(t, \mathbf{m}, \mathbf{k}; P(t, \mathbf{m}, \mathbf{k}) - \vartheta) \\ &- \gamma(t, \mathbf{m}, \mathbf{k})\vartheta. \end{aligned} \tag{3.12}$$

To summarize, the assumption (A) holds for the measure $\mathbb{P}$ defined above in (A.2) and for the Markovian specification of the coefficient $\widetilde{f}_t(\vartheta) := \widetilde{f}(t, X_t, \vartheta)$, where the function $\widetilde{f} = \widetilde{f}^{gc}(t, \mathbf{m}, \widetilde{\mathbf{k}}, \vartheta)$ is given by (3.12), in (A.3).

4. Dynamic Marshall-Olkin Copula TVA Model

The above dynamic Gaussian copula (DGC) model can be sufficient to deal with CVA on portfolios of CDSs. If CDOs are also present in the reference portfolio, a Gaussian copula dependence structure is not rich enough. Instead, one can use the dynamic Marshall-Olkin (DMO) common shock model of [3] and [4] (see Sect. 4.1).

4.1. *Model of Default Times*

We define a certain number m (typically small: a few units) of groups $I_j \subseteq N$ of obligors who are likely to default simultaneously, for $l = 1, \ldots, m$. The idea is that at every time t, there is a positive probability that the survivors of the group of obligors I_j (obligors of group I_j still alive at time t) default simultaneously. Let $\mathcal{I} = \{I_1, \ldots, I_m\}$, $\mathcal{Y} = \{\{-1\}, \{0\}, \{1\}, \ldots, \{n\}, I_1, \ldots, I_m\}$. Let shock intensity processes X^Y be given in the form of extended CIR processes as, for every $Y \in \mathcal{Y}$,

$$dX_t^Y = a(b_Y(t) - X_t^Y)dt + c\sqrt{X_t^Y}\,dW_t^Y \tag{4.1}$$

for non-negative constants a, c, non-negative functions $b_Y(t)$ and independent Brownian motions W^Y (in their own filtration $\mathcal{F}^{\mathbf{W}}$), under a pricing measure $\mathbb{Q}$.

Remark 4.1. The case of deterministic intensities $X_t^Y = b_Y(t)$ can be embedded in this framework as the limiting case of an "infinite mean-reversion speed" a (see [1]).

For every $Y \in \mathcal{Y}$, we define

$$\widehat{\tau}_Y = \inf\{t > 0; \int_0^t X_s^Y\,ds > \mathcal{E}_Y\},$$

where the $\mathcal{E}_Y$ are i.i.d. standard exponential random variables. Then, for every obligor $i \in N$, we set

$$\tau_i = \min_{\{Y \in \mathcal{Y};\, i \in Y\}} \widehat{\tau}_Y\,, \quad H_t^i = \mathbb{1}_{\tau_i \le t}. \tag{4.2}$$

We consider the dynamic model $(\mathbf{X}_t, \mathbf{H}_t) = ((X_t^Y)_{Y \in \mathcal{Y}}, (H_t^i)_{i \in N})$ relatively to the filtration $\mathbb{G} = \mathbb{G}^{mo}$ such that for every t

$$\mathcal{G}_t^{mo} = \mathcal{F}_t^{\mathbf{W}} \vee \Big(\bigvee_{i \in N} (\tau_i \wedge t)\Big).$$

This model can be viewed as a doubly stochastic (via the stochastic intensities X^Y) and dynamized (via the introduction of the filtration $\mathbb{G}$) generalization of the model of [22]. The purpose of the factor process $\mathcal{X}$ is to more realistically model diffusive randomness of credit spreads. As developed in [2], $(\mathbf{X}, \mathbf{H})$ is a Markov process, with infinitesimal generator $\mathcal{L} = \mathcal{L}^{mo}$ acting as follows on every function $u = u(t, \mathbf{x}, \mathbf{k})$ with $t \in \mathbb{R}_+, \mathbf{x} = (x_Y)_{Y \in \mathcal{Y}}$ and $\mathbf{k} = (k_i)_{i \in N}$:

$$\mathcal{L}^{mo} u(t, \mathbf{x}, \mathbf{k}) = \mathcal{A}_x u(t, \mathbf{x}, \mathbf{k}) + \sum_{Y \in \mathcal{Y}} x_Y \delta_Y u(t, \mathbf{x}, \mathbf{k}) \tag{4.3}$$

with

$$\mathcal{A}_x u = \sum_{Y \in \mathcal{Y}} \left(a(b_Y(t) - x_Y)\partial_{x_Y} u + \frac{1}{2} c^2 \; x_Y \partial^2_{x_Y^2} \right) u$$

and where we denote, for $Y \in \mathcal{Y}$,

$$\delta_Y u(t, \mathbf{x}, \mathbf{k}) = u(t, \mathbf{x}, \mathbf{k}^Y) - u(t, \mathbf{x}, \mathbf{k})$$

in which $\mathbf{k}^Y$ denotes the vector obtained from $\mathbf{k} = (k_i)_{i \in N}$ by replacing the components k_i, $i \in Y$, by numbers one. For every $Z \subseteq N$, we have the following expression for the predictable intensity λ_t^Z of the indicator process of the event of a joint default of names in set Z and only in Z:

$$\lambda_t^Z = \lambda_Z(t, \mathbf{X}_t, \mathbf{H}_{t-}) = \sum_{Y \in \mathcal{Y};\, Y_t = Z} X_t^Y, \tag{4.4}$$

where Y_t stands for the set of survivors of set Y "right before time t," for every $Y \in \mathcal{Y}$. So $Y_t = Y \cap \mathrm{supp}^c(\mathbf{H}_{t-})$. One denotes by M^Z the corresponding compensated set-event martingale, so for $t \in [0, T]$,

$$dM_t^Z = d\mathbb{1}_{\tau_Z \leq t} - \lambda_t^Z dt \tag{4.5}$$

where τ_Z stands for the time of a joint default of names in set Z and only in Z. The W^Y and the M^Z have the $(\mathbb{G}, \mathbb{Q})$-martingale representation property.

4.2. *TVA Model*

A DMO setup can also be used as a credit portfolio TVA model, for a mark space of τ given as

$$E = \{Z \subseteq N;\ -1 \text{ or } 0 \in Z\},$$

under the assumption, enforcing (2.3) in the present setup, that there exists a function $\widetilde{\xi}$ satisfying

$$\xi(\pi) = \widetilde{\xi}(\tau, \mathbf{X}_\tau, \mathbf{H}_{\tau-}, \mathbf{H}^Z_{\tau-}; \pi) =: \widetilde{\xi}^Z_\tau(\pi) \tag{4.6}$$

on every event $\{\tau = \tau_Z\}$. Note that in order to meet (4.8) below, we really need "$\widetilde{\xi}(\tau, \mathbf{X}_\tau, \mathbf{H}_{\tau-}, \mathbf{H}^Z_{\tau-}; \pi)$" in (4.6), as opposed to what would be a less specific "$\widetilde{\xi}_Z(\tau, \mathbf{X}_\tau, \mathbf{H}_{\tau-}; \pi)$" (compare (3.5)) . But this is fine since, as discussed in [11], "$\widetilde{\xi}(\tau, \mathbf{X}_\tau, \mathbf{H}_{\tau-}, \mathbf{H}^Z_{\tau-}; \pi)$" holds in the targeted TVA applications.

Let also

$$\mathcal{Z}_t = \{Z \subseteq N; Z = Y_t \text{ for at least one } Y \in \mathcal{Y}\} \setminus \emptyset$$

denote the set of all non-empty sets of survivors of groups Y in $\mathcal{Y}$ right before time t and let

$$\mathcal{Z}_t^\bullet = \{Z \in \mathcal{Z}_t;\ -1 \text{ or } 0 \in Z\} = \mathcal{Z}_t \cap E.$$

Observe that λ_t^Z in (4.4) vanishes for $Z \notin \mathcal{Z}_t$. Therefore, the coefficient $f_t(P_t - \vartheta)$ of (2.5) is given, for $t \in [0, \bar{\tau}]$, through:

$$f_t(P_t - \vartheta) + r_t\vartheta = g_t(P_t - \vartheta) + \sum_{Z \in \mathcal{Z}_t^\bullet} \lambda_t^Z \widetilde{\xi}_t^Z (P_t - \vartheta) - \gamma_t\vartheta. \tag{4.7}$$

Moreover, letting

$$\mathcal{Y}_t^\bullet = \{Y \in \mathcal{Y};\ -1 \text{ or } 0 \in Y_t\},$$

for every process of the form

$$V(\tau, \mathbf{X}_\tau, \mathbf{H}_{\tau-}, \mathbf{H}_{\tau-}^Z), \tag{4.8}$$

we have through (4.4):

$$\begin{aligned}
\sum_{Z \in \mathcal{Z}_t^\bullet} \lambda_t^Z V(t, \mathbf{X}_t, \mathbf{H}_{t-}, \mathbf{H}_{t-}^Z) &= \sum_{Z \in \mathcal{Z}_t^\bullet} \left(\sum_{Y \in \mathcal{Y}_t^\bullet; Y_t = Z} X_t^Y \right) V(t, \mathbf{X}_t, \mathbf{H}_{t-}, \mathbf{H}_{t-}^Z) \\
&= \sum_{Z \in \mathcal{Z}_t^\bullet} \left(\sum_{Y \in \mathcal{Y}_t^\bullet; Y_t = Z} X_t^Y V(t, \mathbf{X}_t, \mathbf{H}_{t-}, \mathbf{H}_{t-}^Y) \right) \\
&= \sum_{Y \in \mathcal{Y}_t^\bullet} X_t^Y V(t, \mathbf{X}_t, \mathbf{H}_{t-}, \mathbf{H}_{t-}^Y),
\end{aligned}$$

where the middle equality follows from the fact that $\mathbf{H}_{t-}^Z = \mathbf{H}_{t-}^Y$, for every $Y \in \mathcal{Y}$ such that $Y_t = Z$. In particular, we have (in view of (4.6), which fits (4.8), regarding the second equality):

$$\gamma_t = \sum_{Z \in \mathcal{Z}_t^\bullet} \lambda_t^Z = \sum_{Y \in \mathcal{Y}_t^\bullet} X_t^Y, \quad \sum_{\mathcal{Z}_t^\bullet} \lambda_t^Z \widetilde{\xi}_t^Z (\pi) = \sum_{Y \in \mathcal{Y}_t^\bullet} X_t^Y \widetilde{\xi}_t^Y (\pi), \tag{4.9}$$

so that by (4.7):

$$f_t(P_t - \vartheta) + r_t\vartheta = g_t(P_t - \vartheta) + \sum_{Y \in \mathcal{Y}_t^\bullet} X_t^Y \left(\widetilde{\xi}_t^Y (P_t - \vartheta) - \vartheta \right). \tag{4.10}$$

4.3. *Reduced-Form TVA Approach*

Next, we want to use a reduced DMO model in a pre-default Markovian TVA approach. Let $\widetilde{\mathcal{Y}} = \{\{1\}, \ldots, \{n\}\} \cup \widetilde{\mathcal{I}}$, where $\widetilde{\mathcal{I}}$ consists of those I_j in $\mathcal{I}$ which do not contain -1 or 0. Defining for every obligor $i \in N^\star$

$$\widetilde{\tau}_i = \min_{\{Y \in \widetilde{\mathcal{Y}};\, i \in Y\}} \widehat{\tau}_Y, \quad \widetilde{H}_t^i = \mathbb{1}_{\widetilde{\tau}_i \leq t},$$

we shall use the reduced model $X_t = (\mathbf{X}_t, \widetilde{\mathbf{H}}_t)$, where $\widetilde{\mathbf{H}} = (\widetilde{H}^i)_{i \in N^\star}$, relatively to the filtration $\mathbb{F} = \mathbb{F}^{mo}$ such that for every t

$$\mathcal{F}_t^{mo} = \mathcal{F}_t^{\mathbf{W}} \vee \Big(\bigvee_{i \in N^\star} (\widetilde{\tau}_i \wedge t) \Big)$$

and to the unchanged probability measure $\mathbb{P} = \mathbb{Q}$. By virtue of the Markov copula DMO features (see [2]), the reduced model is $(\mathbb{F}, \mathbb{Q})$- (in fact, even $(\mathbb{G}, \mathbb{Q})$-) Markov, with the following generator $\mathcal{A} = \mathcal{A}^{mo}$ in a notation similar to (4.3):

$$\mathcal{A}^{mo} u(t, \mathbf{x}, \widetilde{\mathbf{k}}) = \mathcal{A}_x u(t, \mathbf{x}, \widetilde{\mathbf{k}}) + \sum_{Y \in \widetilde{\mathcal{Y}}} x_Y \delta_Y u(t, \mathbf{x}, \widetilde{\mathbf{k}}). \tag{4.11}$$

Moreover, let, for $Z \subseteq N^\star$,

$$d\widetilde{M}_t^Z = d\mathbb{1}_{\tau_Z \leq t} - \widetilde{\lambda}_t^Z dt \text{ with } \widetilde{\lambda}_t^Z = \sum_{Y \in \widetilde{\mathcal{Y}};\, Y_t = Z} X_t^Y. \tag{4.12}$$

The $W^Y, Y \in \mathcal{Y}$ and the $\widetilde{M}^Z, Z \subseteq N^\star$ have the $(\mathbb{F}, \mathbb{Q})$-martingale representation property. Also note that, for $Z \subseteq N^\star$, we have $\widetilde{M}_t^Z = M_t^Z$ on $\{t \leq \bar{\tau}\}$, since by (4.4) and (4.12):

$$\mathbb{1}_{\{t \leq \bar{\tau}\}} \lambda_t^Z dt = \mathbb{1}_{\{t \leq \bar{\tau}\}} \widetilde{\lambda}_t^Z dt.$$

Of course before τ every dependence in $\mathbf{H}_t$ reduces to a dependence in $\widetilde{\mathbf{H}}_t$. Let

$$\mathcal{Y}^\bullet = \{Y \in \mathcal{Y};\, -1 \text{ or } 0 \in Y\}.$$

Under the additional assumption that the processes $r_t, g_t(\pi)$ (for every real π) and P_t are given before τ as continuous functions of (t, X_t) (functions denoted below by the same letters as the related processes), then, in view of (4.10), we have:

$$f_t(P_t - \vartheta)dt = \widetilde{f}(t, X_t, \vartheta)dt, \quad t \in [0, \bar{\tau}],$$

for the function $\widetilde{f} = \widetilde{f}^{mo}$ given, letting $\mathbf{k} = (0, 0, \widetilde{\mathbf{k}})$ for every $\widetilde{\mathbf{k}} \in \{0, 1\}^n$, through:

$$\begin{aligned} \widetilde{f}^{mo}(t, \mathbf{x}, \widetilde{\mathbf{k}}, \vartheta) + r(t, \mathbf{x}, \mathbf{k})\vartheta &= g(t, \mathbf{x}, \mathbf{k}, P(s, \mathbf{m}, \mathbf{k}) - \vartheta) \\ &+ \sum_{Y \in \mathcal{Y}^{\bullet}} x_Y \Big(\widetilde{\xi}(t, \mathbf{x}, \mathbf{k}, \mathbf{k}^Y; P(t, \mathbf{x}, \mathbf{k}) - \vartheta) - \vartheta \Big) . \end{aligned} \tag{4.13}$$

Note that an $\mathbb{F}$-martingale does not jump at τ and an $\mathbb{F}$-martingale stopped at τ is a $\mathbb{G}$-martingale, by the martingale representation properties which are valid in the full and in the reduced DMO models (all under $\mathbb{Q}$). So the assumption (A) holds with $\mathbb{P} = \mathbb{Q}$ in (A.2) and for the Markovian specification of the coefficient $\widetilde{f}_t(\vartheta) := \widetilde{f}(t, X_t, \vartheta)$, where the function $\widetilde{f} = \widetilde{f}^{mo}(t, \mathbf{x}, \widetilde{\mathbf{k}}, \vartheta)$ is given by (4.13), in (A.3).

Conclusion In the dynamic Gaussian copula (DGC) model as in the dynamic Marshall-Olkin (DMO) common shock model, one ends up with a pre-default TVA equation over $[0, T]$ amenable to purely forward (nonlinear as necessary) simulation schemes. Using suitable specifications for the exposure at default ξ and the funding coefficient g_t, one can then develop (2.5) and the ensuing expressions (3.12) (in the DGC model) or (4.13) (in the DMO model) in the form of CVA, DVA, LVA and RC decompositions such as the ones of [10, 15]. See [11] for further developments in this direction, using default times with marks as a unifying modeling tool allowing one not only to deal with counterparty risk and funding on credit derivatives as shown above, but more broadly to account for various possible wrong-way and gap risks scenarios and features.

Acknowledgements

This research benefited from the support of the "Chair Markets in Transition" under the aegis of Louis Bachelier laboratory, a joint initiative of École Polytechnique, Université d'Évry Val d'Essonne and Fédération Bancaire Française.

References

[1] Bielecki, T. R., Cousin, A., Crépey, S. and Herbertsson, A. (2013a). A bottom-up dynamic model of portfolio credit risk with stochastic intensities and random recoveries, Communications in Statistics – Theory and Methods Forthcoming.

[2] Bielecki, T. R., Cousin, A., Crépey, S. and Herbertsson, A. (2013b). A bottom-up dynamic model of portfolio credit risk – Part I: Markov copula perspective, in Recent Advances in Financial Engineering 2012, World Scientific, forthcoming.
[3] Bielecki, T. R., Cousin, A., Crépey, S. and Herbertsson, A. (2013c). Dynamic hedging of portfolio credit risk in a Markov copula model, Journal of Optimization Theory and Applications Forthcoming.
[4] Bielecki, T. R. and Crépey, S. (2011). Dynamic Hedging of Counterparty Exposure, in T. Zariphopoulou, M. Rutkowski and Y. Kabanov. eds., The Musiela Festschrift, Springer, forthcoming.
[5] Bielecki, T. R., Crépey, S. and Jeanblanc, M. (2010). Up and down credit risk, Quantitative Finance **10**, 10, pp. 1469–7696.
[6] Brigo, D. and Capponi, A. (2010). Bilateral counterparty risk with application to CDSs, Risk Magazine , March, pp. 85–90.
[7] Brigo, D., Morini, M. and Pallavicini, A. (2013). Counterparty Credit Risk, Collateral and Funding with pricing cases for all asset classes, Wiley.
[8] Crépey, S. (2011). A BSDE approach to counterparty risk under funding constraints, LAP Preprint n° 326, June 2011 (available at http://www.maths.univ-evry.fr/prepubli/index.html).
[9] Crépey, S. (2012a). Bilateral Counterparty risk under funding constraints – Part I: Pricing, Mathematical Finance DOI 10.1111/mafi.12004.
[10] Crépey, S. (2012b). Bilateral Counterparty risk under funding constraints – Part II: CVA, Mathematical Finance DOI 10.1111/mafi.12005.
[11] Crépey, S. (2013a). Counterparty risk modeling: Beyond immersion, LAP Preprint n° 403, October 2013 (available at http://www.maths.univ-evry.fr/prepubli/index.html).
[12] Crépey, S. (2013b). Financial Modeling: A Backward Stochastic Differential Equations Perspective, Springer Finance Textbooks, Springer.
[13] Crépey, S. (2013c). Preface to the special issue "Frontiers of counterparty risk", International Journal of Theoretical and Applied Finance **16**, 2, p. 1302001 (9 pages).
[14] Crépey, S., Bielecki, T. R. and Brigo, D. (2014). Counterparty Risk and Funding – A Tale of Two Puzzles, Taylor & Francis, forthcoming.
[15] Crépey, S., Gerboud, R., Grbac, Z. and Ngor, N. (2013a). Counterparty risk and funding: The four wings of the TVA, International Journal of Theoretical and Applied Finance **16**, 2, p. 1350006 (31 pages).
[16] Crépey, S., Jeanblanc, M. and Wu, D. L. (2013b). Informationally dynamized Gaussian copula, International Journal of Theoretical and Applied Finance **16**, 2, p. 1350008 (29 pages).
[17] Crépey, S. and Song, S. (2014). General counterparty risk modeling, In preparation.
[18] Fermanian, J.-D. and Vigneron, O. (2010). Pricing and hedging basket credit derivatives in the Gaussian copula, Risk Magazine , February, pp. 92–96.
[19] Fujii, M. and Takahashi, A. (2012). Perturbative expansion technique for non-linear FBSDEs with interacting particle method, ArXiv:1204.2638.
[20] Henry-Labordère, P. (2012). Cutting CVA's complexity, Risk Magazine,

July, pp. 67–73.
[21] Li, D. (2000). On default correlation: A copula function approach, Journal of Fixed Income **9**, 4, pp. 43–54.
[22] Marshall, A. W. and Olkin, I. (1967). A multivariate exponential distribution, J. Amer. Statist. Assoc. **2**, pp. 84–98.
[23] Pallavicini, A., Perini, D. and Brigo, D. (2011). Funding valuation adjustment: a consistent framework including CVA, DVA, collateral, netting rules and re-hypothecation, SSRN.com and arXiv.org.
[24] Pallavicini, A., Perini, D. and Brigo, D. (2012). Funding, collateral and hedging: uncovering the mechanics and the subtleties of funding valuation adjustments, http://dx.doi.org/10.2139/ssrn.2161528.

Dynamic One-default Model

Shiqi Song*

Abstract. A dynamic one-default model consists of a random time whose conditional law with respect to a reference filtration is determined by a stochastic differential equation. This model has been introduced in Jeanblanc et al. [14] under a continuity assumption. This paper extends the result to general cases with jumps. We then study two essential properties : the enlargement of filtration formula and the differentiability of the conditional distribution function of the random time.

1. Introduction

We consider a one-default model, i.e., a model consisting of a random time τ combined with a filtration $\mathbb{F}$ under a probability measure $\mathbb{Q}$. The one-default models are widely applied in modeling financial risk and in valuation of financial products such as CDS. The usefulness of a one-default model is conditional upon the way that the conditional laws of τ can be computed with respect to the filtration $\mathbb{F}$. The most used examples of random times are the independent time, the Cox time, the honest time, the pseudo stopping time, the initial time, etc (cf., for example, the references [4, 8, 12, 13, 21, 24, 25]). In the paper Jeanblanc et al. [14] a new class of random times has been introduced. Precisely, it is proved that, for any continuous increasing process Λ null at the origin, for any continuous non-negative local martingale N such that $0 < N_t e^{-\Lambda_t} < 1$, $t > 0$, for any continuous local martingale Y, for any C^1 function f on $\mathbb{R}$ null at the origin, there exists a random variable τ such that the family of conditional expectations $M_t^u = \mathbb{Q}[\tau \leq u|\mathcal{F}_t]$, $0 < u, t < \infty$, satisfies the following stochastic

*Université d'Évry Val d'Essonne, Laboratoire Analyse et Probabilités,
Email: shiqi.song@univ-evry.fr

differential equation:

$$(\natural_u) \begin{cases} dM_t^u = M_t^u \left(-\frac{e^{-\Lambda_t}}{1-N_t e^{-\Lambda_t}} dN_t + f(M_t^u - (1 - Z_t))dY_t \right), \ t \in [u, \infty), \\ M_u^u = 1 - N_u e^{-\Lambda_u}. \end{cases}$$

We will call such a setting a dynamic model (which was called $\natural$-model in Jeanblanc et al. [14]). Notice that the equation $(\natural_u)$ implies that the Azéma supermartingale of τ is $\mathbb{Q}[u < \tau | \mathcal{F}_u] = N_u e^{-\Lambda_u}$. One aim of the work Jeanblanc et al. [14] is to show that, for any given Azéma supermartingale with continuity condition, there exist an infinity of one-default models with the same Azéma supermartingale.

There are two remarkable properties about a dynamic model. It is the only one in which the conditional laws of τ with respect to $\mathbb{F}$ are defined by a system of dynamic equations. The $\natural$-equation displays the evolution of the defaultable market. The knowledge of market evolution is a valuable property. This evolution equation had allowed Jeanblanc et al. [14] to establish the so-called enlargement of filtration formula, i.e. the $\mathbb{G}$ semimartingale decomposition formula for the $\mathbb{F}$ martingales. It also is proved in Jeanblanc et al. [14] that, reciprocally, the $\natural$-equation can be recovered from the enlargement of filtration formula in a way similar to that a differentiable function can be deduced from its derivative.

We recall that the formula of enlargement of filtration is essential, when the no-arbitrage price valuation is considered in a one-default model (cf. the references [2, 9, 29]). Much as the enlargement of filtration formula is universally valid before the default time τ, for a long time, the part of the enlargement of filtration formula after τ was merely proved for the honest time model or the initial time model. The dynamic models constitute the third family of models where the enlargement of filtration formula is valid on the whole $\mathbb{R}_+$. In addition, the enlargement of filtration formula in a dynamic model has a richer structure than that of honest time model, and has a more accurate expression than that of the initial time model. We recall also how widely the financial models are defined by stochastic differential equations, because it is one of the best ways to represent the evolution of a financial market. Usually, in a one-default model, there is no such a possibility to design the evolution. Now with the dynamic model, this becomes available.

The second remarkable property of the dynamic model is its rich and flexible system of inputs (Z, Y, f). The input Z corresponds to the Azéma supermartingale of the dynamic model and determines the default intensity.

The inputs Y and f describe the evolution of the market after the default time τ. Such a system of inputs sets up a propitious framework for inferring the market behavior and for calibrating the financial data.

We believe that the dynamic model can be a useful instrument to model a financial market. The purpose of this paper is to extend the dynamic model of Jeanblanc et al. [14] to the case of a general Azéma supermartingale.

Here is an overview of the main results.

(1) The dynamic model is founded on the martingale point of view of the one-default models. It was introduced in Jeanblanc et al. [14] with help of the notion of $i\mathcal{M}_Z$ family. In Section 2, the definition and results about $i\mathcal{M}_Z$ are recalled.

(2) The dynamic model with jumps will be defined in Section 3 for suitable parameters $Z, \mathbf{F}, \mathbf{Y}$, where Z is an Azéma supermartingale, $\mathbf{F}$ is a Lipschitz functional in the sense of Protter [26], and $\mathbf{Y}$ is a local martingale.

(3) The first problem when we tried to extend the $\natural$-model from continuous case to jump case, was how the equation $(\natural)_u$ should be changed to take account of the jumps. The correct change is to replace the term $-\frac{e^{-\Lambda_t}dN_t}{1-N_t e^{-\Lambda_t}}$ by $-\frac{dM_t}{^{\mathrm{p}}(1-Z_t)}$, where Z is the given Azéma supermartingale, M is the martingale part of Z, $^{\mathrm{p}}(1-Z_t)$ denotes the $\mathbb{F}$ predictable projection of $(1-Z)$. Our reflection is greatly influenced by Meyer et al. [22] and Yoeurp [30] and by the fact that $-\frac{\Delta_t M}{^{\mathrm{p}}(1-Z_t)} > -1$, where $\Delta_t M$ denotes the jump of M at time t.

(4) The stochastic differential equation of the form $dX_t = {^{\mathrm{p}}X_t} dW_t + dV_t$, where W is a local martingale, V is an increasing predictable process, plays an important role in the study of the dynamic model. This equation had been considered in a different form in the references [22, 30]. We obtain an explicit formula in Section 3 for the solution of such an equation, which is different from that of the references [22, 30]. To the best of our knowledge this formula seems not exist in the literature.

(5) To establish the dynamic model in the continuous case, we needed to look at the equation $(\natural)_u$ in its general form

$$(\natural_u)\begin{cases} dX_t = X_t\Big(-\frac{e^{-\Lambda_t}}{1-N_t e^{-\Lambda_t}}dN_t + f(X_t-(1-Z_t))dY_t\Big), & t\in[u,\infty),\\ X_u = x. \end{cases}$$

We proved three properties on the solutions X^x of the equation $(\natural)_u$: $X_t^x \geq 0, X_t^x \leq 1 - Z_t$ and $X_t^x \leq X_t^y$ if $0 \leq x \leq y \leq 1 - Z_u$. For the first property, it was the consequence of the Doléans-Dade exponential formula. The second property was obtained by the local time technique (cf. Revuz et al. [27]). Finally the third property is issued from the one-dimensional comparison theorem. It happens that the local time technique and the comparison theorem become inefficient in the jump case. This technical difficulty is overcome by introducing the notion of $\natural$-pair in Section 3.

(6) In Section 4 we prove the enlargement of filtration formula for the dynamic models with Markovian coefficients. A striking point in this section is the use of the prediction process (cf. the references [3, 19, 23, 31]). The notion of the prediction process was introduced to represent the filtrations as Markov processes, which would give a pleasant way to make calculus on the filtrations. For long, the prediction process was assumed to contribute to the theory of enlargement of filtrations. See the study given in Song [28]. Nevertheless, the prediction process had not been widely applied in the literature, because likely the prediction process seemed not indispensable in the known examples. (For example, it was not used in the study of the continuous dynamic model of Jeanblanc et al. [14].) However, with the presence of the jumps, the prediction process appears unavoidable in the establishment of the enlargement of filtration formula in the dynamic models.

(7) When a one-default model is applied for a practical purpose, an explicit formula for the conditional laws $\mathbb{Q}[\tau \in du|\mathcal{F}_t]$ will be needed. In Section 5 we explain how to compute these conditional laws in dynamic models. Actually in the case of Markovian coefficient with regularity, we will be able to compute the derivatives

$$\lim_{v \to u} \frac{\mathbb{Q}[u < \tau \leq v|\mathcal{F}_t]}{A_v - A_u},$$

where A is the drift part of the submartingale $1 - Z$. This yields in particular the absolute continuity of $\mathbb{Q}[\tau \in du|\mathcal{F}_t]$ (for finite t) with respect to dA_u.

(8) The dynamic model is not a particular case of known models such as honest model or initial models. However, many links exist between dynamic model and the other models. Roughly speaking,

when $Z = e^{-\Lambda}$ for some continuous increasing predictable process Λ with $\Lambda_0 = 0$ and when $\mathbf{F} \equiv 0$, the dynamic model yields a Cox time. When the drift part of $1 - Z$ is absolutely continuous with respect to Lebesgue measure and when $\mathbf{F}$ is of Markovian form, the dynamic model yields an initial time. When Z is a predictable decreasing process, the dynamic model yields a pseudo stopping time. Certainly, if non of these conditions are satisfied, the dynamic model can produce random times of a new kind. Moreover, the dynamic model studied in this paper can not be honest time because of the assumption $\mathbf{Hy}(Z)$ below.

(9) A short discussion is given in the last section on the case where $1 - Z$ may be null.

2. Increasing Family of Bounded and Positive Martingales

This section is borrowed from Section 2 of Jeanblanc et al. [14] with some modifications.

2.1. *Product probability space*

As explained in Jeanblanc et al. [14] an one-default model can always be imbedded isomorphically into a product probability space. Henceforth, only models on product spaces will be studied in the present paper. Precisely, consider a measurable space $(\Omega, \mathcal{A})$ ($\mathcal{A}$ being a σ-algebra) equipped with a filtration $\mathbb{F} = (\mathcal{F}_t)_{t\geq 0}$ of sub-σ-algebras of $\mathcal{A}$, and the product space $[0,\infty]\times\Omega$ equipped with the product σ-algebra $\mathcal{B}[0,\infty]\otimes\mathcal{A}$. Introduce the two maps π, τ defined as follows: $\pi(s,\omega) = \omega$ and $\tau(s,\omega) = s$ and the filtration $\widehat{\mathbb{F}} = \pi^{-1}(\mathbb{F})$ on $[0,\infty]\times\Omega$. Provided with the pair $(\tau, \widehat{\mathbb{F}})$, constructing a one-default models on the product space amounts to constructing a probability measure on $\mathcal{B}[0,\infty]\otimes\mathcal{A}$. Recall that, working with the product space, it is usual to identify ω with the map π. In this way the filtration $\mathbb{F}$ is identified with $\widehat{\mathbb{F}}$ and the functions on Ω become functions on $[0,\infty]\times\Omega$.

2.2. *iM family associated with a probability measure* $\mathbb{Q}$

There exists a variety of ways to construct probability measures on the product space. Our approach in this paper is based on the following observation. A probability measure $\mathbb{Q}$ on $\mathcal{B}[0,\infty]\otimes\mathcal{A}$ is determined by its disintegration into its restriction on $\mathcal{A}$ and the conditional law of τ given $\mathcal{A}$. We will consider this conditional law $\mathbb{Q}[\tau \in du|\mathcal{A}]$ as the terminal value of the measure-valued martingale $(\mathbb{Q}[\tau \in du|\mathcal{F}_t] : 0 \le t \le \infty)$ and we will define this measure-valued martingale by a stochastic differential equation. For this purpose, we introduce the following notion.

Let $\mathbb{P}$ be a probability measure on $\mathcal{A}$. An increasing family of positive $(\mathbb{P}, \mathbb{F})$ martingales bounded by 1 (in short $i\!M(\mathbb{P}, \mathbb{F})$ or simply $i\!M$) is a family of processes $(M^u : 0 \le u \le \infty)$ satisfying the following conditions:

(1) Every M^u is a càdlàg non-negative $(\mathbb{P}, \mathbb{F})$ martingale on $[u, \infty]$, bounded by 1, and closed by M^u_∞.

(2) For every $0 \le t \le \infty$, the random map $u \in [0, t] \to M^u_t$ is a right continuous non-decreasing function.

(3) $M^\infty_\infty = 1$.

The theorem below gives the relationship between an $i\!M$ family and the construction of a probability measure on the product space. We recall that we identify the elements of $(\Omega, \mathcal{A})$ with elements of the product space.

Theorem 2.1. *Let $\mathbb{P}$ be a probability measure on $\mathcal{A}$. Suppose that the filtration $\mathbb{F}$ is right continuous and contains the $(\mathbb{P}, \mathcal{F}_\infty)$ null sets.*

(1) For any probability measure $\mathbb{Q}$ on the σ-algebra $\mathcal{B}[0, \infty] \otimes \mathcal{A}$ which coincides with $\mathbb{P}$ on $\mathcal{F}_\infty$, there exists a unique $i\!M(\mathbb{P}, \mathbb{F}) = (M^u : 0 \le u \le \infty)$ such that, for $0 \le u \le t \le \infty$,

$$M^u_t = \mathbb{Q}[\tau \le u|\mathcal{F}_t].$$

We shall say that this $i\!M$ is associated with $\mathbb{Q}$.

(2) Let $(M^u : 0 \le u \le \infty)$ be an $i\!M(\mathbb{P}, \mathbb{F})$. There is a unique probability measure $\mathbb{Q}$ on the σ-algebra $\mathcal{B}[0, \infty] \otimes \mathcal{A}$ which coincides with $\mathbb{P}$ on $\mathcal{A}$ and satisfies $\mathbb{Q}[\tau \le u|\mathcal{F}_t] = M^u_t$ for $0 \le u \le t \le \infty$, and $\mathbb{Q}[\tau \le u|\mathcal{A}] = M^u_\infty$. We shall say that $\mathbb{Q}$ is associated with the $i\!M$ and with $\mathbb{P}$.

Proof. Consider the first assertion. The uniqueness is clear. For each $0 \le u \le \infty$, let $(G^u_t : 0 \le t \le \infty)$ be a càdlàg version of the $(\mathbb{P}, \mathbb{F})$ martingale $\mathbb{Q}[\tau \le u|\mathcal{F}_t], t \in [0, \infty]$. For $u < v$, for any $\mathbb{F}$ stopping time T, $\mathbb{P}$ almost surely, we have $0 \le G^u_T \le G^v_T \le 1$. Set

$$M^u_t = \inf_{w \in \mathbb{Q}_+, w > u} (G^w_t \wedge 1)^+, \ 0 \le u < \infty, 0 \le t \le \infty.$$

Then,

- For $0 \le u < \infty$, the process $M^u = (M_t^u)_{0 \le t \le \infty}$ is $\mathbb{F}$-optional.
- For $0 \le t \le \infty$, $u \in [0, \infty) \to M_t^u$ is right continuous.
- For $0 \le t \le \infty, 0 \le u < v < \infty$, $0 \le M_t^u \le M_t^v \le 1$ everywhere.
- For $0 \le u < \infty$, for any $\mathbb{F}$ stopping time T, $\mathbb{P}$ almost surely, $G_T^u \le M_T^u$.

Let $0 \le u < \infty$ and T be an $\mathbb{F}$ stopping time. We write

$$\mathbb{Q}[\tau \le u] = \mathbb{Q}[G_T^u] \le \mathbb{Q}[M_T^u] = \inf_{v \in \mathbb{Q}_+, u < v} \mathbb{Q}[G_T^v]$$
$$= \inf_{v \in \mathbb{Q}_+, u < v} \mathbb{Q}[\tau \le v] = \mathbb{Q}[\tau \le u].$$

This shows that $M_T^u = G_T^u$, $\mathbb{P}$ almost surely. Consequently G^u and M^u are $\mathbb{P}$ indistinguishable (cf. He et al. [10]). Define finally $M_t^\infty = 1$ for $0 \le t \le \infty$. The family of processes $(M^u : 0 \le u \le \infty)$ thus defined satisfies the conditions of an $\mathcal{M}$.

Consider the second assertion. The uniqueness is clear. We denote by $d_u M_\infty^u$ the random measure on $[0, \infty]$ associated with the càdlàg non-decreasing map $u \in [0, \infty] \to M_\infty^u$ (noting that at 0 this measure can have a mass equal to M_∞^0). Define a probability measure on $([0, \infty] \times \Omega, \mathcal{B} \otimes \mathcal{A})$ by

$$\mathbb{Q}[F] = \mathbb{P}[\int_{[0,\infty]} F(s, \cdot) d_s M_\infty^s]$$

where $F(s, \omega) \in \mathcal{B}[0, \infty] \otimes \mathcal{A}$, $F(s, \omega) \ge 0$. Note that, for $A \in \mathcal{A}$:

$$\mathbb{Q}[A] = \mathbb{Q}[A \cap \{0 \le \tau \le \infty\}] = \mathbb{P}[\mathbb{I}_A \int_{[0,\infty]} d_s M_\infty^s] = \mathbb{P}[A].$$

Let $0 \le u \le t \le \infty, A \in \mathcal{F}_t$. We have

$$\mathbb{Q}[A \cap \{\tau \le u\}] = \mathbb{P}[\mathbb{I}_A \int_{[0,u]} d_s M_\infty^s] = \mathbb{P}[\mathbb{I}_A M_\infty^u] = \mathbb{P}[\mathbb{I}_A M_t^u] = \mathbb{Q}[\mathbb{I}_A M_t^u].$$

This implies that $\mathbb{Q}[\tau \le u | \mathcal{F}_t] = M_t^u$. If we rewrite the above computation for $A \in \mathcal{A}$, we will obtain $\mathbb{Q}[\tau \le u | \mathcal{A}] = M_\infty^u$. □

2.3. $\mathcal{M}_Z$ *family*

In this subsection, besides the given probability structure $(\mathbb{P}, \mathbb{F})$, we consider a $(\mathbb{P}, \mathbb{F})$ supermartingale Z such that $0 \le Z \le 1$. (Such a supermartingale is called Azéma supermartingale.) We introduce the following definition.

An increasing family of positive martingales bounded by $1-Z$ (in short $\mathrm{i}M_Z(\mathbb{P},\mathbb{F})$ or simply $\mathrm{i}M_Z$) is an $\mathrm{i}M = (M^u : 0 \le u \le \infty)$ satisfying the following conditions: for any $0 \le u \le t < \infty$, $M^u_u = 1-Z_u$ and $M^u_t \le 1-Z_t$.

The theorem below is an immediate consequence of Theorem 2.1.

Theorem 2.2. *Let $\mathbb{P}$ be a probability measure on $\mathcal{A}$. Suppose that the filtration $\mathbb{F}$ is right continuous and contains the $(\mathbb{P},\mathcal{F}_\infty)$ null sets. Let $(M^u : 0 \le u \le \infty)$ be an $\mathrm{i}M(\mathbb{P},\mathbb{F})$ associated with a probability measure $\mathbb{Q}$ on $\mathcal{B}[0,\infty] \otimes \mathcal{A}$ which coincides with $\mathbb{P}$ on $\mathcal{F}_\infty$. Then, $\mathrm{i}M$ is a $\mathrm{i}M_Z$ if and only if $\mathbb{Q}[t < \tau|\mathcal{F}_t] = Z_t$ for $t \ge 0$.*

3. The Dynamic Model

From now on, we fix a stochastic structure $(\mathbb{P},\mathbb{F})$ satisfying the usual condition and a $(\mathbb{P},\mathbb{F})$ supermartingale Z satisfying the following assumption.

Hy(Z): For $0 < t < \infty$, $1 - Z_t > 0$ and $1 - Z_{t-} > 0$.

3.1. *An affine stochastic differential equation for positive submartingales*

Lemma 3.1. *Let $u \ge 0$, W be a $(\mathbb{P},\mathbb{F})$ local martingale whose jump process satisfies $\Delta W > -1$, V be a non-decreasing $\mathbb{F}$ predictable process and $a \ge 0$. Consider the stochastic differential equation*

$$\begin{cases} d\Theta_t &= {}^{\mathfrak{p}}\Theta_t dW_t + dV_t, \ \ t \in [u,\infty), \\ \Theta_u &= a, \end{cases}$$

where the superscript ${}^{\cdot\mathfrak{p}}$ denotes the predictable projection in $\mathbb{F}$. Then, the solution Θ^a of this equation is given by

$$\Theta^a_t = \mathcal{E}(\mathbb{1}_{(u,\infty)} \centerdot W)_t \left(a + \int_u^t \frac{1}{\mathcal{E}(\mathbb{1}_{(u,\infty)} \centerdot W)_{s-}} dV_s \right), \ \ t \in [u,\infty).$$

Proof. The stochastic differential equation in this lemma has uniqueness of the solution. Let X to be the right hand term in the above formula. We apply the integration by parts formula to check that X is the solution of the equation:

$$
\begin{aligned}
&dX_t \\
&= \mathcal{E}(\mathbb{1}_{(u,\infty)} \centerdot W)_{t-} \left(a + \int_{(u,t)} \frac{1}{\mathcal{E}(\mathbb{1}_{(u,\infty)} \centerdot W)_{s-}} dV_s \right) \mathbb{1}_{(u,\infty)} dW_t \\
&\qquad + \mathbb{1}_{(u,\infty)} dV_t + d\left[\mathcal{E}(\mathbb{1}_{(u,\infty)} \centerdot W), \mathbb{1}_{(u,\infty)} \frac{1}{\mathcal{E}(\mathbb{1}_{(u,\infty)} \centerdot W)_-} \centerdot V \right]_t \\
&= \mathcal{E}(\mathbb{1}_{(u,\infty)} \centerdot W)_{t-} \left(a + \int_{(u,t)} \frac{1}{\mathcal{E}(\mathbb{1}_{(u,\infty)} \centerdot W)_{s-}} dV_s \right) \mathbb{1}_{(u,\infty)} dW_t \\
&\qquad + \mathbb{1}_{(u,\infty)} dV_t + \mathcal{E}(\mathbb{1}_{(u,\infty)} \centerdot W)_{t-} \frac{1}{\mathcal{E}(\mathbb{1}_{(u,\infty)} \centerdot W)_{t-}} \mathbb{1}_{(u,\infty)} d\,[W, V]_t \\
&= \mathcal{E}(\mathbb{1}_{(u,\infty)} \centerdot W)_{t-} \left(a + \int_{(u,t)} \frac{1}{\mathcal{E}(\mathbb{1}_{(u,\infty)} \centerdot W)_{s-}} dV_s \right) \mathbb{1}_{(u,\infty)} dW_t \\
&\qquad + \mathbb{1}_{(u,\infty)} dV_t + \mathcal{E}(\mathbb{1}_{(u,\infty)} \centerdot W)_{t-} \frac{1}{\mathcal{E}(\mathbb{1}_{(u,\infty)} \centerdot W)_{t-}} \mathbb{1}_{(u,\infty)} \Delta_t V dW_t \\
&= \mathcal{E}(\mathbb{1}_{(u,\infty)} \centerdot W)_{t-} \left(a + \int_u^t \frac{1}{\mathcal{E}(\mathbb{1}_{(u,\infty)} \centerdot W)_{s-}} dV_s \right) \mathbb{1}_{(u,\infty)} dW_t + \mathbb{1}_{(u,\infty)} dV_t \\
&= {}^{\mathfrak{p}}X_t \mathbb{1}_{(u,\infty)} dW_t + \mathbb{1}_{(u,\infty)} dV_t.
\end{aligned}
$$

□

As a corollary we have:

Lemma 3.2. *Suppose the same setting as in the previous lemma. Then, the solution Θ^a is non-negative for all $t \in [u, \infty)$. If $a > 0$, then $\Theta^a > 0$ and $\Theta^a_- > 0$ on $[u, \infty)$. Let $T = \inf\{s \geq u : V_s - V_u > 0\}$. Then $\Theta^0 > 0, \Theta^0_- > 0$ on (T, ∞).*

We have a partially inverse result, which is a direct consequence of Théorème 6.31 of Jacod [11] and of Meyer et al. [22].

Lemma 3.3. *Let $u \geq 0$. Let X be a $(\mathbb{P}, \mathbb{F})$ submartingale such that $X > 0$ and $\frac{1}{{}^{\mathfrak{p}}X}$ is locally bounded on $[u, \infty)$. Then there exists a $(\mathbb{P}, \mathbb{F})$ local martingale W with $\Delta W > -1$ and a non-decreasing $\mathbb{F}$ predictable process V such that*

$$dX_t = {}^{\mathfrak{p}}X_t dW_t + dV_t, \ t \in [u, \infty).$$

3.2. *The positive submartingale* $1 - Z$

We recall $\mathbf{Hy}(Z)$. Let $Z = M - A$ be the $(\mathbb{P}, \mathbb{F})$ canonical decomposition of Z with M a $(\mathbb{P}, \mathbb{F})$ local martingale and A a non-decreasing $\mathbb{F}$ predictable process. Then, ${}^{\mathfrak{p}}(1 - Z)_t = 1 - Z_{t-} + \Delta_t A > 0$ for any $0 < t < \infty$. For any $0 < u < \infty$, the submartingale $1 - Z$ satisfies the stochastic differential equation:

$$d(1 - Z)_t = {}^{\mathfrak{p}}(1 - Z)_t \, \frac{-dM_t}{{}^{\mathfrak{p}}(1 - Z)_t} + dA_t, \; t \in [u, \infty).$$

Note that $-\Delta_t M + {}^{\mathfrak{p}}(1 - Z)_t = 1 - Z_t$ so that

$$\frac{-\Delta_t M}{{}^{\mathfrak{p}}(1 - Z)_t} = \frac{1 - Z_t}{{}^{\mathfrak{p}}(1 - Z_t)} - 1 > -1, \; t \in [u, \infty).$$

We define, for $0 < u < \infty$, the $(\mathbb{P}, \mathbb{F})$ local martingale $\widetilde{m}_t^u = \int_u^t \frac{-dM_s}{{}^{\mathfrak{p}}(1-Z)_s}$. Since obviously $d\widetilde{m}_t^u = d\widetilde{m}_t^v$ for $0 < u < v \leq t < \infty$, we omit the superscripts and we denote simply

$$d\widetilde{m}_t = \frac{-dM_t}{{}^{\mathfrak{p}}(1 - Z)_t}, \; t \in (0, \infty).$$

3.3. *Dynamic equation and* $\natural$*-pair*

Let $\mathbb{D}$ denote the space of all càdlàg $\mathbb{F}$ adapted processes. Let $m > 0$ be an integer. Let $\mathbf{Y} = (Y_1, \ldots, Y_m)$ be an m-dimensional $(\mathbb{P}, \mathbb{F})$ local martingale, and $\mathbf{F} = (F_1, \ldots, F_m)$ be a Lipschitz functional from $\mathbb{D}$ into the set of m-dimensional locally bounded $\mathbb{F}$ predictable processes in the sense of Protter [26]. For $0 < u < \infty$, for any $\mathcal{F}_u$-measurable random variable x, we consider the stochastic differential equation determined by the pair $(\mathbf{F}, \mathbf{Y})$:

$$(\natural_u) \begin{cases} dX_t = X_{t-} d\widetilde{m}_t + \mathbf{F}(X)_t^\top d\mathbf{Y}_t, \; t \in [u, \infty), \\ X_u = x. \end{cases}$$

We will call the pair $(\mathbf{F}, \mathbf{Y})$ a $\natural$-pair if it satisfies the following conditions, for any $1 \leq j \leq m$, for any $u > 0$ and for any $X, X' \in \mathbb{D}$:

(i) The process $t \in [u, \infty) \to \frac{F_j(X)_t}{{}^{\mathfrak{p}}(1-Z)_t - X_{t-}} \mathbb{1}_{\{{}^{\mathfrak{p}}(1-Z)_t - X_{t-} \neq 0\}}$ is integrable with respect to Y_j, and satisfies the inequality:

$$\Delta_t \widetilde{m} - \frac{1}{{}^{\mathfrak{p}}(1 - Z)_t - X_{t-}} \mathbb{1}_{\{{}^{\mathfrak{p}}(1-Z)_t - X_{t-} \neq 0\}} \mathbf{F}(X)_t^\top \Delta_t \mathbf{Y} > -1, \quad t \in [u, \infty).$$

(ii) The process $t \in [u, \infty) \to \frac{F_j(X)_t}{X_{t-}} \mathbb{1}_{\{X_{t-} \neq 0\}}$ is integrable with respect to Y_j, and satisfies the inequality:

$$\Delta_t \tilde{m} + \frac{1}{X_{t-}} \mathbb{1}_{\{X_{t-} \neq 0\}} \mathbf{F}(X)_t^\top \Delta_t \mathbf{Y} \geq -1, \ t \in [u, \infty).$$

(iii) If $0 \leq X, X' \leq 1$, the process

$$t \in [u, \infty) \to \frac{F_j(X)_t - F_j(X')_t}{X_{t-} - X'_{t-}} \mathbb{1}_{\{X_{t-} - X'_{t-} \neq 0\}}$$

is integrable with respect to Y_j, and satisfies the inequality:

$$\Delta_t \tilde{m} + \frac{1}{X_{t-} - X'_{t-}} \mathbb{1}_{\{X_{t-} - X'_{t-} \neq 0\}} (\mathbf{F}(X)_t - \mathbf{F}(X')_t)^\top \Delta_t \mathbf{Y} \geq -1,$$
$$t \in [u, \infty).$$

Remark 3.4. Note that the inequality in the condition (iii) is equivalent to

$$X_{t-} + X_{t-} \Delta_t \tilde{m} + \mathbf{F}(X)_t^\top \Delta_t \mathbf{Y} \geq X'_{t-} + X'_{t-} \Delta_t \tilde{m} + \mathbf{F}(X')_t^\top \Delta_t \mathbf{Y},$$

whenever $X_{t-} > X'_{t-}$. This condition looks like very much that one in Chapter V Section 10 Theorem 65 of Protter [26] which ensures that the solutions of a stochastic differential equation form a diffeomorphism.

3.4. *The basic properties of the $\natural$-equation*

Consider an m-dimensional $(\mathbb{P}, \mathbb{F})$ local martingale $\mathbf{Y} = (Y_1, \ldots, Y_m)$ and a Lipschitz functional $\mathbf{F} = (F_1, \ldots, F_m)$ from $\mathbb{D}$ into the set of m-dimensional locally bounded $\mathbb{F}$ predictable processes in the sense of Protter [26].

Lemma 3.5. *For $0 < u < \infty$ let X^x be the solution of the equation $(\natural_u)$ associated with the pair $(\mathbf{F}, \mathbf{Y})$.*

(1) Suppose that $X^x_u = x \leq 1 - Z_u$. Then, if $(\mathbf{F}, \mathbf{Y})$ satisfies the above condition (i) on $[u, \infty)$, we have $(1 - Z)_t - X^x_t \geq 0$ for $t \in [u, \infty)$ and $(1 - Z)_t - X^x_t > 0$ for $t \in (T, \infty)$ where $T = \inf\{s \geq u : (1 - Z_u) - x + A_s - A_u > 0\}$. Conversely, for any pair of $\mathbb{F}$ stopping times $u \leq S \leq T$, if $(1 - Z)_t - X^x_t > 0$ and $(1 - Z)_{t-} - X^x_{t-} > 0$ on $(S, T]$, the above condition (i) is satisfied on $(S, T]$ (instead of $[u, \infty)$) by $(\mathbf{F}, \mathbf{Y})$ with $X = X^x$.

(2) *Suppose that* $X^x_u = x \geq 0$. *Then,* $X^x \geq 0$ *and* $\frac{1}{X^x_-}1\!\!1_{\{X^x_->0\}}$ *is* X^x *integrable on* $[u,\infty)$, *if and only if the above condition (ii) is satisfied by* $(\mathbf{F},\mathbf{Y})$ *with* $X = X^x$. *In addition,* $X^x > 0, X^x_- > 0$ *on* $[u,\infty)$ *if and only if* $(\mathbf{F},\mathbf{Y})$ *and* $X = X^x$ *satisfies the condition (ii) with the strict inequality* > -1 *instead of* ≥ -1.

(3) *Suppose that* $X^x_u = x < y = X^y_u$ *and* $0 \leq X^x, X^y \leq 1$. *Then,* $X^y - X^x \geq 0$ *and* $\frac{1}{(X^y-X^x)_-}1\!\!1_{\{(X^y-X^x)_->0\}}$ *is* $X^y - X^x$ *integrable on* $[u,\infty)$ *if and only if the above condition (iii) is satisfied by* $(\mathbf{F},\mathbf{Y})$ *with* $X = X^y, X' = X^x$. *In addition,* $X^y - X^x > 0, (X^y - X^x)_- > 0$ *on* $[u,\infty)$ *if and only if* $(\mathbf{F},\mathbf{Y})$ *and* $X = X^y, X' = X^x$ *satisfies the condition (iii) with the strict inequality* > -1 *instead of* ≥ -1.

Proof.

(1) Note that $d(1-Z) = {}^{\mathsf{p}}(1-Z)d\widetilde{m} + dA$, also ${}^{\mathsf{p}}X^x = X^x_-$ (X being necessarily a local martingale). From this, we deduce

$$\begin{aligned} & d((1-Z)_t - X^x_t) \\ & = ({}^{\mathsf{p}}(1-Z)_t - X^x_{t-})d\widetilde{m}_t - \mathbf{F}(X^x)^\top_t d\mathbf{Y}_t + dA_t \\ & = ({}^{\mathsf{p}}(1-Z)_t - X^x_{t-})\left(d\widetilde{m}_t - \frac{1\!\!1_{\{{}^{\mathsf{p}}(1-Z)_t - X^x_{t-} \neq 0\}}}{{}^{\mathsf{p}}(1-Z)_t - X^x_{t-}}\mathbf{F}(X^x)^\top_t d\mathbf{Y}_t\right) + dA_t \\ & = {}^{\mathsf{p}}((1-Z) - X^x)_t\left(d\widetilde{m}_t - \frac{1\!\!1_{\{{}^{\mathsf{p}}(1-Z)_t - X^x_{t-} \neq 0\}}}{{}^{\mathsf{p}}(1-Z)_t - X^x_{t-}}\mathbf{F}(X^x)^\top_t d\mathbf{Y}_t\right) + dA_t. \end{aligned}$$

If $(\mathbf{F},\mathbf{Y})$ satisfies the condition (i), Lemma 3.2 is applicable, which yields the positivity of $(1-Z) - X^x$. Conversely, on the random interval $(S,T]$, the condition $(1-Z)_{t-} - X^x_{t-} > 0$ implies the integrability condition in (i). By a direct computation from the identity ${}^{\mathsf{p}}((1-Z) - X^x) = 1 - M_- + A - X^x_-$, we obtain

$$\frac{\Delta_t(-M - X^x)}{{}^{\mathsf{p}}((1-Z) - X^x)_t} > -1, \ t \in (S,T],$$

which is equivalent to say that the inequality of condition (i) is satisfied on $(S,T]$ by $(\mathbf{F},\mathbf{Y})$ with $X = X^x$.

(2) The first assertion is the consequence of the Doléans-Dade exponential formula (cf. Theorem 9.39 of He et al. [10]). The second assertion is the consequence of the same formula together with Theorem 2.62 of He et al. [10].

(3) We make the same argument as in (2).

As a consequence, we obtain the following theorem.

Theorem 3.6. *Let $0 < u < \infty$ and $0 \leq x \leq 1 - Z_u$. Then, the equation ($\natural_u$) associated with a $\natural$-pair $(\mathbf{F}, \mathbf{Y})$ has a unique solution X^x on $[u, \infty)$ such that $0 \leq X^x \leq 1 - Z$. In particular, X^x is a uniformly integrable $\mathbb{F}$ martingale on $[u, \infty)$. If $0 \leq x \leq y \leq 1$, $X^x \leq X^y$ on $[u, \infty)$.*

The following theorem proves that the family of $\natural$-pair is not empty.

Theorem 3.7. *Let $m > 0$ be an integer. Let $g(t, x)$ be any bounded continuously differentiable function defined on $\mathbb{R}_+ \times \mathbb{R}$ taking values in $\mathbb{R}^m$ with bounded derivative. Let φ be a C^1 increasing function on $\mathbb{R}_+$ such that $|\varphi(x)| \leq 2$ and $|\frac{\varphi(x)}{x}| \leq 1$ and φ' is bounded. For $t \in \mathbb{R}_+$, we introduce the set G_t of $\mathbf{z} \in \mathbb{R}^m$ satisfying the two conditions:*

$$\circ : 2\left|g(t,x)^\top \mathbf{z}\right| < 1 + \Delta_t \tilde{m}, \text{ for any } x \in \mathbb{R},$$

$$\begin{aligned}\circ\circ : [&-\varphi'({}^{\mathsf{p}}(1-Z)_t - x)\varphi(x)g(t,x) + \varphi({}^{\mathsf{p}}(1-Z)_t - x)\varphi'(x)g(t,x)\\ &+ \varphi({}^{\mathsf{p}}(1-Z)_t - x)\varphi(x)g'(t,x)]^\top \mathbf{z} > -(1 + \Delta_t \tilde{m}), \text{ for any } x \in \mathbb{R}.\end{aligned}$$

(Here $g'(t, x)$ denotes the derivative with respect to x.) Then, for any $t \in \mathbb{R}_+$, the random set G_t is not empty, and the set-valued process G is $\mathbb{F}$ optional. There exists an m-dimensional $\mathbb{F}$ local martingale $\mathbf{Y} = (Y_1, \ldots, Y_m)$ whose jump at $t \in \mathbb{R}_+$, if it exists, is contained in G_t. Let

$$\mathbf{F}(X)_t = f(t, X_{t-}) = \varphi({}^{\mathsf{p}}(1-Z)_t - X_{t-})\varphi(X_{t-})g(t, X_{t-}), \ X \in \mathbb{D}.$$

Then, the above conditions (i), (ii) and (iii) with strict inequality > -1 instead of ≥ -1 are satisfied for the pair $(\mathbf{F}, \mathbf{Y})$.

Proof. We recall that $\Delta_t \tilde{m} > -1$. Consequently, G_t contains always a no-empty neighbourhood of the origin. The optionality of G with respect to $\mathbb{F}$ can be proved in a usual way (cf. Karatzas et al. [18]). The existence of $\mathbf{Y}$ is deduced from the measurable selection theorem. We now compute

$$\begin{aligned}&\Delta_t \tilde{m} - \frac{1}{{}^{\mathsf{p}}(1-Z)_t - X_{t-}} 1\!\!1_{\{{}^{\mathsf{p}}(1-Z)_t - X_{t-} \neq 0\}} f(t, X_{t-})^\top \Delta_t \mathbf{Y}\\ &\geq \Delta_t \tilde{m} - 1\!\!1_{\{{}^{\mathsf{p}}(1-Z)_t - X_{t-} \neq 0\}} 2|g(t, X_{t-})^\top \Delta_t \mathbf{Y}| > -1,\end{aligned}$$

and

$$\begin{aligned}&\Delta_t \tilde{m} + \frac{1}{X_{t-}} 1\!\!1_{\{X_{t-} \neq 0\}} f(t, X_{t-})^\top \Delta_t \mathbf{Y}\\ &\geq \Delta_t \tilde{m} - 1\!\!1_{\{X_{t-} \neq 0\}} 2|g(t, X_{t-})^\top \Delta_t \mathbf{Y}| > -1.\end{aligned}$$

These computations proves the conditions (i) and (ii) with strict inequality > -1. The condition (iii) with strict inequality is the sequence of the assumption $\circ\circ$, because then the map

$$x \longrightarrow x + x\Delta_t\tilde{m} + f(t,x)^\top \Delta_t \mathbf{Y}$$

is strictly increasing. □

3.5. *The iM_Z associated with the $\natural$-equation*

We assume $\mathbf{Hy}(Z)$.

Theorem 3.8. *Suppose $\mathcal{F}_\infty = \vee_{t\in\mathbb{R}_+}\mathcal{F}_t$. For $0 < u < \infty$, let $(L^u_t : t \in [u,\infty))$ denote the solution of the equation $(\natural_u)$ associated with a $\natural$-pair with the initial condition $L^u_u = 1 - Z_u$. Set $L^u_\infty = \lim_{t\to\infty} L^u_t$. Set*

$$M^u_u = (1 - Z_u),$$
$$M^u_t = \inf_{v\in\mathbb{Q},u<v\leq t}(L^v_t)^+ \wedge (1 - Z_t),\ t \in (u,\infty].$$

Set finally

$$M^0_t = \inf_{u\in\mathbb{Q},0<u\leq t} M^u_t,\ t \in (0,\infty],$$
$$M^0_0 = \lim_{t\downarrow 0} M^0_t \quad \textit{(which exists)},$$
$$M^\infty_t = 1, \quad \textit{for } t \in [0,\infty].$$

Then, for $0 < u < \infty$, M^u is $\mathbb{P}$ indistinguishable from L^u on $[u,\infty]$, and $(M^u : 0 \leq u \leq \infty)$ is an iM_Z.

The above iM_Z will be said to be associated with the $\natural$-equation and the probability measure $\mathbb{Q}^\natural$ constructed in Theorem 2.1 with this iM_Z and with $\mathbb{P}$ will be said to be associated with the $\natural$-equation.

Proof As it has been proved in Theorem 3.6, $0 \leq L^u_t \leq (1 - Z_t)$ for $0 < u \leq t \leq \infty$, and $L^u_t \leq L^v_t$ for $0 < u < v \leq t \leq \infty$. The comparison relation holds because L^u, L^v satisfy the same equation $(\natural_v)$ on $[v,\infty)$ and $L^u_v \leq (1 - Z_v) = L^v_v$. Let T be an $\mathbb{F}$ stopping time with $0 < u \leq T \leq \infty$. We have

$$\begin{aligned}
\mathbb{E}[M^u_T - L^u_T] &= \mathbb{E}[\mathbb{I}_{\{u<T\}} \inf_{v\in\mathbb{Q},u<v\leq T} L^v_T - \mathbb{I}_{\{u<T\}} L^u_T] \\
&= \inf_{v\in\mathbb{Q},u<v} \mathbb{P}[\mathbb{I}_{\{u<T\}} L^{v\wedge T}_T] - \mathbb{E}[\mathbb{I}_{\{u<T\}} L^u_T] \\
&= \inf_{v\in\mathbb{Q},u<v} \mathbb{P}[\mathbb{I}_{\{u<T\}}(1 - Z_{v\wedge T})] - \mathbb{E}[\mathbb{I}_{\{u<T\}}(1 - Z_u)] = 0.
\end{aligned}$$

This shows that M^u, L^u are indistinguishable on $[u,\infty]$ and in particular M^u is càdlàg on $[u,\infty]$.

Let T be an $\mathbb{F}$ stopping time with $0 < T \leq \infty$. We have

$$\begin{aligned}\mathbb{E}[M_T^0] &= \mathbb{E}[\inf_{v\in\mathbb{Q},0<v\leq T} L_T^v] = \mathbb{E}[\inf_{v\in\mathbb{Q},v>0} L_T^{v\wedge T}] \\ &= \inf_{v\in\mathbb{Q},v>0} \mathbb{E}[L_T^{v\wedge T}] = \inf_{v\in\mathbb{Q},v>0} \mathbb{E}[(1-Z_{v\wedge T})] = \mathbb{E}[(1-Z_0)].\end{aligned}$$

The value $\mathbb{E}[M_T^0]$ does not depend on the stopping time T. According to Theorem 4.40 of He et al. [10], M^0 is a càdlàg $\mathbb{F}$ martingale on $(0,\infty]$. From this property we deduce that $M_0^0 = \lim_{t\downarrow 0} M_t^0$ exists.

By definition, for $0 \leq t \leq \infty$, the map $u \in [0,t] \to M_t^u$ is a right continuous non decreasing function. The theorem is proved. □

4. Formula of Enlargement of Filtration in the Case of Markovian Coefficients

4.1. *The problem*

We consider the dynamic model associated with an m-dimensional $\natural$-pair $(\mathbf{F},\mathbf{Y})$. We suppose that the operator $\mathbf{F}(X)$ takes the particular form $\mathbf{F}(X) = f(\omega,t,X_{t-})$, where $f(\omega,t,x)$ is a map from $\Omega\times\mathbb{R}_+\times\mathbb{R}$ into $\mathbb{R}^m$, $\mathcal{P}(\mathbb{F})\otimes\mathcal{B}(\mathbb{R}_+)$ measurable, such that:

(1) For $1\leq j\leq m$, for $\omega\in\Omega$ and $t\geq 0$, the partial derivative $\frac{\partial f_j}{\partial x}(\omega,t,x)$ exists and is bounded, and, for fixed ω, is uniformly continuous with respect to t in every compact set.

(2) $f(\omega,t,0) = 0$.

We know by Theorem 3.7 that such a $\natural$-pair exists. It is usual to call such $\mathbf{F}$ a Markovian coefficient.

Let $\mathbb{Q}^\natural$ be the probability measure on the product space associated with the $\natural$-equation by Theorem 2.1. Let $\mathbb{G} = (\mathcal{G}_t)_{t\geq 0}$ be the filtration defined by $\cap_{s>t}(\mathcal{F}_s \vee \sigma(\tau\wedge s))$ completed with the $\mathbb{Q}^\natural$ negligible sets. In this section, we consider the problem of enlargement of filtration : whether the bounded $\mathbb{F}$ martingales are semimartingales in the filtration $\mathbb{G}$.

In the computations below, the expectations are all taken under $\mathbb{Q}^\natural$ (recalling that $\mathbb{Q}^\natural$ coincides with $\mathbb{P}$ on $\mathcal{A}$).

4.2. *Preliminary results*

We need the following results.

Lemma 4.1. *For $0 \leq u < \infty$, for h a bounded Borel function on $[0, \infty]$, the process :*

$$t \in [u, \infty] \rightarrow \int_0^u h(v) d_v M_t^v$$

is a bounded $(\mathbb{P}, \mathbb{F})$ martingale on $[u, \infty]$.

Proof. Consider the family of bounded Borel functions h on $[0, \infty]$ such that

$$\mathbb{E}[\int_0^u h(v) d_v M_T^v] = \mathbb{E}[\int_0^u h(v) d_v M_u^v]$$

for any $\mathbb{F}$ stopping time $T \geq u$. By monotone class theorem, we see that this family actually contains all bounded Borel functions on $[0, \infty]$. According to Theorem 4.40 of He et al. [10], $\int_0^u h(v) d_v M_t^v, t \in [u, \infty)$ is a $(\mathbb{P}, \mathbb{F})$ martingale. □

Lemma 4.2. *Let $V_s(\omega), 0 \leq s < \infty, \omega \in \Omega$, be a function such that, for fixed s, V_s is $\mathcal{F}_\infty$ measurable and for fixed ω, $s \rightarrow V_s(\omega)$ is càdlàg and non decreasing. We denote by dV_s the induced measure on $[0, \infty)$. Let $F_s(t, \omega), 0 \leq s < \infty, 0 \leq t \leq \infty, \omega \in \Omega$, be a positive function measurable with respect to $\mathcal{B}[0, \infty] \otimes \mathcal{B}[0, \infty] \otimes \mathcal{F}_\infty$. Suppose $\mathbb{E}[\int_0^\infty dV_s] < \infty$. Then,*

$$\mathbb{E}[\int_0^\infty F_s dV_s] = \mathbb{E}[\int_0^\infty \left(\int_{[0,\infty]} F_s(v, \cdot) d_v M_\infty^v \right) dV_s].$$

Proof. By monotone class theorem, we need only to check the relation for a function of the form $F_s(t, \omega) = h(s) H(t, \omega)$, where h is a bounded Borel function on $[0, \infty]$, H is bounded $\mathcal{B}[0, \infty] \otimes \mathcal{F}_\infty$ measurable. The following (obvious) equalities

$$\begin{aligned}
&\mathbb{E}[\int_0^\infty F_s dV_s] = \mathbb{E}[H \int_0^\infty h(s) dV_s] \\
&= \mathbb{E}[\int_{[0,\infty]} H(v, \cdot) d_v M_\infty^v \int_0^\infty h(s) dV_s] = \mathbb{E}[\int_0^\infty \left(\int_{[0,\infty]} F_s(v, \cdot) d_v M_\infty^v \right) dV_s]
\end{aligned}$$

yield to the result. □

We need the following result, consequence of Theorem 13.1 of Aldous [3].

Theorem 4.3. *Let $\sqcap$ denote the space of all probability measures on the measurable space $([0,\infty],\mathcal{B}[0,\infty])$ equipped with the topology of weak convergence (cf. Billingsley [5]). There exists a Skorohod process χ adapted to $\mathbb{F}$ taking values in $\sqcap$ such that, for any $\mathbb{F}$ stopping time T, χ_T is the $\mathbb{Q}^\natural$ regular conditional distribution of τ given $\mathcal{F}_T$.*

Notice that the result in Aldous [3] is proved for a càdlàg adapted process. But actually the proof of the result is valid for any Polish space valued random variable. The process χ is called in the literature the prediction process of the conditional law of τ with respect to the filtration $\mathbb{F}$ under $\mathbb{Q}^\natural$. Notice that χ_{t-} is a probability measure on $\mathcal{B}[0,\infty]$.

Lemma 4.4. *For any $\mathbb{F}$ stopping time $0 < T \leq \infty$, $\mathbb{Q}^\natural$ almost surely, $M^u_T = \chi_T[\tau \leq u]$ for all $u \in [0,T)$. On the other hand, for fixed $u \in (0,\infty)$, for bounded function $h(\omega,s,v)$ measurable with respect to $\mathcal{P}(\mathbb{F}) \times \mathcal{B}(\mathbb{R}_+)$,*

$$^{\mathfrak{p}}\left(\int_{[0,u]} h(\cdot,v)\chi_\infty(dv)\right) = {}^{\mathfrak{p}}\left(\int_{[0,u]} h(\cdot,v)d_vM^v_\infty\right) = \int_{[0,u]} h(\cdot,v)\chi_-(dv) \tag{4.1}$$

on $(u,\infty]$. In particular, $M^u_- = \chi_-[\tau \leq u]$ on $(u,\infty]$.

Proof. We note that, $\mathbb{Q}^\natural$ almost surely,

$$M^u_T \mathbb{1}_{\{u<T\}} = \mathbb{P}[\tau \leq u|\mathcal{F}_T]\mathbb{1}_{\{u<T\}} = \chi_T[\tau \leq u]\mathbb{1}_{\{u<T\}}$$

for any rational number u. By the right continuity in u of M^u_t and of $\chi_t[\tau \leq u]$, we prove the first assertion. For the identity (4.1), let $\epsilon > 0$ and φ be a non negative bounded continuous function on $\mathbb{R}_+$ whose support is contained in $[0,u+\epsilon]$. Consider $h(s,v) = H_s g(v)$, where H is a bounded $\mathbb{F}$ predictable process and g is a bounded continuous function on $\mathbb{R}_+$. Let T be an $\mathbb{F}$ predictable stopping time such that $u + 2\epsilon < T \leq \infty$ and $(T_n)_{n\geq 1}$ a non decreasing sequence of $\mathbb{F}$ stopping times such that $u+\epsilon < T_n < T$ and $T_n \uparrow T$. Using basic properties of martingales, we obtain

$$\begin{aligned}
&^{\mathfrak{p}}\left(\int_{[0,u+\epsilon]} h(\cdot,v)\varphi(v)d_vM^v_\infty\right)_T = \mathbb{E}[\int_{[0,u+\epsilon]} h(T,v)\varphi(v)d_vM^v_\infty|\mathcal{F}_{T-}] \\
&= H_T\mathbb{E}[\int_{[0,u+\epsilon]} g(v)\varphi(v)d_vM^v_\infty|\mathcal{F}_{T-}] \\
&= H_T \lim_{n\uparrow\infty} \mathbb{E}[\int_{[0,u+\epsilon]} g(v)\varphi(v)d_vM^v_\infty|\mathcal{F}_{T_n}] = H_T \lim_{n\uparrow\infty} \int_{[0,u+\epsilon]} g(v)\varphi(v)d_vM^v_{T_n} \\
&= H_T \lim_{n\uparrow\infty} \int_{[0,u+\epsilon]} g(v)\varphi(v)\chi_{T_n}(dv) = \int_{[0,u+\epsilon]} h(T,v)\varphi(v)\chi_{T-}(dv).
\end{aligned}$$

By the monotone class theorem, this identity remains valid for any bounded function $h(\omega, s, v)$ measurable with respect to $\mathcal{P}(\mathbb{F})\times\mathcal{B}(\mathbb{R}_+)$. Now, to finish the proof, let φ decrease to $1\!\!1_{[0,u]}$. □

Remark 4.5. The reason to introduce χ is that $u \to \chi_{t-}[\tau \leq u]$ is a true distribution function, whilst the behavior of $u \to M^u_{t-}$ was undetermined from its definition.

Let us denote the $\mathbb{F}$ predictable bracket of two $(\mathbb{P}, \mathbb{F})$ local martingales X, X' by $\langle X, X'\rangle$ (when it exists).

Lemma 4.6. *Let $0 \leq s, u < \infty$ and $A \in \mathcal{F}_s$. Let X be a bounded $\mathbb{F}$ martingale such that $X_{s\vee u} = 0$ and $\int_{s\vee u}^{\infty} |d\langle \tilde{m}, X\rangle| + \int_{s\vee u}^{\infty} |d\langle \mathbf{Y}, X\rangle|$ is integrable. Let $\mathrm{p}(\omega, w, v)$, $(\omega, w, v) \in \Omega \times \mathbb{R}_+ \times \mathbb{R}_+$, denote the function*

$$\begin{aligned}&\frac{\partial f}{\partial x}(w, \chi_{w-}[\tau \leq v])1\!\!1_{\{\chi_{w-}[\tau=v]=0\}}\\ &\qquad + \frac{f(w, \chi_{w-}[\tau \leq v]) - f(w, \chi_{w-}[\tau < v])}{\chi_{w-}[\tau = v]}1\!\!1_{\{\chi_{w-}[\tau=v]>0\}}.\end{aligned}$$

Then,

$$\mathbb{E}[1\!\!1_A 1\!\!1_{\{\tau\leq u\}}X_\infty]=\mathbb{E}[1\!\!1_A 1\!\!1_{\{\tau\leq u\}}\left(\int_{s\vee u}^{\infty} d\langle \tilde{m}, X\rangle_w + \int_{s\vee u}^{\infty} \mathrm{p}(w,\tau)^\top d\langle \mathbf{Y}, X\rangle_w\right)].$$

Proof. Suppose firstly $u > 0$. We compute

$$\begin{aligned}\mathbb{E}[1\!\!1_A 1\!\!1_{\{\tau\leq u\}}X_\infty] &= \mathbb{E}[1\!\!1_A M^u_\infty X_\infty] = \mathbb{E}[1\!\!1_A \int_{s\vee u}^{\infty} d\langle M^u, X\rangle_w]\\ &= \mathbb{E}[1\!\!1_A \left(\int_{s\vee u}^{\infty} M^u_{w-} d\langle \tilde{m}, X\rangle_w + \int_{s\vee u}^{\infty} f(w, M^u_{w-})^\top d\langle \mathbf{Y}, X\rangle_w\right)].\end{aligned}$$

By Lemma 4.4, we can write, for all $w > s \vee u$,

$$\begin{aligned}&f(w, M^u_{w-})\\ &= f(w, M^0_{w-}) + f(w, M^u_{w-}) - f(w, M^0_{w-})\\ &= f(w, \chi_{w-}[\tau \leq 0]) + f(w, \chi_{w-}[\tau \leq u]) - f(w, \chi_{w-}[\tau \leq 0])\\ &= \frac{f(w, \chi_{w-}[\tau = 0])}{\chi_{w-}[\tau = 0]}1\!\!1_{\{\chi_{w-}[\tau=0]>0\}} + \int_{(0,u]} \frac{\partial f}{\partial x}(w, \chi_{w-}[\tau < v])\chi_{w-}(dv)\\ &\quad + \sum_{0<v\leq u} (f(w, \chi_{w-}[\tau \leq v]) - f(w, \chi_{w-}[\tau < v])\\ &\qquad\qquad - \frac{\partial f}{\partial x}(w, \chi_{w-}[\tau < v])\chi_{w-}[\tau = v])\end{aligned}$$

$$= \int_{[0,u]} \left(\frac{\partial f}{\partial x}(w, \chi_{w-}[\tau \le v]) 1\!\!1_{\{\chi_{w-}[\tau=v]=0\}} \right.$$
$$\left. + \frac{f(w, \chi_{w-}[\tau \le v]) - f(w, \chi_{w-}[\tau < v])}{\chi_{w-}[\tau = v]} 1\!\!1_{\{\chi_{w-}[\tau=v]>0\}} \right) \chi_{w-}(dv)$$
$$= \int_{[0,u]} \mathsf{p}(w, v)\chi_{w-}(dv).$$

We note that p is bounded and is $\mathcal{P}(\mathbb{F}) \times \mathcal{B}(\mathbb{R}_+)$ measurable. Applying Lemma 4.2 and Lemma 4.4 we continue the computation:

$$\mathbb{E}[1\!\!1_A 1\!\!1_{\{\tau \le u\}} X_\infty]$$
$$= \mathbb{E}[1\!\!1_A \left(\int_{s\vee u}^{\infty} M^u_{w-} d\langle \tilde{m}, X\rangle_w + \int_{s\vee u}^{\infty} \int_{[0,u]} \mathsf{p}(w, v)^\top \chi_{w-}(dv) d\langle \mathbf{Y}, X\rangle_w \right)]$$
$$= \mathbb{E}[1\!\!1_A \left(\int_{s\vee u}^{\infty} M^u_{\infty} d\langle \tilde{m}, X\rangle_w + \int_{s\vee u}^{\infty} \int_{[0,u]} \mathsf{p}(w, v)^\top \chi_{\infty}(dv) d\langle \mathbf{Y}, X\rangle_w \right)]$$
$$= \mathbb{E}[1\!\!1_A \left(\int_{s\vee u}^{\infty} 1\!\!1_{\{\tau \le u\}} d\langle \tilde{m}, X\rangle_w + \int_{s\vee u}^{\infty} \int_{[0,u]} \mathsf{p}(w, v)^\top d_v M^v_{\infty} d\langle \mathbf{Y}, X\rangle_w \right)]$$
$$= \mathbb{E}[1\!\!1_A \left(\int_{s\vee u}^{\infty} 1\!\!1_{\{\tau \le u\}} d\langle \tilde{m}, X\rangle_w + \int_{s\vee u}^{\infty} 1\!\!1_{\{\tau \le u\}} \mathsf{p}(w, \tau)^\top d\langle \mathbf{Y}, X\rangle_w \right)]$$

This proves the lemma for $u > 0$. Now let $u \downarrow 0$ to finish the proof of the lemma. □

4.3. *Enlargement of filtration formula*

Recall that the canonical decomposition of Z is given by $Z = M - A$. The following result, which is a direct consequence of Jeulin et al. [15, 17], gives, for any bounded $(\mathbb{P}, \mathbb{F})$ martingale, its $(\mathbb{Q}^\natural, \mathbb{G})$ semimartingale decomposition before τ.

Lemma 4.7. *Let X be a bounded $(\mathbb{P}, \mathbb{F})$ martingale and B^X the $(\mathbb{P}, \mathbb{F})$ predictable dual projection of the jump process $t \to \Delta X_\tau 1\!\!1_{\{0<\tau\le t\}}$. Then,*

$$X_{\cdot\wedge\tau} - \int_0^{\cdot\wedge\tau} \frac{1}{Z_{s-}}(d\langle M, X\rangle_s + dB^X_s)$$

is a $(\mathbb{Q}^\natural, \mathbb{G})$ local martingale.

We complete this result in our model, providing the $(\mathbb{Q}^\natural, \mathbb{G})$ semimartingale decomposition after τ.

Theorem 4.8. *Let X be a bounded $(\mathbb{P}, \mathbb{F})$ martingale. Then the process*

$$X_t - X_0 - \int_0^t \mathbb{1}_{\{s\leq\tau\}} \frac{1}{Z_{s-}}(d\langle M, X\rangle_s + dB_s^X)$$

$$+ \int_0^t \mathbb{1}_{\{\tau<w\}} \frac{1}{\mathrm{p}(1-Z)_w} d\langle M, X\rangle_w - \int_0^t \mathbb{1}_{\{\tau<w\}} \mathrm{p}(w,\tau)^\top d\langle \mathbf{Y}, X\rangle_w,$$

$t \in \mathbb{R}_+$, *is a $(\mathbb{Q}^\natural, \mathbb{G})$ local martingale.*

Proof. We write

$$X_t - X_0 = X_{\tau\vee t} - X_\tau + X_{\tau\wedge t} - X_0.$$

The $\mathbb{G}$ semimartingale decomposition for $X_{\tau\wedge t} - X_0$ is given in Lemma 4.7. We need only to prove the formula for $X_{\tau\vee t} - X_\tau$. Without loss of generality we suppose that X is stopped so that everything in the computations below is integrable. Let $0 < s < t < \infty, 0 < u < \infty$ and $A \in \mathcal{F}_s$.

$$\begin{aligned}
\mathbb{E}[\mathbb{1}_A \mathbb{1}_{\{\tau\leq u\}}(X_{\tau\vee t} - X_{\tau\vee s})] &= \mathbb{E}[\mathbb{1}_A \mathbb{1}_{\{\tau=0\}}(X_t - X_s)] \\
&\quad + \lim_{n\to\infty} \mathbb{E}[\mathbb{1}_A \sum_{k=1}^n \mathbb{1}_{\{\frac{(k-1)u}{n}<\tau\leq\frac{ku}{n}\}}(X_{\tau\vee t} - X_{\tau\vee s})] \\
&= \mathbb{E}[\mathbb{1}_A \mathbb{1}_{\{\tau=0\}}(X_t - X_s)] \\
&\quad + \lim_{n\to\infty} \mathbb{E}[\mathbb{1}_A \sum_{k=1}^n \mathbb{1}_{\{\frac{(k-1)u}{n}<\tau\leq\frac{ku}{n}\}}(X_{\frac{ku}{n}\vee t} - X_{\frac{ku}{n}\vee s})].
\end{aligned}$$

Set $t_n = \frac{ku}{n} \vee t$ and $s_n = \frac{ku}{n} \vee s$. By Lemma 4.6, we have

$$\begin{aligned}
&\mathbb{E}[\mathbb{1}_A \mathbb{1}_{\{\frac{(k-1)u}{n}<\tau\leq\frac{ku}{n}\}}(X_{\frac{ku}{n}\vee t} - X_{\frac{ku}{n}\vee s})] \\
&= \mathbb{E}[\mathbb{1}_A \mathbb{1}_{\{\tau\leq\frac{ku}{n}\}}(X_{\frac{ku}{n}\vee t} - X_{\frac{ku}{n}\vee s})] - \mathbb{E}[\mathbb{1}_A \mathbb{1}_{\{\tau\leq\frac{(k-1)u}{n}\}}(X_{\frac{ku}{n}\vee t} - X_{\frac{ku}{n}\vee s})] \\
&= \mathbb{E}[\mathbb{1}_A \mathbb{1}_{\{\tau\leq\frac{ku}{n}\}}\left(\int_{s_n}^{t_n} d\langle\tilde{m}, X\rangle_w + \int_{s_n}^{t_n} \mathrm{p}(w,\tau)^\top d\langle \mathbf{Y}, X\rangle_w\right)] \\
&\quad -\mathbb{E}[\mathbb{1}_A \mathbb{1}_{\{\tau\leq\frac{(k-1)u}{n}\}}\left(\int_{s_n}^{t_n} d\langle\tilde{m}, X\rangle_w + \int_{s_n}^{t_n} \mathrm{p}(w,\tau)^\top d\langle \mathbf{Y}, X\rangle_w\right)] \\
&= \mathbb{E}[\mathbb{1}_A \mathbb{1}_{\{\frac{(k-1)u}{n}<\tau\leq\frac{ku}{n}\}}\left(\int_{s_n}^{t_n} d\langle\tilde{m}, X\rangle_w + \int_{s_n}^{t_n} \mathrm{p}(w,\tau)^\top d\langle \mathbf{Y}, X\rangle_w\right)].
\end{aligned}$$

Also,

$$\begin{aligned}
&\mathbb{E}[\mathbb{1}_A \mathbb{1}_{\{\tau=0\}}(X_t - X_s)] \\
&\quad = \mathbb{E}[\mathbb{1}_A \mathbb{1}_{\{\tau=0\}}\left(\int_s^t d\langle\tilde{m}, X\rangle_w + \int_s^t \mathrm{p}(w,\tau)^\top d\langle \mathbf{Y}, X\rangle_w\right)]
\end{aligned}$$

We turn back to $\mathbb{E}[1\!\!1_A 1\!\!1_{\{\tau\leq u\}}(X_{\tau\vee t} - X_{\tau\vee s})]$:

$$\begin{aligned}
&\mathbb{E}[1\!\!1_A 1\!\!1_{\{\tau\leq u\}}(X_{\tau\vee t} - X_{\tau\vee s})] \\
&= \mathbb{E}[1\!\!1_A 1\!\!1_{\{\tau=0\}}(X_t - X_s)] \\
&\quad + \lim_{n\to\infty} \mathbb{E}[1\!\!1_A \sum_{k=1}^{n} 1\!\!1_{\{\frac{(k-1)u}{n}<\tau\leq\frac{ku}{n}\}}(X_{\frac{ku}{n}\vee t} - X_{\frac{ku}{n}\vee s})] \\
&= \mathbb{E}[1\!\!1_A 1\!\!1_{\{\tau=0\}}\left(\int_s^t d\langle\tilde{m}, X\rangle_w + \int_s^t \mathsf{p}(w,\tau)^\top d\langle \mathbf{Y}, X\rangle_w\right)] \\
&+ \quad \lim_{n\to\infty} \mathbb{E}[1\!\!1_A \sum_{k=1}^{n} 1\!\!1_{\{\frac{(k-1)u}{n}<\tau\leq\frac{ku}{n}\}}\left(\int_{s_n}^{t_n} d\langle\tilde{m}, X\rangle_w + \int_{s_n}^{t_n} \mathsf{p}(w,\tau)^\top d\langle \mathbf{Y}, X\rangle_w\right)] \\
&= \mathbb{E}[1\!\!1_A 1\!\!1_{\{\tau=0\}}\left(\int_s^t d\langle\tilde{m}, X\rangle_w + \int_s^t \mathsf{p}(w,\tau)^\top d\langle \mathbf{Y}, X\rangle_w\right)] \\
&\quad + \mathbb{E}[1\!\!1_A 1\!\!1_{\{0<\tau\leq u\}}\left(\int_{\tau\vee s}^{\tau\vee t} d\langle\tilde{m}, X\rangle_w + \int_{\tau\vee s}^{\tau\vee t} \mathsf{p}(w,\tau)^\top d\langle \mathbf{Y}, X\rangle_w\right)] \\
&= \mathbb{E}[1\!\!1_A 1\!\!1_{\{\tau\leq u\}}\left(\int_s^t 1\!\!1_{\{\tau<w\}} d\langle\tilde{m}, X\rangle_w + \int_s^t 1\!\!1_{\{\tau<w\}}\mathsf{p}(w,\tau)^\top d\langle \mathbf{Y}, X\rangle_w\right)].
\end{aligned}$$

Note that, for $0 < s < s'$, $\mathcal{G}_s \subset \mathcal{F}_{s'} \otimes \sigma(\tau)$. The above formula implies

$$\begin{aligned}
&\mathbb{E}[1\!\!1_B (X_{\tau\vee t} - X_{\tau\vee s'})] \\
&= \mathbb{E}[1\!\!1_B \left(\int_{s'}^t 1\!\!1_{\{\tau<w\}} d\langle\tilde{m}, X\rangle_w + \int_{s'}^t 1\!\!1_{\{\tau<w\}}\mathsf{p}(w,\tau)^\top d\langle \mathbf{Y}, X\rangle_w\right)]
\end{aligned}$$

for any $B \in \mathcal{G}_s$. Taking the limit when $s' \downarrow s$, we prove the theorem. □

5. Regularity of $u \to M^u$ in the Case of Markovian Coefficients

Consider the $\natural$-equation associated with a $\natural$-pair $(\mathbf{F}, \mathbf{Y})$ of the type presented in Theorem 3.7:

$$\begin{aligned}
\Delta_t Y &\in \mathtt{G}_t, \\
\mathbf{F}(X)_t &= \varphi({}^{\mathfrak{p}}(1-Z)_t - X_{t-})\varphi(X_{t-})g(t, X_{t-}), \ X \in \mathbb{D}.
\end{aligned}$$

We suppose moreover that $\varphi(x) = x$ for $x \in [0,1]$, and the function g is autonomous, i.e. $g(t,x) = g(x)$ (in the sense of Protter [26]), and g is C^∞ with bounded derivatives of all order, and finally g has a compact support. Consider the $\mathtt{i}M_Z = (M^u : 0 \leq u \leq \infty)$ associated with $(\mathbf{F}, \mathbf{Y})$. Since $0 \leq M^u \leq 1 - Z$, we have

$$\varphi({}^{\mathfrak{p}}(1-Z) - M^u_-)\varphi(M^u_-) = ({}^{\mathfrak{p}}(1-Z) - M^u_-)M^u_-.$$

Hence, M^u is a solution of the stochastic differential equation (the explosion time for M^u being ∞):

$$\begin{cases} dX_t = X_{t-}d\widetilde{m}_t + ({}^{\mathfrak{p}}(1-Z)_t - X_{t-})X_{t-}g(X_{t-})^\top d\mathbf{Y}_t, \ u \le t < \infty, \\ X_u = x. \end{cases} \tag{5.1}$$

Theorem 39 and Theorem 65 in Chapter V Section 10 of Protter [26] are applicable to such an equation.

Theorem 5.1. *Let* $0 < u \le t < \infty$ *and denote by* $x \to \Xi^u_t(x)$ *the stochastic differential flow defined by equation (5.1). Let*

$$\kappa_v = (1 + \Delta_v\widetilde{m} - (1 - Z_{v-})g(1 - Z_{v-})^\top \Delta_v\mathbf{Y}), \quad \text{for } v \in (0,t].$$

Then, when $\Delta_v A > 0, v \in (0,t]$,

$$M^v_t - M^{v-}_t = \Xi^v_t(1 - Z_v) - \Xi^v_t(1 - Z_v - \kappa_v\Delta_v A).$$

When $\Delta_v A = 0$,

$$\lim_{u\uparrow v} \tfrac{M^v_t - M^u_t}{A_v - A_u} = \tfrac{d\Xi^v_t}{dx}(1 - Z_v)\kappa_v, \quad \text{for } v \in (0,t],$$

$$\lim_{u\downarrow v} \tfrac{M^v_t - M^u_t}{A_v - A_u} = \tfrac{d\Xi^v_t}{dx}(1 - Z_v), \qquad \text{for } v \in (0,t).$$

Proof. For any $0 < t < \infty$, we consider the difference $M^v_t - M^u_t$ for $u, v \in (0,t], u < v$. By the uniqueness of stochastic differential equation,

$$M^v_t - M^u_t = \Xi^v_t(1 - Z_v) - \Xi^v_t(M^u_v) = \tfrac{d\Xi^v_t}{dx}(\xi)((1 - Z_v) - M^u_v),$$

where ξ is some random variable such that $M^u_v \le \xi \le 1 - Z_v$. The process $(1 - Z_v) - M^u_v$ satisfies the equation

$$\begin{cases} d((1-Z) - M^u)_t = {}^{\mathfrak{p}}((1-Z) - M^u)_t dW^u_t + dA_t, \ t \in [u,\infty), \\ ((1-Z) - M^u)_u \ = 0, \end{cases}$$

with

$$dW^u_t = d\widetilde{m}_t - M^u_{t-}g(M^u_{t-})^\top d\mathbf{Y}_t.$$

According to Lemma 3.1,

$$(1 - Z_t) - M^u_t = \mathcal{E}(\mathbb{1}_{(u,\infty)} \centerdot W^u)_t \int_u^t \frac{1}{\mathcal{E}(\mathbb{1}_{(u,\infty)} \centerdot W^u)_{s-}} dA_s, \ t \in [u,\infty).$$

By the property of the iM_Z, $\lim_{u\uparrow t} M^u_{t-} = 1 - Z_{t-}$. By Doléans-Dade formula,

$$\lim_{u\uparrow t} \mathcal{E}(\mathbb{1}_{(u,\infty)} \centerdot W^u)_{t-} = 1.$$

Consequently,

$$\begin{aligned}
&\lim_{u\uparrow t}\mathcal{E}(\mathbb{1}_{(u,\infty)}\centerdot W^u)_t \\
&=\lim_{u\uparrow t}\mathcal{E}(\mathbb{1}_{(u,\infty)}\centerdot W^u)_{t-}(1+\Delta_t\widetilde{m}-M^u_{t-}g(M^u_{t-})^\top\Delta_t\mathbf{Y}) \\
&=(1+\Delta_t\widetilde{m}-(1-Z_{t-})g(1-Z_{t-})^\top\Delta_t\mathbf{Y})=\kappa_t.
\end{aligned}$$

We also need to estimate the difference, for $0<u<v\le t$,

$$\begin{aligned}
\frac{d\Xi^v_t}{dx}(y)-\frac{d\Xi^u_t}{dx}(y)&=\frac{d\Xi^v_t}{dx}(y)-\frac{d\Xi^v_t}{dx}(\Xi^u_v(y))\frac{d\Xi^u_v}{dx}(y) \\
&=\frac{d\Xi^v_t}{dx}(y)-\frac{d\Xi^v_t}{dx}(\Xi^u_v(y))+\frac{d\Xi^v_t}{dx}(\Xi^u_v(y))(1-\frac{d\Xi^u_v}{dx}(y)).
\end{aligned}$$

When u is fixed and $v\downarrow u$, by Chapter V Section 7 Theorem 39 of Protter [26], we have $\lim_{v\downarrow u}\Xi^u_v(y)=y$ and $\lim_{v\downarrow u}\frac{d\Xi^u_v}{dx}(y)=1$. According to the same reference, by Kolmogorov continuity criterion, for any $a\in(0,t)$, the random map $y\to\frac{d\Xi^v_t}{dx}(y)$ is continuous uniformly with respect to $v\in[a,t]$. This uniform continuity implies

$$\lim_{v\downarrow u}\frac{d\Xi^v_t}{dx}(y)-\frac{d\Xi^u_t}{dx}(y)=0.$$

Now we write

$$\begin{aligned}
&\frac{M^v_t-M^u_t}{A_v-A_u}=\frac{d\Xi^v_t}{dx}(\xi)((1-Z_v)-M^u_v)\frac{1}{A_v-A_u} \\
&=\frac{d\Xi^v_t}{dx}(1-Z_u)\frac{(1-Z_v)-M^u_v}{A_v-A_u}+(\frac{d\Xi^v_t}{dx}(\xi)-\frac{d\Xi^v_t}{dx}(1-Z_u))\frac{(1-Z_v)-M^u_v}{A_v-A_u} \\
&=\frac{d\Xi^v_t}{dx}(1-Z_v)\frac{(1-Z_v)-M^u_v}{A_v-A_u}+(\frac{d\Xi^v_t}{dx}(\xi)-\frac{d\Xi^v_t}{dx}(1-Z_v))\frac{(1-Z_v)-M^u_v}{A_v-A_u}.
\end{aligned}$$

We consider separately limits when v is fixed and $u\uparrow v$ or when u is fixed and $v\downarrow u$, we have

$$\begin{aligned}
&\lim_{u\uparrow v}((1-Z_v)-M^u_v)=\kappa_v\Delta_vA, \\
&\lim_{u\uparrow v}\frac{(1-Z_v)-M^u_v}{A_v-A_u} &&=\kappa_v, \\
&\lim_{v\downarrow u}((1-Z_v)-M^u_v) &&=0, \\
&\lim_{v\downarrow u}\frac{(1-Z_v)-M^u_v}{A_v-A_u} &&=1.
\end{aligned}$$

From these computations we deduce that

$$\begin{array}{lll}
\lim_{u\uparrow v} \dfrac{M_t^v - M_t^u}{A_v - A_u} = \frac{\Xi_t^v(1-Z_v)-\Xi_t^v(1-Z_v-\kappa_v\Delta_v A)}{\Delta_v A}, & \text{if } \Delta_v A > 0,\ v \in (0,t], \\
\lim_{v\downarrow u} \dfrac{M_t^v - M_t^u}{A_v - A_u} = 0, & \text{if } \Delta_u A > 0,\ u \in (0,t), \\
\lim_{u\uparrow v} \dfrac{M_t^v - M_t^u}{A_v - A_u} = \frac{d\Xi_t^v}{dx}(1 - Z_v)\kappa_v, & \text{if } \Delta_v A = 0,\ v \in (0,t], \\
\lim_{v\downarrow u} \dfrac{M_t^v - M_t^u}{A_v - A_u} = \frac{d\Xi_t^u}{dx}(1 - Z_u), & \text{if } \Delta_u A = 0,\ u \in (0,t).
\end{array}$$

□

The knowledge on the regularity of the random map $u \to M_t^u$ for $u \in [0,t]$ will be very helpful in the study of the non-arbitrage property or of the optimization problem. In the immediate term, this regularity already gives answers to several questions about the dynamic model.

In Jeanblanc et al. [14] we proved results under the assumption that $u \in (0,t] \to M_t^u$ was continuous. We asked if this continuity effectively hold in the continuous dynamic model. Now we can say:

Corollary 5.2. *For the dynamic model in Theorem 5.1, for any* $0 < t < \infty$, *the map* $u \in (0,t] \to M_t^u$ *is continuous if and only if* A *is continuous.*

Another question was how the random measure $d_u M_t^u$ was linked with dA_u. If we refine slightly the proof of Theorem 5.1 we see that, for any $0 < a < b < \infty$, for almost all ω, the quotient $\frac{M_t^v(\omega)-M_t^u(\omega)}{A_v(\omega)-A_u(\omega)}$ is uniformly bounded for $v, u \in [a,b]$. Theorem 4.3.4 of Krantz et al. [20] together with Theorem 5.1 yield to the following result. (Note that $\kappa \neq 1$ only on at most countably many points.)

Corollary 5.3. *For the dynamic model in Theorem 5.1, for any* $0 < t < \infty$, $d_s M_t^s$ *is absolutely continuous with respect to* dA_s *on* $(0,t]$ *with the density function*

$$\frac{d\Xi_t^s}{dx}(1-Z_s)1\!\!1_{\{\Delta_s A=0\}} + \frac{\Xi_t^s(1Z_s) - \Xi_t^s(1-Z_s-\kappa_s\Delta_s A)}{\Delta_s A}1\!\!1_{\{\Delta_s A>0\}},\ s\in(0,t].$$

The absolute continuity property of $d_u M_t^u$ with respect to dA_u is established for finite interval $(0,t]$. The situation when $t = \infty$ can be very different. We have the following lemma from Aksamit [1]:

Lemma 5.4. *Suppose that $Z_t = e^{-t}, t \in [0,\infty)$. Let B be a linear Brownian motion. Consider the $\natural$-equation:*

$$(\natural_u) : \begin{cases} dX_t = X_t(1 - e^{-t} - X_t)dB_t, \ t \in [u,\infty), \\ \\ X_u \ = 1 - e^{-u}, \end{cases}$$

for $0 < u < \infty$, and the associated $\mathcal{M}_Z = (M^u : 0 \leq u \leq \infty)$. Then, the map $u \in [0,\infty] \to M^u_\infty$ takes only two values 0 or 1.

In this lemma, clearly $d_u M^u_\infty$ is not absolutely continuous with respect to $dA_u = e^{-u}du$.

To end this section, notice that, if a one-default model satisfies the density hypothesis in the sense of El Karoui et al. [7] with respect to a determinist measure μ, the random map $u \to M^u_t$ for $u \in [0,t]$ is absolutely continuous with respect to μ. Consequently, we have:

Corollary 5.5. *There exists a dynamic model which does not satisfies the density hypothesis.*

6. One-default Model without Hy(Z)

Until now we study the dynamic one-default model under the assumption **Hy**(Z). The case where $1-Z$ can be null was studied in Jeanblanc et al. [14] under technical conditions. In this section we show a different technic to deal with the problem.

Given a stochastic structure $(\mathbb{P},\mathbb{F})$, let Z be a $(\mathbb{P},\mathbb{F})$ supermartingale with $0 \leq Z \leq 1$. For any $\alpha > 0$ set $Z^{(\alpha)}_t = e^{-\alpha t}Z_t, t \in \mathbb{R}_+$. The supermartingale $Z^{(\alpha)}$ satisfies clearly **Hy**(Z). Choose a $\natural$-pair and consider the corresponding $\natural$-equation associated with $Z^{(\alpha)}$. Denote by $\mathbb{Q}^{(\alpha)}$ the probability measure generated by this $\natural$-equation via Theorem 3.8.

We suppose that Ω is a Polish space with its Borel σ-algebra $\mathcal{A}$. We consider $[0,\infty]$ as a compact metric space. Under these assumptions the product space $[0,\infty] \times \Omega$ is a Polish space. We suppose that, for $t \in \mathbb{R}_+$, the σ-algebra $\mathcal{F}_{t-}$ is generated by the null sets and by the $\mathcal{F}_{t-}$ measurable bounded continuous functions on Ω.

Lemma 6.1. *The family of probability measures $\mathbb{Q}^{(\alpha)}, \alpha > 0$, on $[0,\infty] \times \Omega$ is tight.*

Proof. For any $\epsilon > 0$ let $K \subset \Omega$ be a compact set such that $\mathbb{P}[K^c] < \epsilon$. Then, $[0,\infty] \times K$ is a compact subset in the product space and, as the restriction of $\mathbb{Q}^{(\alpha)}$ on $\mathcal{A}$ is $\mathbb{P}$, $\mathbb{Q}^{(\alpha)}[([0,\infty]\times K)^c] < \epsilon$. □

Theorem 6.2. *There exists a probability measure $\mathbb{Q}$ on the product space such that $\mathbb{Q} = \mathbb{P}$ on $\mathcal{F}_\infty$ and $\mathbb{Q}[t < \tau | \mathcal{F}_t] = Z_t$ for any $t \in \mathbb{R}_+$.*

Proof. Let $\mathbb{Q}^{(0)}$ be the limit point of a sequence of $\mathbb{Q}^{(\alpha_n)}, n \in \mathbb{N}, \alpha_n > 0, \alpha_n \to 0$. Note that, for any bounded continuous function g on Ω,

$$\mathbb{E}^{(0)}[g] = \lim_{n\to\infty} \mathbb{E}^{(\alpha_n)}[g] = \mathbb{E}^{\mathbb{P}}[g].$$

This shows that $\mathbb{Q}^{(0)} = \mathbb{P}$ on $\mathcal{F}_\infty$.

Let $s \in \mathbb{R}_+$ be such that $\mathbb{Q}^{(0)}[\tau = s] = 0$. Let g be a $\mathcal{F}_{s-}$ measurable bounded continuous function on Ω. Let h and h' be bounded continuous functions on $[0,\infty]$ such that $0 \le h(x) \le \mathbb{1}_{\{s\le x\}} \le h'(x)$. We have

$$\begin{aligned}
&\mathbb{E}^{(0)}[gh(\tau)] = \lim_{n\to\infty} \mathbb{E}^{(\alpha_n)}[gh(\tau)] \\
&\le \liminf_{n\to\infty} \mathbb{E}^{(\alpha_n)}[g\mathbb{1}_{\{s\le\tau\}}] = \liminf_{n\to\infty} \mathbb{E}^{(\alpha_n)}[ge^{-\alpha_n s} Z_{s-}] = \mathbb{E}^{\mathbb{P}}[gZ_{s-}] \\
&\le \lim_{n\to\infty} \mathbb{E}^{(\alpha_n)}[gh'(\tau)] = \mathbb{E}^{(0)}[gh'(\tau)].
\end{aligned}$$

This being true for all suitable g, h, h', we obtain

$$\mathbb{Q}^{(0)}[s \le \tau | \mathcal{F}_{s-}] = Z_{s-}.$$

Taking the right limit, we prove the theorem. □

Acknowledgements

This research benefited from the support of the "Chair Markets in Transition" under the aegis of Louis Bachelier laboratory, a joint initiative of École polytechnique, Université d'Évry Val d'Essonne and Fédération Bancaire Française.

References

[1] Aksamit A. *Random times, enlargement of filtration and arbitrages* Thesis University Evry (2014)

[2] Aksamit A., Choulli T., Deng J. and Jeanblanc M. "Non-arbitrage up to random horizon and after honest times for semimartingale models" arXiv:1310.1142 (2013)

[3] Aldous D. *Weak convergence and the general theory of processes* University of California (1981)

[4] Bielecki T., Jeanblanc M. and Rutkowski M. *Credit Risk Modelling* Osaka University Press (2009)

[5] Billingsley P. *Convergence of probability measures* John Wiley & Sons (1999)

[6] Callegaro G., Jeanblanc M. and Zargari B. "Carthaginian enlargement of filtrations" *ESAIM: Probability and Statistics* **17** 550-566 (2013)
[7] El Karoui N., Jeanblanc M. and Ying J. "What happens after a default : the conditional density approach" *Stochastic Processes and their Applications* **120** 1011-1032 (2010)
[8] Elliot R., Jeanblanc M. and Yor M. "On models of default risk" *Mathematical Finance* **10** 179-195 (2000)
[9] Fontana C., Jeanblanc M. and Song S. "On arbitrages arising from honest times" arxiv:1207.1759 (2013)
[10] He S.W., Wang J.G. and Yan J.A. *Semimartingale Theory And Stochastic Calculus* Science Press CRC Press Inc (1992)
[11] Jacod J. *Calcul Stochastique et Problèmes de Martingales* Lecture Notes in Mathematics **714** Springer (1979)
[12] Jeanblanc M. and LeCam Y. "Progressive enlargement of filtrations with initial times" *Stochastic Processes and their Applications* **119** 2523-2543 (2009)
[13] Jeanblanc M. and Song S. "An explicit model of default time with given survival probability" *Stochastic Processes and their Applications* **121**(8) 1678-1704 (2010)
[14] Jeanblanc M. and Song S. "Random times with given survival probability and their $\mathbb{F}$-martingale decomposition formula" *Stochastic Processes and their Applications* **121**(6) 1389-1410 (2010)
[15] Jeulin T. *Semi-martingales et grossissement d'une filtration* Lecture Notes in Mathematics **833** Springer (1980)
[16] Jeulin T. and Yor M. "Nouveaux résultats sur le grossissement des tribus". *Ann. Scient. Ec. Norm. Sup.* **4** t.11 429-443 (1978)
[17] Jeulin T. and Yor M. "Grossissement d'une filtration et semi-martingales: formules explicites" *Séminaire de Probabilités* **12** 78-97 (1978)
[18] Karatzas I. and Kardaras C. "The numéraire portfolio in semimartingale financial models" *Finance and Stochastics* **11**(4) 447-493 (2007)
[19] Knight F. *Essays on the prediction process* IMS Lecture Notes Monograph Series (1981)
[20] Krantz S. and Parks H. *Geometric Integration Theory* Springer (2008)
[21] Li L. and Rutkowski M. "Random times and multiplicative systems" *Stochastic Processes and their Applications* **122**(5) 2053-2077 (2012)
[22] Meyer P. and Yoeurp C. "Sur la décomposition multiplicative des sous-martingales positives" *Séminaire de Probabilités* **10** 501-504 (1976)
[23] Meyer P. and Yor M. "Sur la théorie de la prédiction" *Séminaire de Probabilités* **10** 104-117 (1976)
[24] Nikeghbali A. and Platen E. "On honest times in financial modeling" arxiv:0808.2892 (2008)
[25] Nikeghbali A. and Yor M. "A definition and some characteristic properties of pseudo-stopping times" *The Annals of Probability* **33**(5) 1804-1824 (2005)

[26] Protter P. *Stochastic integration and differential equations* Springer (2005)
[27] Revuz D. and Yor M. *Continuous martingales and Brownian motion* Springer (2004)
[28] Song S. *Grossissement d'une filtration et problèmes connexes* Thesis Université Paris VI (1987)
[29] Song S. "Drift operator in a market affected by the expansion of information flow: a case study" arxiv:1207.1662 (2012)
[30] Yoeurp C. "Décomposition des martingales locales et formules exponentielles" *Séminaire de Probabilités* **10** 432-480 (1976)
[31] Yor M. "Sur les théories du filtrage et de la prédiction" *Séminaire de Probabilités* **11** 257-297 (1977)

Stochastic Sensitivity Study for Optimal Credit Allocation

Laurence Carassus*, Simone Scotti†

Abstract. In this paper we propose a short overview of error calculus theory using Dirichlet forms which was developed by Bouleau. Then we apply this method to an optimal credit allocation problem.

1. Introduction

The study of uncertainties propagation from the parameters to the model outputs is an old topic which is generally treated following two alternative approaches. The uncertainties can be seen as infinitely small deterministic quantities. This allows to calculate the uncertainties propagation using classical differential calculus. The main drawback of this method is that it forgets the uncertainties random nature: in particular the uncertainties correlation can not be addressed at all.

In the second approach, the uncertainties are represented as random variables. However, the knowledge of uncertainties probability law is out of reach in almost all real cases: only the bias and the mean squared error can be generally estimated. Moreover, the explicit computation of a probability measure through a non-affine transformation is an extremely complex problem. This second drawback is well-known and has forced the probability community to develop different ways to construct complex but treatable objects. In his seminal work, Bouleau [5] shows that the Dirichlet forms theory offers a general framework to treat the infinitesimal random uncertainties propagation using a second order differential calculus. In a first part, we propose an overview of this theory. We present the intuition as well as the axiomatic of Dirichlet forms theory applied to error calculus. We also develop a concrete example of error structure.

*LMR, Université Reims Champagne Ardenne and LPMA, Université Paris Diderot. email: laurence.carassus@univ-reims.fr

†LPMA, Université Paris Diderot. email: simone.scotti@univ-paris-diderot.fr

The second part of the paper is devoted to a concrete application of the preceding theory. We adopt the setup proposed by Bernis et al. [2]. We consider the problem of a credit allocation under a hidden regime switching model. The global evolution of the credit market is driven by a benchmark, the drift of which is given by a two-state continuous time hidden Markov chain. The investor performs a simple mean-variance allocation on credit assets. In Bernis et al. [2], the authors apply filtering techniques to obtain the diffusion of the credit assets under partial observation, and show that they have a specific excess return with respect to the benchmark. Then they assume that the views on the excess returns are approximatively accurate and contain uncertainty. At this point it is possible to apply Dirichlet forms theory and to perform a sensitivity analysis with respect to the random excess returns on the optimal solution of the portfolio optimization problem. In this paper, we present the complete proof of this sensitivity analysis (see Theorem 3.12): we compute the bias and the variance of the optimal portfolio estimator as a function of the bias and the variance of the extra drifts. Even if Theorem 3.12 looks quite abstract, it can be applied to concrete case. In Bernis et al. [2], a numerical application to the case of a sector allocation with CDS, fully calibrated with observable data or direct input given by the portfolio manager, is proposed.

The rest of the paper is organized as follows. In Section 2, we propose an overview of error calculus theory. In Section 3, we present an application to credit optimal allocation. In particular, in Section 3.2, we apply the error calculus machinery to determine the variance and bias of the optimal allocation.

2. Error Calculus

2.1. *Intuition*

This subsection is devoted to the presentation of error calculus in the sense of sensitivity with respect to a stochastic perturbation.

We shortly recall Gauss idea for error propagation. All the implied functions are assumed to be smooth enough. Let $(\Omega, \mathcal{F}, \mathbb{P})$ be a probability space. Let $V = F(U_1, \ldots, U_n)$ be a real valued function of $U_1, \ldots, U_n$. The U_i are seen as random variables with values very close to their mean value $\mathbb{E}[U_i]$ and $\mathbb{E}[U_i]$ is interpreted as a measure of U_i. This measure can be erroneous and the possible measure error is given by $\sigma_{i,j} = \mathbb{E}\left[(U_i - \mathbb{E}[U_i])(U_j - \mathbb{E}[U_j])\right]$. Here the error is assumed to be small, this allows us to use a first order approximation and to obtain:

$$F(U_1,\dots,U_n)-F(\mathbb{E}[U_1],\dots,\mathbb{E}[U_n)])$$
$$=\sum_{i=1}^{n}(U_i-\mathbb{E}[U_i])\frac{\partial F}{\partial U_i}(\mathbb{E}[U_1],\dots,\mathbb{E}[U_n)]).$$

If we denote by $\sigma_V^2=\mathbb{E}\left[F(U_1,\dots,U_n)-F(\mathbb{E}[U_1],\dots,\mathbb{E}[U_n])\right]^2$,

$$\sigma_V^2=\sum_{i,j=1}^{n}\sigma_{i,j}\frac{\partial F}{\partial U_i}(\mathbb{E}[U_1],\dots,\mathbb{E}[U_n)])\frac{\partial F}{\partial U_j}(\mathbb{E}[U_1],\dots,\mathbb{E}[U_n)]) \quad (2.1)$$

We try to generalize Gauss equation (2.1) by considering that the measure of U_i is no more the constant value $\mathbb{E}[U_i]$ but a random variable. This random variable is defined on an (independent) copy of $(\Omega,\mathcal{F},\mathbb{P})$ denoted by $(\widetilde{\Omega},\widetilde{\mathcal{F}},\widetilde{\mathbb{P}})$. We called it $\widehat{U_i}$: in statistical language it is the estimator of U_i. Consider another real valued function G. We introduce the following random variables called bias and mean squared error and defined on the space $(\widetilde{\Omega},\widetilde{\mathcal{F}},\widetilde{\mathbb{P}})$:

$$\text{Bias}(U_i)=\mathbb{E}\left(U_i-\widehat{U_i}\right)=\mathbb{E}\left(U_i\right)-\widehat{U_i}$$

$$\text{Bias}(F(U_1,\dots,U_n))=\mathbb{E}\left[F(U_1,\dots,U_n)-F(\widehat{U_1},\dots,\widehat{U_n})\right] \quad (2.2)$$
$$=\mathbb{E}\left[F(U_1,\dots,U_n)\right]-F(\widehat{U_1},\dots,\widehat{U_n})$$

$$\text{MSE}(U_i,U_j)=\mathbb{E}\left[\left(U_i-\widehat{U_i}\right)\left(U_j-\widehat{U_j}\right)\right] \quad (2.3)$$

$$\text{MSE}(F(U_1,\dots,U_n),G(U_1,\dots,U_n))$$
$$=\mathbb{E}\left[\left(F(U_1,\dots,U_n)-F(\widehat{U_1},\dots,\widehat{U_n})\right)\times\left(G(U_1,\dots,U_n)-G(\widehat{U_1},\dots,\widehat{U_n})\right)\right].$$

Let $H\in\{F,G\}$, using a first order approximation we obtain:

$$H(U_1,\dots,U_n)-H(\widehat{U_1},\dots,\widehat{U_n})=\sum_{i=1}^{n}(U_i-\widehat{U_i})\frac{\partial H}{\partial U_i}(\widehat{U_1},\dots,\widehat{U_n}).$$

This leads to:

$$\text{MSE}(F(U_1,\dots,U_n),G(U_1,\dots,U_n))= \quad (2.4)$$
$$\sum_{i,j=1}^{n}\frac{\partial F}{\partial U_i}(\widehat{U_1},\dots,\widehat{U_n})\times\frac{\partial G}{\partial U_i}(\widehat{U_1},\dots,\widehat{U_n})\text{MSE}(U_i,U_j).$$

Thus one has:

$$F(U_1,\ldots,U_n) - F(\widehat{U_1},\ldots,\widehat{U_n}) = \sum_{i=1}^{n}(U_i - \widehat{U_i})\frac{\partial F}{\partial U_i}(\widehat{U_1},\ldots,\widehat{U_n}) + \frac{1}{2}\sum_{i,j=1}^{n}(U_i - \widehat{U_i})(U_j - \widehat{U_j})\frac{\partial^2 F}{\partial U_i U_j}(\widehat{U_1},\ldots,\widehat{U_n})$$

We also get that:

$$\mathrm{Bias}(F(U_1,\ldots,U_n)) = \sum_{i=1}^{n}\mathrm{Bias}(U_i)\frac{\partial F}{\partial U_i}(\widehat{U_1},\ldots,\widehat{U_n}) + \frac{1}{2}\sum_{i,j=1}^{n}\mathrm{MSE}(U_i,U_j)\frac{\partial^2 F}{\partial U_i U_j}(\widehat{U_1},\ldots,\widehat{U_n}). \tag{2.5}$$

The bias follows a second order differential calculus involving the mean square error.

The idea developed by Bouleau is to allow the errors to be random variables and to propose an axiomatization in the spirit of (2.4) and (2.5). The quadratic error on each U_i will be given by $\Gamma(U_i, U_i)$ (shortly denoted by $\Gamma(U_i)$) where Γ is a bilinear operator associated to the mean squared error and the bias by $\mathcal{A}(U_i)$, where $\mathcal{A}$ is a linear operator. This representation would have the nice following property that if the sequence of pairs $(Y_n, \Gamma(Y_n))$ converges (in a certain sense) it should tend to $(Y, \Gamma(Y))$.

2.2. *Error Calculus Axiomatic*

We now formalize the intuition exposed in the preceding section. The axiomatization of random uncertainty propagation was introduced by Bouleau [4] as follows:

Definition 2.1. Error structure An error structure is a term $\left(\widetilde{\Omega}, \widetilde{\mathcal{F}}, \widetilde{\mathbb{P}}, \mathbb{D}, \Gamma\right)$, where

- $\left(\widetilde{\Omega}, \widetilde{\mathcal{F}}, \widetilde{\mathbb{P}}\right)$ is a probability space;
- $\mathbb{D}$ is a dense sub-vector space of $L^2\left(\widetilde{\Omega}, \widetilde{\mathcal{F}}, \widetilde{\mathbb{P}}\right)$ such that for any function F of class C^1 and globally Lipschitz (afterward denoted $C^1 \cap Lip$) and $U \in \mathbb{D}$, one has $F(U) \in \mathbb{D}$.
- Γ is a positive symmetric bilinear function from $\mathbb{D} \times \mathbb{D}$ into $L^1\left(\widetilde{\Omega}, \widetilde{\mathcal{F}}, \widetilde{\mathbb{P}}\right)$ satisfying the following functional calculus inspired

by (2.4): for any functions F and G of class $\mathcal{C}^1 \cap Lip$ and $U, V \in \mathbb{D}^n$

$$\Gamma\left[F(U),\, G(V)\right] = \sum_{i,j=1}^{n} \frac{\partial F}{\partial U_i}(U)\, \frac{\partial G}{\partial U_j}(V)\, \Gamma[U,\, V] \;\; \widetilde{\mathbb{P}}\; a.s.; \tag{2.6}$$

- the bilinear form $\mathcal{E}[U,\, V] = \frac{1}{2}\widetilde{\mathbb{E}}\left[\Gamma[U,\, V]\right]$ is closed, i.e. $\mathbb{D}$ equipped with the norm $|U|_{\mathbb{D}} = \left(\widetilde{\mathbb{E}}\left[U^2\right] + \frac{1}{2}\mathcal{E}[U,\, U]\right)^{1/2}$ is complete;
- the constant 1 belongs to $\mathbb{D}$ and $\mathcal{E}[1, 1] = 0$

An error structure is thus a probability space equipped with a so called "carré du champ operator" Γ. We generally write $\Gamma[U]$ for $\Gamma[U,\, U]$. The Hille-Yosida theorem guarantees that there exist a semigroup and a generator $\mathcal{A}$ that are coherent with the Dirichlet form $\mathcal{E}$, see for instance Albeverio [1] and Fukushima et al. [7]. This generator $\mathcal{A} : \mathcal{DA} \to L^2\left(\widetilde{\Omega},\, \widetilde{\mathcal{F}},\, \widetilde{\mathbb{P}}\right)$ is a self-adjoint operator, its domain $\mathcal{DA}$ is included into $\mathbb{D}$. It is such that for all $U \in \mathcal{DA}$ and $V \in L^2\left(\widetilde{\Omega},\, \widetilde{\mathcal{F}},\, \widetilde{\mathbb{P}}\right)$

$$\mathcal{E}[U, V] = -\widetilde{\mathbb{E}}\left[\mathcal{A}[U]V\right]. \tag{2.7}$$

Moreover this operator satisfies, for $F \in \mathcal{C}^2 \cap Lip$, $U \in \mathcal{DA}$, $F(U) \in \mathcal{DA}$ and $\Gamma[U] \in L^2\left(\widetilde{\Omega},\, \widetilde{\mathcal{F}},\, \widetilde{\mathbb{P}}\right)$:

$$\mathcal{A}\left[F(U)\right] = \sum_{i=1}^{n} \frac{\partial F}{\partial U_i}(U)\mathcal{A}[U] + \frac{1}{2}\sum_{i,j=1}^{n} \frac{\partial^2 F}{\partial U_i U_j}(U)\, \Gamma[U] \;\; \widetilde{\mathbb{P}}\; a.s.. \tag{2.8}$$

This equation is similar to (2.5). From (2.7), $\mathcal{A}$ is a closed operator with respect to the norm $|\;|_{\mathbb{D}}$, in the sense that $\mathcal{DA}$ equipped with the norm $|\;|_{\mathbb{D}}$ is complete.

We present below a basic example of error structure. This shows that the set of error structures is not empty and give some further intuitions about the different operators introduced above.

Example 2.2. Ornstein-Uhlenbeck error structure The Ornstein-Uhlenbeck structure is $(\mathbb{R}, \mathcal{B}(\mathbb{R}), \mu, \mathbb{D}, \Gamma)$, where μ is unidimensional centered Gaussian law and $\mathbb{D} := H^{1,2}(\mu)$, i.e. the first Sobolev space associated to $L^2(\mathbb{R}, \mathcal{B}(\mathbb{R}), \mu)$. Here we choose a particular semi-group and we compute the associated generator and Dirichlet form. To this end we introduce an auxiliary Ornstein-Uhlenbeck process in a probability space $(\Omega^B, \mathcal{F}^B, \mathbb{P}^B)$ equipped with a Brownian motion B:

$$dX_\epsilon = -\frac{1}{2}X_\epsilon d\epsilon + dB_\epsilon.$$

We denote by X^x_ϵ the solution of the preceding equation with $X_0 = x$. We then define the semigroup as follows: for all $\omega \in \mathbb{R}$,

$$P_\epsilon[U](\omega) = \mathbb{E}^B\left[U(X^x_\epsilon)|x = \omega\right],$$

where $\mathbb{E}^B$ denotes the expectation with respect to $\mathbb{P}^B$. Using Itô lemma, we obtain that for $U \in \mathbf{H}^{2,2}(\mu)$:

$$\begin{aligned}
X^x_\epsilon &= xe^{-\frac{1}{2}\epsilon} + \int_0^\epsilon e^{-\frac{1}{2}(\epsilon-s)}dB_s \\
U(X^x_\epsilon) &= U(x) - \frac{1}{2}\int_0^\epsilon U'(X^x_s)X^x_s ds + \frac{1}{2}\int_0^\epsilon U''(X^x_s)ds \\
&\quad + \int_0^\epsilon U'(X^x_s)dB_s \\
< U(X^x) >_\epsilon &= \int_0^\epsilon \left(U'(X^x_s)\right)^2 ds \qquad (2.9)
\end{aligned}$$

where $<>_\epsilon$ denotes the quadratic variation operator. The generator $\mathcal{A}$ and the Dirichlet form $\mathcal{E}$ associated to the semi-group P_ϵ are given by (see Bouleau [5] chapter II):

$$\begin{aligned}
\mathcal{A}[U](\omega) &= \lim_{\epsilon\to 0}\frac{P_\epsilon[U](\omega) - U(\omega)}{\epsilon} = \frac{1}{2}U''(\omega) - \frac{1}{2}\omega U'(\omega) \qquad (2.10) \\
\mathcal{E}[U] &= \langle -\mathcal{A}[U], U\rangle_{L^2(\mu)} = \frac{1}{2}\int(\omega U'(\omega) - U''(\omega))U(\omega)d\mu(\omega) \\
&= \frac{1}{2}\int(U'(\omega))^2 d\mu(\omega) \qquad (2.11)
\end{aligned}$$

where (2.11) comes from integration by parts formula (recall that μ is a centered gaussian law). As $\mathcal{E}[U] = \frac{1}{2}\int\Gamma[U,U](\omega)d\mu(\omega)$, we deduce from (2.11) that $\Gamma(U) = (U')^2$ for all $U \in \mathbb{D}$. From (2.10), the related generator is $\mathcal{A}[U] = \frac{1}{2}U'' - \frac{1}{2}Id\cdot U'$, where Id denotes the identity operator. Here the domain $\mathcal{DA} := \mathbf{H}^{2,2}(\mu)$, i.e. the second Sobolev space.

In this particular case, (2.6) and (2.8) are easy to obtain:

$$\begin{aligned}
\Gamma\left[F(U), G(V)\right](\omega) &= F'(U)(\omega)U'(\omega)G'(V)(\omega)V'(\omega) \\
&= F'(U)(\omega)G'(V)(\omega)\Gamma[U,V](\omega) \\
\mathcal{A}\left[F(U)\right](\omega) &= \frac{1}{2}\left(F''(U)(\omega)\left(U'(\omega)\right)^2 + F'(U)(\omega)U''(\omega)\right) \\
&\quad -\frac{1}{2}\omega F'(U)(\omega)U'(\omega) \\
&= F'(U)(\omega)\mathcal{A}[U](\omega) + \frac{1}{2}F''(U)(\omega)\Gamma[U](\omega)
\end{aligned}$$

We finish this example with the following remark. In section 2.1, we have associated to the statistical notions of bias and mean squared error the operators $\mathcal{A}$ and Γ (see (2.2), (2.3) and (2.4)). In this example we can do the other way around: it is easy to show that $\mathcal{A}$ is associated to a bias and Γ is associated to a mean square error in the following sense: for ϵ small enough, one has:

$$\epsilon\mathcal{A}[U](\omega) = \mathbb{E}^B\left[U(X_\epsilon^x) - U(x)|x = \omega\right]$$

$$\epsilon\Gamma[U](\omega) = \mathbb{E}^B\left[\left(U(X_\epsilon^x) - U(x)\right)^2|x = \omega\right]$$

Afterwards, we will omit the explicit dependency on ω.

Remark 2.3. Statistical interpretation It is possible to push further the analogy between statistics and Dirichlet forms theory applied to errors computation. Bouleau and Chorro [6] show that one can construct confidence interval for some random variable U using $\mathcal{A}[U]$ and $\Gamma[U]$. This is achieved by choosing an error structure linked to Fisher's information matrix.

The main drawback of the "carré du champ" operator Γ is its bilinearity, which makes computations awkward to perform. An easy way to overcome this drawback is to introduce a new operator, called the gradient, see Bouleau and Hirsch [3], section II.6. We recall the definition of gradient operator associated to Γ.

Proposition 2.4. Gradient operator *From now we assume that the space $\mathbb{D}$ is separable. Let $\left(\widetilde{\Omega}, \widetilde{\mathcal{F}}, \widetilde{\mathbb{P}}, \mathbb{D}, \Gamma\right)$ be an error structure. Let $\left(\overline{\Omega}, \overline{\mathcal{F}}, \overline{\mathbb{P}}\right)$ be an (independent) copy of $\left(\widetilde{\Omega}, \widetilde{\mathcal{F}}, \widetilde{\mathbb{P}}\right)$ and $\mathcal{H} = L^2\left(\overline{\Omega}, \overline{\mathcal{F}}, \overline{\mathbb{P}}\right)$ be an auxiliary Hilbert space equipped with scalar product $< X, Y >_{\mathcal{H}} = \overline{\mathbb{E}}(XY)$, where $\overline{\mathbb{E}}$ is the expectation computed under $\overline{\mathbb{P}}$. Let $L^2\left(\left(\overline{\Omega}, \overline{\mathcal{F}}, \overline{\mathbb{P}}\right), \widetilde{\mathbb{P}}\right)$ or shortly $L^2(\widetilde{\mathbb{P}}, \mathcal{H})$ the space of $L^2\left(\overline{\Omega}, \overline{\mathcal{F}}, \overline{\mathbb{P}}\right)$ valued random variables equipped with the scalar product $< A, B >_{L^2(\widetilde{\mathbb{P}},\mathcal{H})} = \widetilde{\mathbb{E}}[< A, B >_{\mathcal{H}}] = \widetilde{\mathbb{E}}[\overline{\mathbb{E}}(AB)]$. Then there exists a linear operator, called gradient and denoted by $(\,)^{\#} : \mathbb{D} \to L^2(\widetilde{\mathbb{P}}, \mathcal{H})$, with the following two properties:*

$$\forall U, V \in \mathbb{D},\ \Gamma[U, V] = < U^{\#}, V^{\#} >_{\mathcal{H}} = \overline{\mathbb{E}}\left(U^{\#}V^{\#}\right) \tag{2.12}$$

$$\forall U \in \mathbb{D}^n,\ F \in \mathcal{C}^1 \cap Lip, (F(U_1, \ldots, U_n))^{\#} = \sum_{i=1}^{n}\left(\frac{\partial F}{\partial x_i} \circ U\right) U_i^{\#} \tag{2.13}$$

The gradient operator is a useful tool when computing Γ because it is linear, whereas the "carré du champ" operator is bilinear.

Let $U \in \mathbb{D}$, by (2.12) and Definition 2.1 (item 4) $\widetilde{\mathbb{E}}[\overline{\mathbb{E}}[(U^{\#})^2]] = \widetilde{\mathbb{E}}[\Gamma[U]] = 2\mathcal{E}(U,U)$ and the gradient operator is closed in the sense that $\mathbb{D}$ equipped with the norm $|\ |_{\mathbb{D}}$ is complete.

In the sequel, the two following lemmas will be in force:

Lemma 2.5. Chain rules for a product of random variables *Let U and $V \in \mathbb{D}$, then one has:*

$$(UV)^{\#} = UV^{\#} + VU^{\#} \tag{2.14}$$

$$\Gamma[UV] = U^2\Gamma[V] + V^2\Gamma[U] + 2UV\Gamma[U,V] \tag{2.15}$$

Moreover, if U and $V \in \mathcal{DA}$, then one has:

$$\mathcal{A}[UV] = U\mathcal{A}[V] + V\mathcal{A}[U] + \Gamma[U,V] \tag{2.16}$$

Proof: (2.14) follows from (2.13), (2.15) follows from (2.6) and (2.16) follows from (2.8). □

Lemma 2.6. Expectation *Let $(\Omega, \mathcal{F}, \mathbb{P})$ be a probability space independent of $\left(\widetilde{\Omega}, \widetilde{\mathcal{F}}, \widetilde{\mathbb{P}}\right)$ and $\left(\overline{\Omega}, \overline{\mathcal{F}}, \overline{\mathbb{P}}\right)$, we will denote $\mathbb{E}$ the conditional expectation w.r.t. $\mathbb{P}$. Let $U \in \mathbb{D} \subset L^2\left(\widetilde{\Omega}, \widetilde{\mathcal{F}}, \widetilde{\mathbb{P}}\right)$ and $V \in \mathcal{DA}$ then one has*

$$\mathbb{E}[U^{\#}] = U^{\#}$$

$$\mathbb{E}[\mathcal{A}[V]] = \mathcal{A}[V].$$

Let $U, V \in \mathbb{D}$ then one has

$$\mathbb{E}[\Gamma[U,V]] = \Gamma[U,V].$$

Let $V \in L^2\left(\Omega \times \widetilde{\Omega}, \mathcal{F} \otimes \widetilde{\mathcal{F}}, \mathbb{P} \otimes \widetilde{\mathbb{P}}\right)$ such that $V \in \mathbb{D}$ and $\mathbb{E}[V] \in \mathbb{D}$, one has:

$$(\mathbb{E}[V])^{\#} = \mathbb{E}[V^{\#}].$$

Let $V \in L^2\left(\Omega \times \widetilde{\Omega}, \mathcal{F} \otimes \widetilde{\mathcal{F}}, \mathbb{P} \otimes \widetilde{\mathbb{P}}\right)$ such that $V \in \mathbb{D}\mathcal{A}$ and $\mathbb{E}[V] \in \mathbb{D}\mathcal{A}$, one has:

$$\mathcal{A}[\mathbb{E}[V]] = \mathbb{E}[\mathcal{A}[V]].$$

Proof: Let $U \in L^1\left(\widetilde{\Omega}, \widetilde{\mathcal{F}}, \widetilde{\mathbb{P}}\right)$ then $\mathbb{E}[U] = U$. So the first three equalities follow from the fact that $U^{\#} \in L^2(\widetilde{\mathbb{P}}, \mathcal{H})$, $\mathcal{A}[V] \in L^2\left(\widetilde{\Omega}, \widetilde{\mathcal{F}}, \widetilde{\mathbb{P}}\right)$ and

$\Gamma[U,V] \in L^1\left(\widetilde{\Omega}, \widetilde{\mathcal{F}}, \widetilde{\mathbb{P}}\right)$. For the fourth equation we use the fact that the gradient operator is closed and we can exchange the gradient operator and the integral sign. The proof of this fact proceeds by an approximation of the integral by a sum, then we apply the gradient operator and finally we take the limit using the gradient operator closeness, see for instance Bouleau [4] section VI.2. Lemma VI.8. The proof of the last equation is similar. □

We observe that the error theory based on Dirichlet forms restricts its analysis to the study of the first two orders of error propagation, i.e. the bias and the variance. This fact is justified by the lack of information on the parameter uncertainties, generally given by the Fischer information matrix, that is often quite limited. The study of higher orders is a very difficult problem for both mathematical and practical reasons. From the mathematical point of view, it would be necessary to study chain rules of higher orders, involving skewness and kurtosis, and to prove that the related operators are closed in a suitable space. However, the crucial problem remains to have sufficiently accurate estimates for the higher order uncertainties. This statistical obstacle cannot be overcome easily. Therefore, it seems reasonable to restrict the study to the two first orders.

3. An Application to Optimal Credit Allocation

3.1. *Financial setup*

This section is taken from the paper of Bernis et al. [2] and reported here for the reader convenience. We consider a continuous time financial model with K credit issuers and one synthetic asset referred to as the benchmark. This synthetic asset represents the global evolution of the credit market. The synthetic asset and each credit issuer k are characterized by their spread over the risk free rate which are respectively denoted by X_0 and X_k (for $k = 1, \dots, K$). The uncertainty is represented by a filtered probability space $(\Omega, \mathcal{F}, \mathbb{F} := \{\mathcal{F}_t\}_{t\geq 0}, \mathbb{P})$ satisfying the usual conditions. First, we define on this space a $\mathbb{F}$-adapted, continuous-time, two-state valued Markov chain, $Y = (Y_t)_{t\geq 0}$. The state space of the Markov chain is equal to $\{g, b\}$.

Let $(W(t))_{t\geq 0} = (W_0(t), \dots, W_K(t))_{t\geq 0}$ be a standard $(\mathbb{F}, \mathbb{P})$-Brownian motion of dimension $K+1$. Let us define the $K+1$-dimensional stochastic process $(X(t))_{t\geq 0}$, with $X(t) := (X_k(t))_{0\leq k\leq K}$. The process X_0 represents the spread of the benchmark, the drift of which μ_0 is assumed to be a measurable function of the Markov chain Y:

$$\frac{dX_0(t)}{X_0(t)} = \mu_0(Y_t)dt + \sigma_0 dW_0(t),$$
$$X_0(0) = x_0$$

For all $k \in \{1, \ldots, K\}$, $\Theta_k = (\Theta_k(t))_{t\geq 0}$ are $\mathbb{F}$-adapted processes (thus influenced by Y), representing the unknown drift of X_k, the spread over the risk free rate of credit issuer:

$$\frac{dX_k(t)}{X_k(t)} = \Theta_k(t)dt + \sigma_k \left(\rho_k dW_0(t) + \sqrt{1-\rho_k^2} \sum_{j=1}^{K} L_{k,j} dW_j(t) \right)$$
$$X_k(0) = x_k$$

where $L := [L_{i,j}]$ is a $K \times K$ lower triangular matrix, such that $C := L \cdot L'$ is a (non-degenerated) correlation matrix. It means that $C := (C_{i,j})_{1\leq i,j\leq K}$ is a symmetric, semi-definite positive matrix with unit diagonal coefficients. We also denote $Z_k^{\perp}(t) = \sum_{j=1}^{K} L_{k,j} W_j(t)$. It is a standard $(\mathbb{F}, \mathbb{P})$-Brownian motion of dimension 1.

We introduce $\mathbb{G} := \{\mathcal{G}_t\}_{t\geq 0}$ and $\mathbb{G}^0 := \{\mathcal{G}_t^0\}_{t\geq 0}$, the right continuous, complete filtrations generated respectively by the following processes:

$$\mathcal{G}_t = \sigma\{X_0(s), X_1(s), \ldots, X_K(s) | s \leq t\}$$
$$\mathcal{G}_t^0 = \sigma\{X_0(s) | s \leq t\}$$

Now we state the following assumptions which will prevail throughout the paper.

Assumption 3.1. We assume that W_0 and Y are independent. We assume that the processes $\mu_0(Y)$ and Θ_k are uniformly bounded and measurable. Finally $\sigma_k > 0$ for $k = 0, \ldots, K$ and $-1 < \rho_k < 1$ for $k = 1, \ldots, K$.

Assumption 3.2. Let $p_t := \mathbb{P}\left[\{Y_t = b\} \mid \mathcal{G}_t^0\right]$, for any $t \geq 0$. We assume that

$$p_t = \mathbb{P}\left[\{Y_t = b\} | \mathcal{G}_t\right].$$

In Bernis et al. [2], the following proposition is proved.

Proposition 3.3. *Under Assumptions 3.1 and 3.2,*

$$\begin{aligned}
\frac{dX_0(t)}{X_0(t)} &= \mathbb{E}\left\{\mu_0(Y_t) \mid \mathcal{G}_t^0\right\} dt + \sigma_0 d\widehat{W}_0(t) \\
\frac{dX_k(t)}{X_k(t)} &= \left(\mathbb{E}\left\{\mu_0(Y_t) \mid \mathcal{G}_t^0\right\} + e_k(t)\right) dt \\
&\quad + \sigma_k \left(\rho_k d\widehat{W}_0(t) + \sqrt{1-\rho_k^2}\, d\widehat{Z}_k^{\perp}(t)\right),
\end{aligned} \tag{3.1}$$

where

$$\begin{aligned}
e_k(t) &= \mathbb{E}\left\{\Theta_k(t)|\mathcal{G}_t\right\} - \mathbb{E}\{\mu_0(Y_t)|\mathcal{G}_t^0\} \\
\widehat{W}_0(t) &= W_0(t) + \frac{1}{\sigma_0}\int_0^t \left(\mu_0(Y_s) - \mathbb{E}\left\{\mu_0(Y_s) \mid \mathcal{G}_s^0\right\}\right) ds \\
\widehat{Z}_k^{\perp}(t) &= Z_k^{\perp}(t) + \frac{1}{\sigma_k\sqrt{1-\rho_k^2}}\int_0^t \left(\Theta_k(s) - \mathbb{E}\left\{\mu_0(Y_s) \mid \mathcal{G}_s^0\right\} - e_k(t)\right) ds \\
&\quad - \frac{\rho_k}{\sqrt{1-\rho_k^2}}\int_0^t \frac{\mu_0(Y_s) - \mathbb{E}\{\mu_0(Y_s)|\mathcal{G}_s^0\}}{\sigma_0} ds.
\end{aligned}$$

$\widehat{W}_0$ *is a* $(\mathbb{G}^0, \mathbb{P})$*-Brownian motion and for* $k = 1, \ldots, K$, $\widehat{Z}_k^{\perp}$ *is a* $(\mathbb{G}, \mathbb{P})$*-Brownian motion. Moreover* $\mathbb{E}\left\{\mu_0(Y_t) \mid \mathcal{G}_t^0\right\} = (\mu_0(b) - \mu_0(g))p_t + \mu_0(g)$, *where* $(p_t)_{t\geq 0}$ *is solution of*

$$dp_t = \left[-\left(\lambda_b + \lambda_g\right) p_t + \lambda_b\right] dt + \frac{\mu_0(b) - \mu_0(g)}{\sigma_0} p_t(1-p_t) d\widehat{W}_0(t).$$

The $e_k(t)$ can be interpreted as the $\mathcal{G}_t$-adapted views of the portfolio manager on the excess return of the spread X_k with respect to the benchmark. Let

$$\epsilon_k = \frac{1}{T}\int_0^T e_k(t)dt. \tag{3.2}$$

Then from (3.1) and (3.2), we obtain that

$$\begin{aligned}
X_k(T) = x_k \exp\Bigg\{ &\int_0^T \left(\mathbb{E}\left\{\mu_0(Y_t) \mid \mathcal{G}_t^0\right\}\right) dt + T\epsilon_k - \frac{\sigma_k}{2} T \\
&+ \sigma_k \int_0^T \left(\rho_k d\widehat{W}_0(t) + \sqrt{1-\rho_k^2}\, d\widehat{Z}_k^{\perp}(t)\right)\Bigg\},
\end{aligned} \tag{3.3}$$

We assume that it is possible to invest in the benchmark and on each credit issuer through a debt product - a bond or a CDS - the price of which at time $t \geq 0$ is given by $P^{(k)}(t, X_k(t))$, $k \in \{0, \ldots, K\}$.

As an example we consider a bond paying at future time $T_1 < \cdots < T_n$ the deterministic cash flows $(C_i)_{1\leq i\leq n}$ (interest and capital). Here we will drop the superscript k as there is only one bond. Consider a recovery rate $Rec \in [0,1[$. Set, for any $t < T_1$,

$$P(t,x) = \sum_{i=1}^{n} C_i \frac{Cap(t)}{Cap(T_i)} e^{-x(T_i-t)} + Rec \int_t^{T_n} \frac{Cap(t)}{Cap(u)} x e^{-xu} du$$

which means that the spread x is a default intensity associated with the recovery rate Rec. When $Rec = 0$, x becomes a (continuous) spread over the risk free rate. Here $Cap(t)$ denotes the deterministic capitalisation factor at time t of the risk free rate.

Assumption 3.4. The mappings $P^{(k)}(\cdot,\cdot)$ from $\mathbb{R}^+ \times \mathbb{R}^+$ to $(0,\infty)$ are, at least, twice continuously differentiable.

We will denote by $\dot{P}_1^{(k)}(\cdot,\cdot)$ its first order derivative with respect to the first variable, by $\dot{P}_2^{(k)}(\cdot,\cdot)$ its first order derivative with respect to the second variable, and by $\ddot{P}_2^{(k)}(\cdot,\cdot)$ its second order derivative with respect to the second variable.

The P&L, at time $t \leq T$, of a buy and hold position on the asset $k \in \{0,\ldots,K\}$ is given by

$$\text{P\&L}_k(t, X_k(t)) = P^{(k)}(t, X_k(t)) - P^{(k)}(0, x_k) \times Cap(t). \tag{3.4}$$

We consider two alternative portfolio representations. The first one uses the allocation on the assets to outperform the benchmark (benchmarked allocation): this corresponds to the case $\zeta = 1$. The second one is a simple allocation on the K assets with no benchmark reference (total return allocation) and corresponds to the case $\zeta = 0$. For any $\pi \in \mathbb{R}^K$:

$$G(\pi, t, X(t)) := \sum_{k=1}^{K} \text{P\&L}_k(t, X_k(t)) \times \pi_k - \zeta \text{P\&L}_0(t, X_0(t)) \tag{3.5}$$

Define for any $(k,j) \in \{0,\ldots,K\}^2$,

$$Cov[k,j] := \mathbb{E}\{\text{P\&L}_k(t)\text{P\&L}_j(t)\} - M_k M_j,$$

where for any $k \in \{0,\ldots,K\}$, $M_k := \mathbb{E}\{\text{P\&L}_k(t)\}$. Note that for ease of notation we drop the time indexation. We also set

$$\begin{aligned} M &:= (M_k)_{1\leq k\leq K}, \\ Cov &:= (Cov[i,j])_{(i,j)\in\{1,\ldots,K\}^2} \text{ and} \\ Cov[0] &:= (Cov[0,k])_{k\in\{1,\ldots,K\}}. \end{aligned}$$

Then $\mathbb{V}\{G(\pi,T)\} = \pi' \cdot Cov \cdot \pi - 2\zeta\pi' \cdot Cov[0] + \zeta^2 Cov[0,0]$ and $\mathbb{E}\{G(\pi,T)\} = \pi' \cdot M - \zeta M_0$. So the mean-variance program $(\mathcal{P})$ solved by the investor can be written

$$(\mathcal{P}): \begin{cases} \min_{\pi \in \mathbb{R}^K} \frac{1}{2} (\pi' \cdot Cov \cdot \pi - \zeta\pi' \cdot Cov[0]\} \\ \\ \text{s.t. } \pi' \cdot M \geq r + \zeta M_0 \\ \\ \text{and } \pi' \cdot \mathbb{I} = 1 \end{cases}$$

where $\mathbb{I}$ is the element of $\mathbb{R}^K$ with all its components equal to 1 and $r > 0$ is the return budget constraint.

In order to solve the optimization program, we define the following quantities:

$$\begin{array}{ll} z_1 := \mathbb{I}' \cdot Cov^{-1} \cdot \mathbb{I} & z_2 := M' \cdot Cov^{-1} \cdot M \\ z_3 := M' \cdot Cov^{-1} \cdot \mathbb{I} & z_4 := \mathbb{I}' \cdot Cov^{-1} \cdot Cov[0] \\ z_5 := M' \cdot Cov^{-1} \cdot Cov[0] & z_6 := 1 - \zeta z_4 \\ z_7 = r + \zeta M_0 - \zeta z_5 & z_{10} := (z_1 z_2 - z_3^2)^{-1} \end{array} \tag{3.6}$$

Proposition 3.5. *Assume that the following condition holds*

$$z_1 z_7 > z_3 z_6, \tag{3.7}$$

and that M is not co-linear to $\mathbb{I}$. Then, the solution of $(\mathcal{P})$ is given by

$$\pi^* = Cov^{-1} \cdot (\zeta Cov[0] + \mu M - \nu \mathbb{I}) . \tag{3.8}$$

where

$$\mu = \frac{z_7 z_1 - z_6 z_3}{z_1 z_2 - z_3^2} \quad \textit{and} \quad \nu = \frac{z_7 z_3 - z_6 z_2}{z_1 z_2 - z_3^2},$$

Proof: see proof of Proposition 2 in Bernis et al. [2]. □

3.2. *Error Calculus on the Optimal Allocation*

Recall that the ϵ_k (see (3.2)) represents the portfolio manager views on the spread excess return. We want to compute the ϵ_k error measure influence on the optimal allocation. The uncertainty is represented by general random variables described by their bias and their covariance. We use the methodology introduced in section 3.2 and developed by Bouleau [5]. In Theorem 3.12, we are able to compute the bias and the covariance of the optimal allocation, and, thus, measure the influence of the error made on the excess return estimation. Note that even if the errors are centered (no bias in the estimation), the optimal allocation will be biased, because the optimal allocation is not a linear mapping of the excess returns.

We define the following quantities, for $1 \leq k \leq K$ and $0 \leq j \leq K$:

$$\begin{aligned}
\phi_k &:= T\mathbb{E}^{\mathbb{P}}\left\{X_k(T)\dot{P}_2^{(k)}(T, X_k(T))\right\} \\
\psi_k &:= T^2\mathbb{E}^{\mathbb{P}}\left\{X_k^2(T)\ddot{P}_2^{(k)}(T, X_k(T))\right\} \\
\Phi_{k,j} &:= T\mathbb{E}^{\mathbb{P}}\left\{X_k(T)\dot{P}_2^{(k)}(T, X_k(T))P\&L_j(T, X_j(T))\right\} \\
\Psi_{k,j} &:= T^2\mathbb{E}^{\mathbb{P}}\left\{X_k^2(T)\ddot{P}_2^{(k)}(T, X_k(T))P\&L_j(T, X_j(T))\right\} \\
\Upsilon_{kj} &:= T^2\mathbb{E}^{\mathbb{P}}\left\{X_k(T)X_j(T)\dot{P}_2^{(k)}(T, X_k(T))\dot{P}_2^{(j)}(T, X_j(T))\right\}
\end{aligned}$$

The first result provides the sensitivity analysis for the expected returns M.

Proposition 3.6. Sensitivity of the expected value of P&L *For any* $1 \leq k, j \leq K$, *we have*

$$P\&L_k^{\#}(T, X_k) = \dot{P}_2^{(k)}(T, X_k(T))TX_k(T)\epsilon_k^{\#} \tag{3.9a}$$

$$M_k^{\#} = \phi_k \epsilon_k^{\#} \tag{3.9b}$$

$$\Gamma[M_k, M_j] = \phi_k\phi_j\Gamma[\epsilon_k, \epsilon_j] \tag{3.9c}$$

$$\begin{aligned}\Gamma[P\&L_k(T, X_k), P\&L_j(T, X_j)] = {}& \dot{P}_2^{(k)}(T, X_k(T))\dot{P}_2^{(j)}(T, X_j(T)) \\ &\times T^2X_k(T)X_j(T)\Gamma[\epsilon_k, \epsilon_j]\end{aligned} \tag{3.9d}$$

$$\begin{aligned}\mathcal{A}[P\&L_k(T, X_k)] = {}& \dot{P}_2^{(k)}(T, X_k(T))TX_k(T)\mathcal{A}[\epsilon_k] \\ &+\frac{1}{2}\dot{P}_2^{(k)}(T, X_k(T))T^2X_k(T)\Gamma[\epsilon_k] \\ &+\frac{1}{2}\ddot{P}_2^{(k)}(T, X_k(T))T^2X_k^2(T)\Gamma[\epsilon_k]\end{aligned} \tag{3.9e}$$

$$\mathcal{A}[M_k] = \phi_k\mathcal{A}[\epsilon_k] + \left(\frac{1}{2}T\phi_k + \frac{1}{2}\psi_k\right)\Gamma[\epsilon_k] \tag{3.9f}$$

Proof: From (3.3) and (2.13), we have $X_k^{\#}(T) = TX_k(T)\epsilon_k^{\#}$. Then (2.13) again gives (3.9a). Equation (3.9b) comes from Lemma 2.6. For (3.9c) we use (3.9b) and (2.12). Moreover, (3.9d) follows from (2.12) and (3.9a). Finally a direct application of bias chain rule (2.8) gives (3.9e) and applying Lemma 2.6 we find Equation (3.9f). □

We state the following Proposition about the sensitivity of the variance covariance matrix.

Proposition 3.7. Sensitivity of the variance covariance matrix *For any $1 \le k, j \le K$, we have*

$$Cov_{kj}^{\#} = \delta_{k,j}^{C}\epsilon_k^{\#} + \delta_{j,k}^{C}\epsilon_j^{\#} \tag{3.10}$$

$$Cov[0]_j^{\#} = \delta_{j,0}^{C}\epsilon_j^{\#} \tag{3.11}$$

$$\begin{aligned}\Gamma[Cov_{kj}, Cov_{il}] &= \delta_{k,j}^{C}\delta_{i,l}^{C}\Gamma[\epsilon_k, \epsilon_i] + \delta_{j,k}^{C}\delta_{i,l}^{C}\Gamma[\epsilon_j, \epsilon_i] \\ &\quad + \delta_{k,j}^{C}\delta_{l,i}^{C}\Gamma[\epsilon_k, \epsilon_l] + \delta_{j,k}^{C}\delta_{l,i}^{C}\Gamma[\epsilon_j, \epsilon_l]\end{aligned} \tag{3.12}$$

$$\begin{aligned}\mathcal{A}[Cov_{kj}] &= \delta_{k,j}^{C}\mathcal{A}[\epsilon_k] + \delta_{j,k}^{C}\mathcal{A}[\epsilon_j] \\ &\quad + \alpha_{k,j}^{C}\Gamma[\epsilon_k] + \alpha_{j,k}^{C}\Gamma[\epsilon_j] + \beta_{kj}^{C}\Gamma[\epsilon_k, \epsilon_j]\end{aligned} \tag{3.13}$$

$$\mathcal{A}[Cov[0]_k] = \delta_{k,0}^{C}\mathcal{A}[\epsilon_k] + \alpha_{k,0}^{C}\Gamma[\epsilon_k] \tag{3.14}$$

where we set:

$$\begin{aligned}\delta_{k,j}^{C} &:= \Phi_{k,j} - \phi_k M_j \\ \alpha_{k,j}^{C} &:= \frac{1}{2}\left[T\Phi_{k,j} + \Psi_{k,j} - (T\phi_k + \psi_k) M_j\right] \\ \beta_{kj}^{C} &:= \Upsilon_{kj} - \phi_k\phi_j\end{aligned}$$

with $1 \le k \le K$ and $0 \le j \le K$.

Proof: This result is a direct consequence of Lemmas 2.5, 2.6 and Proposition 3.6. We recall that as X_0 does not depend on $\{\epsilon_k\}_{k=1,\dots K}$ it is unaffected by drift estimation uncertainty. □

Our optimal allocation depends on the inverse of the variance covariance matrix, see (3.8).

Proposition 3.8. Sensitivity of the inverse variance covariance matrix *For any $1 \le i, l, m, n \le K$, we have*

$$\left(Cov_{il}^{-1}\right)^{\#} = \sum_k \delta_{k,il}^{Cov^{-1}} \epsilon_k^{\#} \tag{3.15a}$$

$$\Gamma[Cov_{il}^{-1}, Cov_{mn}^{-1}] = \sum_{kj} \delta_{k,il}^{Cov^{-1}} \delta_{j,mn}^{Cov^{-1}} \Gamma[\epsilon_k, \epsilon_j] \tag{3.15b}$$

$$\begin{aligned}\mathcal{A}[Cov_{il}^{-1}] &= \sum_k \left(\delta_{k,il}^{Cov^{-1}} \mathcal{A}[\epsilon_k] + \alpha_{k,il}^{Cov^{-1}} \Gamma[\epsilon_k]\right) \\ &\quad + \sum_{kj} \beta_{kj,il}^{Cov^{-1}} \Gamma[\epsilon_k, \epsilon_j]\end{aligned}$$

where

$$\delta_{k,il}^{Cov^{-1}} := -\sum_m \left(Cov_{ik}^{-1} Cov_{ml}^{-1} + Cov_{im}^{-1} Cov_{kl}^{-1}\right) \delta_{k,m}^{C}$$

$$\alpha_{k,il}^{Cov^{-1}} := -\sum_m \left(Cov_{ik}^{-1} Cov_{ml}^{-1} + Cov_{im}^{-1} Cov_{kl}^{-1}\right) \alpha_{k,m}^{C}$$

$$\beta_{kj,il}^{Cov^{-1}} := -Cov_{ik}^{-1} Cov_{jl}^{-1} \beta_{kj}^{C} - \sum_m \left(Cov_{ik}^{-1} \delta_{k,m}^{C} \delta_{j,ml}^{Cov^{-1}} + Cov_{im}^{-1} \delta_{j,m}^{C} \delta_{k,jl}^{Cov^{-1}}\right)$$

Proof: Note that $Cov\ Cov^{-1} = \mathbb{I}$, where $\mathbb{I}$ (the identity matrix) is unaffected by the uncertainty on the coefficients of Cov. Using (2.14), we find $0 = \mathbb{I}^{\#} = Cov^{\#} Cov^{-1} + Cov(Cov^{-1})^{\#}$. Using (3.10), (3.15a) follows. (3.15b) follows from (3.15a) and (2.12). From (2.16), we get that

$$0 = \mathcal{A}[\mathbb{I}] = \mathcal{A}[Cov Cov^{-1}] = \mathcal{A}[Cov]\, Cov^{-1} + Cov\, \mathcal{A}[Cov^{-1}] + \Gamma[Cov, Cov^{-1}].$$

Thus, using (2.12), we get that

$$\mathcal{A}[Cov^{-1}] = -Cov^{-1}\, \mathcal{A}[Cov]\, Cov^{-1} - Cov^{-1}\, \mathbb{E}\left(Cov^{\#}(Cov^{-1})^{\#}\right)$$

and we conclude using (3.10), and (3.15a). □

We turn now to study the sensitivity of z_i, see relations (3.6), we remark that only the sensitivities of z_1 to z_5 need to be computed since z_6 and z_7 are linear combination of z_1 to z_5.

Proposition 3.9. Sensitivity of z_i *For* $a = 1, \ldots, 5$, *we have*

$$z_a^{\#} = \sum_k \delta_k^{z_a} \epsilon_k^{\#}$$

$$\mathcal{A}[z_a] = \sum_k \delta_k^{z_a} \mathcal{A}[\epsilon_k] + \sum_k \alpha_k^{z_a} \Gamma[\epsilon_k] + \sum_{kj} \beta_{kj}^{z_a} \Gamma[\epsilon_k, \epsilon_j]$$

where all sums are taken from 1 *to* K *and*

$$\delta_k^{z_1} := \sum_{il} \delta_{k,il}^{Cov^{-1}} \qquad \alpha_k^{z_1} := \sum_{il} \alpha_{k,il}^{Cov^{-1}} \qquad \beta_{kj}^{z_1} := \sum_{il} \beta_{kj,il}^{Cov^{-1}}$$

$$\delta_k^{z_2} := \sum_{il} M_i \delta_{k,il}^{Cov^{-1}} M_l + 2 \sum_i \phi_k\, Cov_{ki}^{-1} M_i$$

$$\alpha_k^{z_2} := \sum_{il} M_i \alpha_{k,il}^{Cov^{-1}} M_l + (T\phi_k + \psi_k) \sum_i M_i Cov_{ik}^{-1}$$

$$\beta_{kj}^{z_2} := \sum_{il} M_i \beta_{kj,il}^{Cov^{-1}} M_l + 2 \sum_i \phi_j M_i \delta_{k,ij}^{Cov^{-1}} + Cov_{jk}^{-1} \phi_j \phi_k$$

$$\delta_k^{z_3} := \sum_i \phi_k\, Cov_{ki}^{-1} + \sum_{il} \delta_{k,il}^{Cov^{-1}} M_i$$

$$\alpha_k^{z_3} := \sum_{il} M_i \alpha_{k,il}^{Cov^{-1}} + \sum_i \frac{1}{2}(T\phi_k + \psi_k) Cov_{ki}^{-1}$$

$$\beta_{kj}^{z_3} := \sum_{il} M_i \beta_{kj,il}^{Cov^{-1}} + \sum_i \phi_k \delta_{j,ki}^{Cov^{-1}}$$

$$\delta_k^{z_4} := \sum_{il} \delta_{k,il}^{Cov^{-1}} Cov[0]_l + \sum_i \delta_{k,0}^C Cov_{ik}^{-1}$$

$$\alpha_k^{z_4} := \sum_{il} \alpha_{k,il}^{Cov^{-1}} Cov[0]_l + \sum_i \alpha_{k,0}^C Cov_{ik}^{-1}$$

$$\beta_{kj}^{z_4} := \sum_{il} \beta_{kj,il}^{Cov^{-1}} Cov[0]_l + \sum_i \delta_{j,0}^C \delta_{k,ij}^{Cov^{-1}}$$

$$\delta_k^{z_5} := \sum_{il} M_i \delta_{k,il}^{Cov^{-1}} Cov[0]_l + \sum_i M_i Cov_{ik}^{-1} \delta_{k,0}^C + \sum_i \phi_k Cov_{ki}^{-1} Cov[0]_i$$

$$\alpha_k^{z_5} := \sum_{il} M_i \alpha_{k,il}^{Cov^{-1}} Cov[0]_l + \sum_i M_i Cov_{ik}^{-1} \alpha_{k,0}^C + \frac{1}{2}(T\phi_k + \psi_k) \sum_i Cov_{ki}^{-1} Cov[0]_i$$

$$\beta_{kj}^{z_5} := \sum_{il} M_i \beta_{kj,il}^{Cov^{-1}} Cov[0]_l + \sum_i M_i \delta_{j,0}^C \delta_{k,ij}^{Cov^{-1}} + \sum_i \delta_{k,ji}^{Cov^{-1}} \phi_j Cov[0]_i + Cov_{jk}^{-1} \phi_j \delta_{k,0}^C$$

Proof: The proof is similar to the proof of Proposition 3.8 and is based on Propositions 3.6, 3.7, 3.8 and Lemma 2.5 (recall also definitions (3.6)). □

Remark 3.10. From (3.6), $z_6^{\#} = -\zeta z_4^{\#}$ and $\mathcal{A}[z_6] = -\zeta \mathcal{A}[z_4]$ and we set

$$\delta_k^{z_6} := -\zeta \delta_k^{z_4} \qquad \alpha_k^{z_6} := -\zeta \alpha_k^{z_4} \qquad \beta_{kj}^{z_6} := -\zeta \beta_{kj}^{z_4}$$

Similarly, $z_7^{\#} = -\zeta z_5^{\#}$ and $\mathcal{A}[z_7] = -\zeta \mathcal{A}[z_5]$ and we set

$$\delta_k^{z_7} := -\zeta \delta_k^{z_5} \qquad \alpha_k^{z_7} := -\zeta \alpha_k^{z_5} \qquad \beta_{kj}^{z_7} := -\zeta \beta_{kj}^{z_5}$$

The following corollary comes directly from Proposition 3.9 together with (2.8).

Corollary 3.11. Sensitivity of μ and ν *We have*

$$
\begin{aligned}
\mu^{\#} &= \sum_k \delta_k^{\mu} \epsilon_k^{\#} \\
\nu^{\#} &= \sum_k \delta_k^{\nu} \epsilon_k^{\#} \\
\mathcal{A}[\mu] &= \sum_k \delta_k^{\mu} \mathcal{A}[\epsilon_k] + \sum_k \alpha_k^{\mu} \Gamma[\epsilon_k] + \sum_{k,j} (\beta_{kj}^{\mu} + \chi_{kj}^{\mu}) \Gamma[\epsilon_k, \epsilon_j] \\
\mathcal{A}[\nu] &= \sum_k \delta_k^{\nu} \mathcal{A}[\epsilon_k] + \sum_k \alpha_k^{\nu} \Gamma[\epsilon_k] + \sum_{kj} (\beta_{kj}^{\nu} + \chi_{kj}^{\nu}) \Gamma[\epsilon_k, \epsilon_j]
\end{aligned}
$$

where

$$
\delta_k^{\mu} := \sum_{a=1,\dots,5} \frac{\partial \mu}{\partial z_a} \delta_k^{z_a} \qquad \alpha_k^{\mu} := \sum_{a=1,\dots,5} \frac{\partial \mu}{\partial z_a} \alpha_k^{z_a} \qquad \beta_{kj}^{\mu} := \sum_{a=1,\dots,5} \frac{\partial \mu}{\partial z_a} \beta_{kj}^{z_a}
$$

$$
\delta_k^{\nu} := \sum_{a=1,\dots,5} \frac{\partial \nu}{\partial z_a} \delta_k^{z_a} \qquad \alpha_k^{\nu} := \sum_{a=1,\dots,5} \frac{\partial \nu}{\partial z_a} \alpha_k^{z_a} \qquad \beta_{kj}^{\nu} := \sum_{a=1,\dots,5} \frac{\partial \nu}{\partial z_a} \beta_{kj}^{z_a}
$$

$$
\chi_{kj}^{\mu} := \frac{1}{2} \sum_{a,b=1,\dots,5} \frac{\partial^2 \mu}{\partial z_a \partial z_b} \delta_k^{z^a} \delta_j^{z^b} \qquad \chi_{kj}^{\nu} := \frac{1}{2} \sum_{a,b=1,\dots,5} \frac{\partial^2 \nu}{\partial z_a \partial z_b} \delta_k^{z^a} \delta_j^{z^b}
$$

All derivatives are listed in appendix for sake of completeness.

We are now in a position to state the result giving the sensitivity analysis of the optimal allocation π^*.

Theorem 3.12. Sensitivity of the optimal strategy *Let π^* be the solution of Program ($\mathcal{P}$), given by Proposition 3.5. We get that:*

$$
\begin{aligned}
(\pi_i^*)^{\#} &= \sum_k \delta_{k,i}^{\pi} \epsilon_k^{\#} \\
\Gamma[\pi_i^*, \pi_l^*] &= \sum_{kj} \delta_{k,i}^{\pi} \delta_{j,l}^{\pi} \Gamma[\epsilon_k, \epsilon_j] \\
\Gamma[\pi_i^*, M_l] &= \sum_k \delta_{k,i}^{\pi} \phi_l \Gamma[\epsilon_k, \epsilon_l] \\
\mathcal{A}[\pi_i^*] &= \sum_k \delta_{k,i}^{\pi} \mathcal{A}[\epsilon_k] + \sum_k \alpha_{k,i}^{\pi} \Gamma[\epsilon_k] + \sum_{kj} \beta_{kj,i}^{\pi} \Gamma[\epsilon_k, \epsilon_j]
\end{aligned}
$$

where

$$\begin{aligned}
\delta^{\pi}_{k,i} :=& \sum_l \delta^{Cov^{-1}}_{k,il} \left(\zeta Cov[0]_l + \mu M_l - \nu\right) \\
&+ \sum_l Cov^{-1}_{il} \left(\delta^{\mu}_k M_l - \delta^{\nu}_k\right) + Cov^{-1}_{ik} \left(\zeta \delta^{C}_{k,0} + \mu \phi_k\right) \\
\alpha^{\pi}_{k,i} :=& \sum_l \alpha^{Cov^{-1}}_{k,il} \left(\zeta Cov[0]_l + \mu M_l - \nu\right) + \sum_l Cov^{-1}_{il} \left(M_l \alpha^{\mu}_k - \alpha^{\nu}_k\right) \\
&+ Cov^{-1}_{ik} \left[\zeta \alpha^{C}_{k,0} + \frac{1}{2}\mu \left(T\phi_k + \psi_k\right)\right] \\
\beta^{\pi}_{kj,i} :=& \sum_l \beta^{Cov^{-1}}_{kj,il} \left(\zeta Cov[0]_l + \mu M_l - \nu\right) + \sum_l \delta^{Cov^{-1}}_{k,il} \left(\delta^{\mu}_j M_l - \delta^{\nu}_j\right) \\
&+ \sum_l Cov^{-1}_{il} \left(M_l \chi^{\mu}_{kj} + M_l \beta^{\mu}_{kj} - \chi^{\nu}_{kj} - \beta^{\nu}_{kj}\right) + \delta^{Cov^{-1}}_{k,ij} \left(\zeta \delta^{C}_{j,0} + \mu \phi_j\right) \\
&+ Cov^{-1}_{ik} \phi_k \delta^{\mu}_j
\end{aligned}$$

Proof: Again the proof is similar to the previous ones. We apply Propositions 3.6, 3.7, 3.8, 3.9, Corollary 3.11, Lemma 2.5 together with (2.8) and (2.12). □

We conclude with a final result for the optimal return sensitivity analysis.

Corollary 3.13. *With the notations of Theorem 3.12, let R^* be the optimal return defined by $R^* := M'\pi^*$. Then we get that:*

$$\begin{aligned}
(R^*)^{\#} &= \sum_k \left(\phi_k \pi^*_k + \sum_i \delta^{\pi}_{k,i} M_i\right) \epsilon^{\#}_k \\
\Gamma[R^*] &= \sum_{k,j} \left(\phi_k \pi^*_k + \sum_i \delta^{\pi}_{k,i} M_i\right) \left(\phi_j \pi^*_j + \sum_i \delta^{\pi}_{j,i} M_i\right) \Gamma[\epsilon_k, \epsilon_j] \\
\mathcal{A}[R^*] &= \sum_k \left(\phi_k \pi^*_k + \sum_i \delta^{\pi}_{k,i} M_i\right) \mathcal{A}[\epsilon_k] \\
&\quad + \sum_k \left(\sum_i \alpha^{\pi}_{k,i} M_i + \frac{1}{2}(T\phi_k + \psi_k)\pi^*_k\right) \Gamma[\epsilon_k] \\
&\quad + \sum_{k,j} \left(\delta^{\pi}_{k,j} \phi_j + \sum_i \beta^{\pi}_{kj,i} M_i\right) \Gamma[\epsilon_k, \epsilon_j]
\end{aligned}$$

A.1. Explicit Derivatives of ν and μ

$$\frac{\partial \nu}{\partial z_1} = -\nu z_2 z_{10} \qquad \frac{\partial \nu}{\partial z_2} = -\mu z_3 z_{10}$$

$$\frac{\partial \nu}{\partial z_3} = \nu z_3 z_{10} + \mu z_2 z_{10} \qquad \frac{\partial \nu}{\partial z_4} = \zeta z_2 z_{10}$$

$$\frac{\partial \nu}{\partial z_5} = -\zeta z_3 z_{10} \qquad \frac{\partial \mu}{\partial z_1} = -\nu z_3 z_{10}$$

$$\frac{\partial \mu}{\partial z_2} = -\mu z_1 z_{10} \qquad \frac{\partial \mu}{\partial z_3} = \nu z_1 z_{10} + \mu z_3 z_{10}$$

$$\frac{\partial \mu}{\partial z_4} = \zeta z_3 z_{10} \qquad \frac{\partial \mu}{\partial z_5} = -\zeta z_1 z_{10}$$

$$\frac{\partial^2 \nu}{\partial z_1^2} = 2\nu z_2^2 z_{10}^2 \qquad \frac{\partial^2 \nu}{\partial z_1 \partial z_2} = (\nu z_1 + \mu z_3) z_2 z_{10}^2 - \nu z_{10}$$

$$\frac{\partial^2 \nu}{\partial z_1 \partial z_3} = -(3\nu z_2 + \mu z_2) z_2 z_{10}^2 \qquad \frac{\partial^2 \nu}{\partial z_1 \partial z_4} = -\zeta z_2^2 z_{10}^2$$

$$\frac{\partial^2 \nu}{\partial z_1 \partial z_5} = \zeta z_2 z_3 z_{10}^2 \qquad \frac{\partial^2 \nu}{\partial z_2^2} = 2\mu z_1 z_3 z_{10}^2$$

$$\frac{\partial^2 \nu}{\partial z_2 \partial z_3} = -(\nu z_1 + 3\mu z_3) z_3 z_{10}^2 - \mu z_{10} \qquad \frac{\partial^2 \nu}{\partial z_2 \partial z_4} = -\zeta z_3^2 z_{10}^2$$

$$\frac{\partial^2 \nu}{\partial z_2 \partial z_5} = \zeta z_1 z_3 z_{10}^2 \qquad \frac{\partial^2 \nu}{\partial z_3^2} = [\nu(3z_3^2 + z_1 z_2) + 4\mu z_2 z_3] z_{10}^2 + \nu z_{10}$$

$$\frac{\partial^2 \nu}{\partial z_3 \partial z_4} = 2\zeta z_2 z_3 z_{10}^2 \qquad \frac{\partial^2 \nu}{\partial z_3 \partial z_5} = -\zeta(z_3^2 + z_1 z_2) z_{10}^2$$

$$\frac{\partial^2 \nu}{\partial z_4^2} = \frac{\partial^2 \nu}{\partial z_5^2} = \frac{\partial^2 \nu}{\partial z_4 \partial z_5} = 0 \qquad \frac{\partial^2 \mu}{\partial z_1^2} = 2\nu z_2 z_3 z_{10}^2$$

$$\frac{\partial^2 \mu}{\partial z_1 \partial z_2} = (\mu z_3 + \nu z_1) z_3 z_{10}^2 \qquad \frac{\partial^2 \mu}{\partial z_1 \partial z_3} = -(3\nu z_3 + \mu z_2) z_3 z_{10}^2 - \nu z_{10}$$

$$\frac{\partial^2 \mu}{\partial z_1 \partial z_4} = -\zeta z_2 z_3 z_{10}^2 \qquad \frac{\partial^2 \mu}{\partial z_1 \partial z_5} = \zeta z_3^2 z_{10}^2$$

$$\frac{\partial^2 \mu}{\partial z_2^2} = 2\mu z_1^2 z_{10}^2 \qquad \frac{\partial^2 \mu}{\partial z_2 \partial z_3} = -(3\mu z_3 + \nu z_1) z_1 z_{10}^2$$

$$\frac{\partial^2 \mu}{\partial z_2 \partial z_4} = -\zeta z_1 z_3 z_{10}^2 \qquad \frac{\partial^2 \mu}{\partial z_2 \partial z_5} = \zeta z_1^2 z_{10}^2$$

$$\frac{\partial^2 \mu}{\partial z_3^2} = [2\mu(z_3^2 + z_1 z_2) + 4\nu z_1 z_3] z_{10}^2 \qquad \frac{\partial^2 \mu}{\partial z_3 \partial z_4} = \zeta z_{10}(1 + 2z_3^2 z_{10})$$

$$\frac{\partial^2 \mu}{\partial z_3 \partial z_5} = -2\zeta z_1 z_3 z_{10}^2 \qquad \frac{\partial^2 \mu}{\partial z_4^2} = \frac{\partial^2 \mu}{\partial z_5^2} = \frac{\partial^2 \mu}{\partial z_4 \partial z_5} = 0$$

Acknowledgements

We are grateful to G. Bernis and G. Docq for bringing this nice optimal credit allocation problem to our attention. The present version of the paper benefits from the fruitful discussions we had together.

References

[1] Albeverio, S. (2003) Theory of Dirichlet forms and application. Springer-Verlag, Berlin.

[2] Bernis, G.; Carassus, L.; Docq, G. and S. Scotti (2013) *Optimal Credit Allocation under Regime Uncertainty with Sensitivity Analysis.* Preprint.

[3] Bouleau, N. and Hirsch, F. (1991): *Dirichlet Forms and Analysis on Wiener space*, De Gruyter, Berlin.

[4] Bouleau, N. (2001) *Calcul d'erreur complet et Lipschitzien et formes de Dirichlet.* Journal de Mathématiques Pures et Appliquées, 80 (9), 961-976.

[5] Bouleau, N. (2003) Error Calculus for Finance and Physics. De Gryuter.

[6] Chorro, C and N. Bouleau (2004): *Error structures and parameter estimation*, C.R. Acad. Sci. Paris, Ser. I 338, 305-310.

[7] Fukushima, M.; Oshima, Y. and Takeda, M. (1994): *Dirichlet Forms and Markov Process*, De Gruyter, Berlin.

PART 3

Control Problem and Information Risks

Discrete-Time Multi-Player Stopping and Quitting Games with Redistribution of Payoffs

Ivan Guo, Marek Rutkowski*

A novel class of multi-period multi-player competitive stopping games with redistribution of payoffs is constructed. Each player can either exit the game for a fixed payoff, determined a priori, or stay and receive an adjusted payoff depending on the decision of other players. The single-period case is shown to be weakly unilaterally competitive under some assumptions on redistribution of payoffs. We present an explicit construction of the unique value at which Nash and optimal equilibria are attained. The multi-period stochastic extension of the game is also studied and solved by the backward induction. The game has interpretations in economic and financial contexts, for example, as a consumption model with bounded resources or a starting point to the construction of multi-person financial game options. Deterministic multi-period quitting games are also examined as an alternative to stopping games.

1. Introduction

In the seminal paper by Dynkin [7], he introduced the concept of a zero-sum, optimal stopping game between two players, where each player can stop the game for a payoff observable at that time. An abundant research was subsequently done on zero-sum and non-zero sum Dynkin stopping games and related problems; see, e.g., [1, 8, 9, 13–15, 20, 21, 23–25, 28, 29, 32] and [33].

An important application of two-player Dynkin games is in design of *game* (or *Israeli*) *options*, as formally defined in the path-breaking paper by Kifer [18], who also proved the existence and uniqueness of its arbitrage price in the Cox–Ross–Rubinstein and the Black–Scholes models. For further research in this vein, in particular, for financial applications to convertible bonds, the interested reader is referred to [2, 4–6, 15, 16], and a recent review paper by Kifer [19].

*School of Mathematics and Statistics, The University of Sydney, Australia.
E-mail: marek.rutkowski@sydney.edu.au

Several alternative formulations of multi-player Dynkin games can be found in the existing literature. For instance, Solan and Vieille [30] introduced a *quitting game*, which terminates as soon as any player chooses to quit; then each player receives a payoff depending the set of players who decide to quit the game. Under certain payoff conditions, a subgame perfect uniform ϵ-equilibrium using cyclic strategies can be found. In Solan and Vieille [31], another version is examined, in which the players are given the opportunity to stop the game in a turn-based fashion. A subgame perfect ϵ-equilibrium was once again shown to exist and consist of pure strategies when the game is not degenerate.

More recently, Guo and Rutkowski [11] introduced a zero-sum m-player stopping game with a focus on designing the explicit dependencies between the payoffs of all players and their stopping decisions. The goal was to model a multilateral 'contract' where all the players are competing for a fixed total sum of wealth. Each player can either exit (and thus also terminate) the contract for a predetermined benefit, or do nothing and receive an adjusted benefit, which reflects the discrepancies caused by any exiting decisions. These adjustments were judiciously designed to ensure that the total wealth redistributed was fixed.

We extend here results of Guo and Rutkowski [11] to m-player games that are not necessarily zero-sum but still retain, at least in the single-period case, the *weakly unilaterally competitive* (WUC) property introduced by Kats and Thisse [17]. In essence, the WUC property means that if only one of the players changes her decision and enjoys a gain, then none of the other players may improve her payoff as well (for details, see Definition 2.5).

As discussed by Pruzhansky [26], Nash equilibria are not always adequate as a solution concept. This certainly occurs in the valuation of the game options, since the Nash equilibria payoffs cannot be guaranteed by each player. Instead, we deal with the stronger *optimal equilibria*, which induce a unique value for the game. Recall that the WUC property is known to ensure that all Nash equilibria are also optimal (see De Wolf [3] and Kats and Thisse [17]). The main results for single-period nonzero-sum m-player deterministic games are Theorems 3.6 and 3.15, which prove the existence of the value and expresses it as the projection onto some subset of $\mathbb{R}^m$ under an appropriate choice of inner product. The construction also produces a pure strategy optimal equilibrium.

Several practically important extensions are also discussed. First, all single-period results can be immediately applied to the stochastic case where both terminal and exercise payoffs are random, as long as expectations are incorporated into the definitions of a solution and equilibria. Second, a multi-period generalization is also studied and the existence of the value and optimal equilibrium is established (see Theorem 4.5). The recursive stopping games can be readily applied to multi-person financial game options, where the properties of the optimal equilibrium become imperative in the pricing arguments. More details on arbitrage pricing of multi-person game options are presented in the follow-up paper [12].

Apart from multi-person game options, the multi-player competitive games presented here may find interpretation in other economic and financial contexts, for example, as a consumption model with bounded resources (see, for instance, Ramasubramanian [27]). They may serve as a starting point to more exhaustive studies of competitive multi-player games. Continuous-time generalizations are also of interest and they are under further research (see the recent paper by Nie and Rutkowski [22]).

This work is organized as follows. In Section 2, some preliminary results in game theory are recalled. In Section 3, we first construct the single-period m-player *redistribution game* and prove the existence and uniqueness of the value. Then, in Subsection 3.4, we revisit the results of Guo and Rutkowski [11] and we extend them to the nonzero-sum single-period redistribution game. As in [11], the value and an optimal equilibrium of the game is constructed using a suitable projection. In Section 4, we deal with multi-period m-player stochastic *stopping games* and solve them by the backward induction. Finally, in Section 5, we briefly examine multi-period m-player deterministic *quitting games*.

2. Optimal Equilibria in WUC Games

Consider a game $\mathcal{G}$ with m *players*, enumerated by the indices $1, 2, \ldots, m$. The set of all players is denoted by $\mathcal{M}$. Each player k can choose a *strategy* $s_k \in \mathcal{S}_k$, and the m-tuples of strategies $s = [s_1, \ldots, s_m] \in \mathcal{S}$ are called *strategy profiles*.

We emphasize that we only consider here the so-called *pure* strategies, so that *mixed* (that is, randomized) strategies are not allowed. We also make a frequent use of the shorthand notation $s = [s_k, s_{-k}]$ where s_{-k} represents strategies of all players except for player k.

Given an arbitrary strategy profile s, it is possible to compute a vector of *payoff* functions $V(s) = [V_1(s), \ldots, V_m(s)]$ for the players, with higher payoffs being more desirable.

Definition 2.1. A strategy profile $s^* \in \mathcal{S}$ is called a *Nash equilibrium*, or simply an *equilibrium*, if for every $k \in \mathcal{M}$,

$$V_k\left(\left[s_k^*, s_{-k}^*\right]\right) \geq V_k\left(\left[s_k, s_{-k}^*\right]\right), \quad \forall\, s_k \in \mathcal{S}_k.$$

Motivated by applications to multi-person game options (see [10, 12]), we will henceforth focus on the stronger concept of an *optimal equilibrium*.

Definition 2.2. A strategy profile $s^* \in \mathcal{S}$ is called an *optimal equilibrium* if, for every $k \in \mathcal{M}$,

$$V_k\left(\left[s_k^*, s_{-k}\right]\right) \geq V_k\left(\left[s_k^*, s_{-k}^*\right]\right) \geq V_k\left(\left[s_k, s_{-k}^*\right]\right), \; \forall\, s_k \in \mathcal{S}_k, \forall\, s_{-k} \in \mathcal{S}_{-k}.$$

It is clear that an optimal equilibrium is essentially a saddle point. It has the properties of a Nash equilibrium, with the addition that each player can guarantee a lower bound on her payoff without knowing the actions of other players. The optimal equilibrium is one of many ways to strengthen the Nash equilibrium for multi-player games. In particular, it replicates the properties of a Nash equilibrium with maximin strategies, as discussed by Pruzhansky [26]. Recall that the *maximin value* $\underline{V}_k$ of player k is the maximum payoff she can guarantee, specifically,

$$\underline{V}_k = \max_{s_k \in \mathcal{S}_k} \min_{s_{-k} \in \mathcal{S}_{-k}} V_k\left(\left[s_k, s_{-k}\right]\right),$$

whereas the *minimax value* $\overline{V}_k$ of player k is the lowest payoff that the other players can force upon her, that is,

$$\overline{V}_k = \min_{s_{-k} \in \mathcal{S}_{-k}} \max_{s_k \in \mathcal{S}_k} V_k\left(\left[s_k, s_{-k}\right]\right).$$

It is well known that the inequality $\overline{V}_k \geq \underline{V}_k$ is always satisfied.

Definition 2.3. If the equality $\overline{V}_k = \underline{V}_k$ holds then the common value $V_k^* = \overline{V}_k = \underline{V}_k$ is called the *value of the game $\mathcal{G}$ for player k*. We say that the game $\mathcal{G}$ has the *value* when it has the value for every player $k \in \mathcal{M}$.

In general, a multi-player game may fail to have a value. The next result shows that the existence of an optimal equilibrium guarantees the existence of the value (the uniqueness of the value is obvious from Definition 2.3).

Proposition 2.4. *If a multi-player game admits at least one optimal equilibrium then it has the value and every optimal equilibrium s^* attains the value, that is,* $V(s^*) = V^* = \left[V_1^*, \ldots, V_m^*\right]$.

Proof. Let s^* be any optimal equilibrium. Then

$$\begin{aligned} V_k\left([s_k^*, s_{-k}^*]\right) &= \max_{s_k \in \mathcal{S}_k} V_k\left([s_k, s_{-k}^*]\right) \geq \overline{V}_k \\ &\geq \underline{V}_k \geq \min_{s_{-k} \in \mathcal{S}_{-k}} V_k\left([s_k^*, s_{-k}]\right) = V_k\left([s_k^*, s_{-k}^*]\right), \end{aligned}$$

and thus all terms are equal. In particular, $\overline{V}_k = \underline{V}_k$ for every $k \in \mathcal{M}$. □

In a two-player zero-sum game, optimal equilibria are equivalent to Nash equilibria. This should be contrasted with a multi-player zero-sum game, where the payoff of any particular player is not sufficient to determine the individual payoffs of other players, in general. As a consequence, Nash equilibria are not necessarily optimal equilibria and they may not achieve the same payoff. However, a result similar to von Neumann's minimax theorem is still valid if the zero-sum condition is strengthened to *weakly unilaterally competitive*, as introduced and studied by Kats and Thisse [17].

Definition 2.5. A game is said to be *weakly unilaterally competitive* (WUC) if, for every $k, l \in \mathcal{M}$,

$$\begin{aligned} V_k([s_k, s_{-k}]) > V_k([s'_k, s_{-k}]) &\implies V_l([s_k, s_{-k}]) \leq V_l([s'_k, s_{-k}]), \\ V_k([s_k, s_{-k}]) = V_k([s'_k, s_{-k}]) &\implies V_l([s_k, s_{-k}]) = V_l([s'_k, s_{-k}]), \end{aligned}$$

for all strategy profiles $s_k, s'_k \in \mathcal{S}_k$ and $s_{-k} \in \mathcal{S}_{-k}$.

Note that WUC explicitly quantifies the concept of competitiveness: if a player deviates from a strategy profile, any changes to her payoff are opposite in sign to the changes of other payoffs. In Kats and Thisse [17] (see also De Wolf [3]), the authors established the following result.

Proposition 2.6. *Assume that an m-player game is WUC. Then any Nash equilibrium is also an optimal equilibrium.*

3. Single-Period Deterministic Redistribution Games

Throughout this paper, the game option terminology of "exercise" will be utilized when referring to the action of stopping the game by a player and the corresponding payoff from doing so will be called the "exercise payoff". We begin by setting up a single-period deterministic game with m players indexed by the set $\mathcal{M} = \{1, 2, \ldots, m\}$, where early exercising (and thus stopping the game) is only allowed at time 0. If no one exercises at time 0, then all players are assumed to exercise at time 1.

Definition 3.1. A *single-period deterministic m-player redistribution game* $\mathrm{RG}(X, P, \alpha)$ is specified by the following inputs:
(a) the vector $X = [X_1, \dots, X_m]$, where X_k is received by player k if she exercises at time 0,
(b) the vector $P = [P_1, \dots, P_m]$, where P_k is received by player k if no player exercises at time 0,
(c) the vector $\alpha = [\alpha_1, \dots, \alpha_m]$, where $\alpha_i > 0$ for all $i \in \mathcal{M}$ and the inequality $\sum_{i \in \mathcal{M}} \alpha_i \leq 1$ holds,
(d) for any strategy profile $s \in \mathcal{S}$, the *weights* $w_k(\mathcal{E}(s))$ for all $k \in \mathcal{M} \setminus \mathcal{E}(s)$ defined by

$$w_k(\mathcal{E}(s)) = \frac{\alpha_k}{1 - \sum_{i \in \mathcal{E}(s)} \alpha_i},$$

where the *exercise set* $\mathcal{E}(s)$ is the collection of players who exercise at time 0, and the rules of the game:
(i) the strategy $s_k \in \mathcal{S}_k := \{0, 1\}$ of player k specifies whether player k exercises, namely, $s_k = 0$ ($s_k = 1$, resp.) means that player k exercises (does not exercise, resp.) at time 0; given a strategy profile $s \in \mathcal{S} := \prod_{i \in \mathcal{M}} \mathcal{S}_i$,
(ii) for any $s \in \mathcal{S}$, the outcome of the game $\mathrm{RG}(X, P, \alpha)$ is the *payoff vector* $V(s) = [V_1(s), \dots, V_m(s)]$ where the payoff $V_k(s)$ received by player k if a strategy profile s is carried out equals

$$V_k(s) = \begin{cases} X_k, & k \in \mathcal{E}(s), \\ P_k - w_k(\mathcal{E}(s))D(s), & k \in \mathcal{M} \setminus \mathcal{E}(s), \end{cases}$$

where $D(s) = \sum_{i \in \mathcal{E}(s)} (X_i - P_i)$ is the *difference due to exercise* and $w_k(\mathcal{E}(s))$.

Remark 3.2. In general, the sign of the difference due to exercise $D(s)$ is arbitrary. It is known, however, that $D(s^*) \geq 0$ if a strategy profile s^* is a Nash equilibrium (see Proposition 2.1 in Guo and Rutkowski [11]). The inequality $D(s^*) > 0$ means that there is an aggregate gain for the cohort of exercising players. It generates negative effects for the payoffs of non-exercising players. The gain of exercising players should be covered by non-exercising players through the redistribution mechanism specified in Definition 3.1. This mechanism ensures that the single-period redistribution game is zero-sum, unless all players decide to exercise at time 0.

Remark 3.3. The weights specified by (c) and (d) satisfy the properties $w_k(\mathcal{E}(s)) > 0$ and $\sum_{k \in \mathcal{M} \setminus \mathcal{E}(s)} w_k(\mathcal{E}(s)) \leq 1$. The choice of this particular weight function is justified in Section 3.2

Remark 3.4. Note that the payoffs $V_k(s)$ are linear functions of X_i and P_i. Due to this salient feature, most results holding for the deterministic case can be extended to a single-period stochastic setting.

3.1. *Examples*

Consider a company with m employees, whose individual salaries are given by $P_1, \ldots, P_m$. At the end of the year, each employee is offered a new salary for the new year, denoted by $X_1, \ldots, X_m$. Each employee may either choose to take the offer (0) or reject the offer (1), forming a strategy profile s. Let $\mathcal{E}$ be the set of employees taking the offer.

The wage budget of the company is fixed at the quantity $\sum_{i=1}^m P_i$. We assume the total offer is within the budget, that is, $\sum_{i=1}^m X_i \leq \sum_{i=1}^m P_i$. If employee i chooses to take the offer, he will receive X_i. If he rejects the offer, then he will be receiving an adjusted salary $P_i(s)$. To understand this adjustment, first note that the company would require the additional fund of $D(s) = \sum_{i \in \mathcal{E}}(X_i - P_i)$ to cover the salary offer of $\mathcal{E}$. This additional fund will be split up and taken away from the wages of those who did not take the offer, $\mathcal{M} \setminus \mathcal{E}$. So the adjusted salary is given by

$$P_i(s) = P_i - w_i(s)D(s)$$

where the weight functions $w_i(s)$ satisfy $w_i(s) > 0$ and $\sum_{i \in \mathcal{M} \setminus \mathcal{E}} w_i(s) = 1$.

Intuitively, the employees with better offers $X_i \geq P_i$ will take their new offer, while for those with worse offers $X_i < P_i$ will have to compare the offer X_i with the adjusted salary $P_i(s)$ in order to make a decision. The problem is equivalent to finding Nash equilibrium in a redistribution game.

If each employee is assigned a *redistribution quotient* α_i, indicating how much he is affected by the adjustment, one could restrict the weight functions by setting $w_i(s)/w_j(s) = \alpha_i/\alpha_j$. After scaling, this is equivalent to the conditions

$$w_i(s) = \frac{\alpha_i}{\sum_{j \in \mathcal{M} \setminus \mathcal{E}} \alpha_j} = \frac{\alpha_i}{1 - \sum_{j \in \mathcal{E}} \alpha_j}, \quad \alpha_i > 0, \quad \sum_{i=1}^m \alpha_i = 1.$$

The game then becomes the zero-sum redistribution game ZRG(X, P, α) in the sense of Definition 3.10.

Let us now suppose that the adjustment $D(s)$ is not entirely redistributed amongst the m employees, but is also partially subsidized by the company boss. In particular, the boss is given a redistribution quotient of

$\alpha_{m+1} = \sum_{i=1}^{m} \alpha_i$. The weights in this new game are defined by

$$w_i(s) = \frac{\alpha_i}{\alpha_{m+1} + \sum_{j \in \mathcal{M} \setminus \mathcal{E}} \alpha_j} = \frac{\alpha_i}{1 - \sum_{j \in \mathcal{E}} \alpha_j}, \quad \alpha_i > 0, \quad \sum_{i=1}^{m} \alpha_i < 1.$$

This game is an example of a general redistribution game $\mathrm{GRG}(X, P, \alpha)$, which is formally specified in Definition 3.13 below.

3.2. *Weights and the WUC Property*

Before proceeding further, we will first formally justify our choice of the weight function. In this section, we begin with a redistribution game $\mathcal{G}$ as described by Definition 3.1 but without the inputs (c) and (d). Instead, a generic weight function $w_k(\mathcal{E})$ will be used as it is narrowed down based on a set of desirable conditions. Consider the class of all redistribution games with some fixed weight functions $w_k(\mathcal{E})$, but with all possible choices of vectors X and P. The goal is to find all weight functions fulfilling the following three conditions:
(C.1) every weight is non-zero,
(C.2) every game in the class is WUC,
(C.3) every game in the class has at least one equilibrium in pure strategies.

The non-zero condition (C.1) will ensure that the decision of any exercising player will always affect the payoffs of all non-exercising players. It also eliminates various degenerate cases. The WUC condition (C.2) further refines the restrictions on the weights, as shown in Proposition 3.5, which deals with the case $m \geq 4$ (for the cases $m = 2$ and $m = 3$, see Remark 6.3).

Proposition 3.5. *Assume that $m \geq 4$. Then the game $\mathcal{G}$ is WUC for all $X, P \in \mathbb{R}^m$ if and only if the weights can be represented as follows: for all $\mathcal{E}$,*

$$w_k(\mathcal{E}) = \frac{\alpha_k}{1 - \sum_{i \in \mathcal{E}} \alpha_i} \quad \textit{where} \quad \alpha_k > 0 \ \textit{and} \ \sum_{i \neq k} \alpha_i < 1 \ \forall\, k \in \mathcal{M}. \tag{3.1}$$

The proof of Proposition 3.5 is postponed to the appendix. For the purpose of this work, redistribution games will only use the weights defined by (3.1) for any number of players. In equation (3.1), the relative sizes of α_ks determine the weights $w_k(\mathcal{E})$ used to redistribute $D(s)$. By Proposition 3.5, the game $\mathcal{G}$ with weights defined by (3.1) is always WUC, so that condition (C.2) holds. For condition (C.3), the further restriction of $\sum_{i \in \mathcal{M}} \alpha_i \leq 1$ is needed, as shown by Proposition 6.4 in the appendix.

3.3. *Existence of the Value*

We henceforth work under the assumptions stated in Definition 3.1. We will now show that they ensure that the redistribution game has the value, which is achieved by all Nash and optimal equilibria.

Theorem 3.6. *The redistribution game $\mathcal{G} = \mathrm{RG}(X, P, \alpha)$ has at least one optimal equilibrium. Hence the game has the value.*

The proof of Theorem 3.6 is preceded by two lemmas. Consider a subset $\mathcal{E} \subset \mathcal{M}$ and assume that every player in $\mathcal{E}$ exercises at time 0, while the players from $\mathcal{M}' = \mathcal{M} \setminus \mathcal{E}$ are still free to make choices. The possible outcomes of the game for the players from $\mathcal{M}'$ define a subgame of the game $\mathcal{G} = \mathrm{RG}(X, P, \alpha)$, which we denote by $\mathcal{G}'$. The next lemma shows that the game $\mathcal{G}'$ has the same features as $\mathrm{RG}(X, P, \alpha)$, but with suitably modified parameters.

Lemma 3.7. *The subgame $\mathcal{G}'$ is equivalent to a game $\mathcal{G}_{\mathcal{M}'} = \mathrm{RG}(X', P', \alpha')$ with:*
(i) the set of players $\mathcal{M}' = \mathcal{M} \setminus \mathcal{E}$,
(ii) $X'_k = X_k$ for $k \in \mathcal{M}'$,
(iii) $P'_k = P_k - w_k(\mathcal{E}) \sum_{i \in \mathcal{E}} (X_i - P_i)$ for $k \in \mathcal{M}'$,
(iv) the weights defined by $\alpha'_k = w_k(\mathcal{E})$for $k \in \mathcal{M}'$,
(v) the strategy profiles $s' = [s'_k,\ k \in \mathcal{M}'] \in \mathcal{S}_{\mathcal{M}'}$.

Proof. Let $s' \in \mathcal{S}_{\mathcal{M}'}$ be any strategy profile of the subgame $\mathcal{G}_{\mathcal{M}'}$ and $\mathcal{E}' = \mathcal{E}(s')$ be the corresponding set of exercising players. Let $s \in \mathcal{S}$ be the matching strategy profile of the original game $\mathcal{G}$, that is,

$$s = [s_k = s'_k, k \in \mathcal{M}'; s_k = 0, k \notin \mathcal{M}']$$

and $\mathcal{E}(s) = \mathcal{E} \cup \mathcal{E}'$. It is sufficient to check that for any $k \in \mathcal{M}'$, the payoff $V'_k(s')$ of the subgame matches the payoff $V_k(s)$ of the original game.

We first note that the weights of $\mathcal{G}_{\mathcal{M}'}$ can be written as

$$w'_k(\mathcal{E}') = \frac{\alpha'_k}{1 - \sum_{i \in \mathcal{E}'} \alpha'_i} = \frac{w_k(\mathcal{E})}{1 - \sum_{i \in \mathcal{E}'} w_i(\mathcal{E})} \tag{3.2}$$

$$= \frac{\frac{\alpha_k}{1 - \sum_{i \in \mathcal{E}} \alpha_i}}{1 - \sum_{i \in \mathcal{E}'} \frac{\alpha_i}{1 - \sum_{j \in \mathcal{E}} \alpha_j}} = \frac{\alpha_k}{1 - \sum_{i \in \mathcal{E} \cup \mathcal{E}'} \alpha_i} = w_k(\mathcal{E}(s)). \tag{3.3}$$

If $k \in \mathcal{E}'$, then

$$V'_k(s') = X'_k = X_k = V_k(s).$$

Otherwise, we obtain

$$V'_k(s') = P'_k - w'_k(\mathcal{E}') \sum_{i \in \mathcal{E}'} (X'_i - P'_i) \tag{3.4}$$

$$= P_k - w_k(\mathcal{E}) \sum_{i \in \mathcal{E}} (X_i - P_i) - w'_k(\mathcal{E}') \sum_{i \in \mathcal{E}'} \Big(X_i - P_i + w_i(\mathcal{E}) \sum_{j \in \mathcal{E}} (X_j - P_j) \Big)$$

$$= P_k - \Big(w_k(\mathcal{E}) + w'_k(\mathcal{E}') \sum_{i \in \mathcal{E}'} w_i(\mathcal{E}) \Big) \sum_{i \in \mathcal{E}} (X_i - P_i) - w'_k(\mathcal{E}') \sum_{i \in \mathcal{E}'} (X_i - P_i).$$

Rearranging (3.2) as

$$w_k(\mathcal{E}) + w'_k(\mathcal{E}') \sum_{i \in \mathcal{E}'} w_i(\mathcal{E}) = w'_k(\mathcal{E}'),$$

we can rewrite (3.4) as follows

$$\begin{aligned} V'_k(s') &= P_k - w'_k(\mathcal{E}') \sum_{i \in \mathcal{E}} (X_i - P_i) - w'_k(\mathcal{E}') \sum_{i \in \mathcal{E}'} (X_i - P_i) \\ &= P_k - w'_k(\mathcal{E}') \sum_{i \in \mathcal{E}(s)} (X_i - P_i) \\ &= P_k - w_k(\mathcal{E}(s)) \sum_{i \in \mathcal{E}(s)} (X_i - P_i) \qquad \text{(by (3.3))} \\ &= V_k(s), \end{aligned}$$

as was required to show. □

We will also need Lemma 3.8, which asserts that the difference due to exercise has to be positive under any Nash equilibrium.

Lemma 3.8. *In the game* $\mathrm{RG}(X, P, \alpha)$, *if* s^* *is an equilibrium, then* $D(s^*) = \sum_{i \in \mathcal{E}(s^*)} (X_i - P_i) \geq 0$.

Proof. Assume the contrary, so $D(s^*) < 0$. Then there must exists a player $k \in \mathcal{E}(s^*)$ for which $X_k - P_k < 0$. Assuming that player k chooses not to exercise at time 0, her payoff will be

$$V_k(s') = P_k - w_k(\mathcal{E}(s')) \sum_{i \in \mathcal{E}(s')} (X_i - P_i)$$

where s' is the modified strategy profile. Note that $s' = [1, s^*_{-k}]$ and $s^* = [0, s^*_{-k}]$ and thus $\mathcal{E}(s') = \mathcal{E}(s^*) \setminus \{k\}$. Since s^* is an equilibrium, we have $V_k(s') \leq V_k(s^*) = X_k$. Consequently, we obtain

$$X_k - V_k(s') = X_k - P_k + w_k(\mathcal{E}(s')) \sum_{i \in \mathcal{E}(s')} (X_i - P_i) \geq 0. \tag{3.5}$$

Since $X_k - P_k < 0$, equality (3.5) implies $\sum_{i\in\mathcal{E}(s')}(X_i - P_i) \geq 0$. Recall that it is postulated that $0 < w_k(\mathcal{E}(s')) \leq 1$. Therefore,

$$D(s^*) = \sum_{i\in\mathcal{E}(s^*)} (X_i - P_i) \geq X_k - P_k + w_k(\mathcal{E}(s')) \sum_{i\in\mathcal{E}(s')} (X_i - P_i) \geq 0,$$

contradicting the assumption of $D(s^*) < 0$. □

Proof of Theorem 3.6. Kats and Thisse [17] and De Wolf [3] have shown that in any WUC game all Nash equilibria are optimal equilibria and attain the same value. It is thus sufficient to construct a Nash equilibrium in pure strategies.

We proceed by induction on the number of players. Note that the game is still well-defined as a single-player game when $m = 1$ and, obviously, all single-player games have at least one equilibrium. In particular, $s^* = [1]$ if $P_1 > X_1$ or $s^* = [0]$ if $P_1 \leq X_1$.

Let us now assume that $m \geq 2$. If $P_i > X_i$ for all i, then $s^* = 1$ is an equilibrium. If $P_k \leq X_k$ for some k, we consider the $m-1$ player subgame $\mathcal{G}_{\{-k\}}$. Let s' be an equilibrium of $\mathcal{G}_{\{-k\}}$, which exists by the induction hypothesis. Consider the strategy profile $s^* = [1, s_{-k} = s']$; we will show it is an equilibrium of $\mathcal{G}$.

By construction, s' is an equilibrium of $\mathcal{G}_{\{-k\}}$, so any player $i \neq k$ cannot improve her payoff by changing her strategy. Hence it is sufficient to check that player k cannot improve by not exercising, or $V_k(s^*) = X_k \geq V_k(s) = P_k - w_k(\mathcal{E})D(s)$ where $s = [0, s_{-k} = s']$ and $\mathcal{E}(s') = \mathcal{E}(s) = \mathcal{E}$. To this end, let us write $D(s)$ in terms of the subgame $\mathcal{G}_{\{-k\}}$ variables P'_i and $D'(s')$ (see Lemma 3.7)

$$D(s) = \sum_{i\in\mathcal{E}}(X_i - P'_i) + (P'_i - P_i) = D'(s') - \sum_{i\in\mathcal{E}} w_i(\{k\})(X_k - P_k). \quad (3.6)$$

Substituting (3.6) back, we want the following expression to be non-negative

$$\begin{aligned}
&X_k - (P_k - w_k(\mathcal{E})D(s)) \\
&= X_k - P_k + w_k(\mathcal{E})D'(s') - w_k(\mathcal{E})(X_k - P_k)\sum_{i\in\mathcal{E}} w_i(\{k\}) \\
&= w_k(\mathcal{E})D'(s') + (X_k - P_k)\Big(1 - w_k(\mathcal{E})\sum_{i\in\mathcal{E}} w_i(\{k\})\Big). \quad (3.7)
\end{aligned}$$

From Lemma 3.8 applied to the subgame $\mathcal{G}_{\{k\}}$, we obtain $D'(s') \geq 0$. Recall that $w_k(\mathcal{E}) \geq 0$ and $X_k - P_k \geq 0$, by assumptions.

It remains to examine the last term in (3.7), which can be represented as follows

$$\begin{aligned} 1 - w_k(\mathcal{E}) \sum_{i \in \mathcal{E}} w_i(\{k\}) &= 1 - \frac{\alpha_k}{1 - \sum_{i \in \mathcal{E}} \alpha_i} \frac{\sum_{i \in \mathcal{E}} \alpha_i}{1 - \alpha_k} \\ &= \frac{1 - \alpha_k - \sum_{i \in \mathcal{E}} \alpha_i}{\left(1 - \sum_{i \in \mathcal{E}} \alpha_i\right)(1 - \alpha_k)} \geq 0 \end{aligned}$$

where the inequality holds since $\alpha_k + \sum_{i \in \mathcal{E}} \alpha_i \leq \sum_{i \in \mathcal{M}} \alpha_i \leq 1$. We have thus shown that $X_k \geq V_k(s)$. Consequently, s^* is an equilibrium and the induction is complete. □

3.4. *Explicit Constructions of Optimal Equilibria*

The main result of this section, Theorem 3.15, furnishes an explicit construction of the value and an optimal equilibrium of the redistribution game. The construction is motivated by results from Guo and Rutkowski [11], who examined the case of the zero-sum redistribution game and expressed its value and an optimal equilibrium in terms of a suitable projection. In Subsection 3.4.1, we briefly revisit the results from [11]. Next, in Subsection 3.4.2, these results will be extended to the case of the general redistribution game.

Let us fix $X \in \mathbb{R}^m$. For any subset $\mathcal{E} \subseteq \mathcal{M}$, we define the hyperplane $\mathbb{H}_{\mathcal{E}}(X)$ as

$$\mathbb{H}_{\mathcal{E}}(X) := \{x \in \mathbb{R}^m : x_i = X_i, \ \forall \, i \in \mathcal{E}\}.$$

In particular, $\mathbb{H}_{\emptyset}(X) = \mathbb{R}^m$ and $\mathbb{H}_{\mathcal{M}}(X) = \{X\}$. The hyperplane $\mathbb{H}_{\mathcal{E}}(X)$ contains all the possible payoffs if all players in $\mathcal{E}$ exercise. We also define the orthant $\mathbb{O}(X)$

$$\mathbb{O}(X) := \{x \in \mathbb{R}^m : x_i \geq X_i, \ \forall \, i \in \mathcal{M}\}.$$

Lemma 3.9. *The value V^* of the redistribution game $\mathcal{G}$ belongs to the orthant $\mathbb{O}(X)$.*

Proof. Given any equilibrium s^*, the payoff of each player should be at least as great as her exercise payoff, so that $V_k(s^*) \geq X_k$ for all $k \in \mathcal{M}$. Hence $V^* \in \mathbb{O}(X)$. □

3.4.1. *Zero-Sum Redistribution Game*

We will only present the main results for the zero-sum (or constant sum) game established by Guo and Rutkowski [11]. Analogous results for the nonzero-sum game will be derived in Subsection 3.4.2 (see Lemma 3.14 and Theorem 3.15). Only in this subsection, we add the constraint $\sum_{i\in\mathcal{M}}\alpha_i = 1$ to Definition 3.1.

Definition 3.10. Assume that the equality $\sum_{i\in\mathcal{M}}\alpha_i = 1$ holds in Definition 3.1. Then the game is called the *zero-sum redistribution game* and it is denoted as $\mathrm{ZRG}(X, P, \alpha)$.

We have that

$$\sum_{i\notin\mathcal{E}} w_i(\mathcal{E}) = \frac{\sum_{i\notin\mathcal{E}}\alpha_i}{1-\sum_{j\in\mathcal{E}}\alpha_j} = 1, \quad \forall\,\mathcal{E}\subset\mathcal{M},\ \mathcal{E}\neq\emptyset. \tag{3.8}$$

If $\mathcal{E}(s) = \mathcal{M}$, then $\sum_{i\in\mathcal{M}} V_i(s) = \sum_{i\in\mathcal{M}} X_i$. Otherwise,

$$\sum_{i\in\mathcal{M}} V_i(s) = \sum_{i\in\mathcal{E}} X_i + \sum_{i\notin\mathcal{E}}\Big(P_i - w_i(\mathcal{E})\sum_{j\in\mathcal{E}}(X_j - P_j)\Big)$$

and thus

$$\sum_{i\in\mathcal{M}} V_i(s) = \sum_{i\in\mathcal{E}} X_i + \sum_{i\notin\mathcal{E}} P_i - \sum_{j\in\mathcal{E}}(X_j - P_j) = \sum_{i\in\mathcal{M}} P_i.$$

This means that the payoffs of the game $\mathrm{ZRG}(X, P, \alpha)$ have a constant sum $c := \sum_{i=1}^m P_i$, except when everyone exercises so that the sum of payoffs equals $\sum_{i=1}^m X_i$.

Let us fix the vectors $X, P \in \mathbb{R}^m$. All possible payoffs, except one, lie on the hyperplane

$$\mathbb{H}(c) := \Big\{x\in\mathbb{R}^m : \sum_{i=1}^m x_i = c\Big\}.$$

In particular, the game $\mathrm{ZRG}(X, P, \alpha)$ is almost zero-sum if the equality $\sum_{i\in\mathcal{M}} P_i = 0$ holds, and thus we decided to use the common term 'zero-sum game', rather than the more precise name 'constant-sum game'.

For any proper subset $\mathcal{E}\subset\mathcal{M}$, we define the hyperplane

$$\mathbb{H}_{\mathcal{E}}(X, c) := \mathbb{H}_{\mathcal{E}}(X)\cap\mathbb{H}(c) = \Big\{x\in\mathbb{R}^m : \sum_{i=1}^m x_i = c \text{ and } x_i = X_i,\ \forall\, i\in\mathcal{E}\Big\}.$$

In particular, $\mathbb{H}_{\emptyset}(X, c) = \mathbb{H}(c)$. By convention, we also set

$$\mathbb{H}_{\mathcal{M}}(X, c) := \mathbb{H}_{\mathcal{M}}(X) = \{X\}.$$

We endow the space $\mathbb{R}^m$ with the inner product $\langle \cdot, \cdot \rangle^0$ given by

$$\langle x, y \rangle^0 = \sum_{i=1}^{m} \left(\frac{x_i y_i}{\alpha_i} \right) \tag{3.9}$$

and the associated norm $\| \cdot \|^0$. This choice of the norm will be justified by Lemma 3.11, which gives a very handy way of representing and computing the payoff vector, provided that the exercise set is known. Given a normed vector space $\mathbb{R}^m$, for any vector P and any closed convex set $\mathbb{K}$, we denote by $\pi_{\mathbb{K}}(P)$ the *projection* of P onto $\mathbb{K}$.

Lemma 3.11. *In the game* ZRG(X, P, α)*, for any strategy profile* $s \in \mathcal{S}$*, the payoff vector* $V(s)$ *equals*

$$V(s) = \pi^0_{\mathbb{H}_{\mathcal{E}(s)}(X,c)}(P)$$

where the mapping

$$\pi^0_{\mathbb{H}_{\mathcal{E}(s)}(X,c)} : \mathbb{R}^m \to \mathbb{H}_{\mathcal{E}(s)}(X, c)$$

is the projection under the norm $\| \cdot \|^0$.

Consider the (possibly empty) simplex $\mathbb{S}(X, c)$ given by the formula

$$\mathbb{S}(X, c) := \mathbb{O}(X) \cap \mathbb{H}(c) = \left\{ x \in \mathbb{R}^m : \sum_{i=1}^{m} x_i = c \text{ and } x_i \geq X_i, \forall i \in \mathcal{M} \right\}.$$

As observed in Lemma 3.9, the value V^* must belong to $\mathbb{O}(X)$. Since all the payoffs of ZRG(X, P, α), but one, lie on $\mathbb{H}(c)$, we expect V^* to belong to $\mathbb{S}(X, c)$ (if it is non-empty).

The following theorem, which is borrowed from [11], shows that the value V^* coincides with the projection of a vector P onto $\mathbb{S}(X, c)$.

Theorem 3.12. *Let us consider the single-period zero-sum redistribution game* ZRG(X, P, α)*. If* $\sum_{i=1}^m X_i \leq c$*, then the value* V^* *equals*

$$V^* = V(s^*) = \pi^0_{\mathbb{S}(X,c)}(P) = \pi^0_{\mathbb{H}_{\mathcal{E}(s^*)}(X,c)}(P)$$

where the projection π^0 *is taken under the norm* $\| \cdot \|^0$ *induced by the inner product* (3.9) *and an optimal equilibrium* $s^* = [s_1^*, \dots, s_m^*]$ *is characterized by*

$$s_i^* = 0 \iff \left[\pi^0_{\mathbb{S}(X,c)}(P) \right]_i = X_i.$$

If $\sum_{i=1}^m X_i > c$*, then the value of the game* ZRG(X, P, α) *equals* $V^* = V(s^*) = X$ *with an optimal equilibrium given by* $s^* = [0, \dots, 0]$.

3.4.2. *Nonzero-Sum Redistribution Game*

This subsection extends the projection representation of the value and an optimal equilibrium to the case of a game that is not zero-sum.

Definition 3.13. Assume that the strict inequality $\sum_{i\in\mathcal{M}} \alpha_i < 1$ holds in Definition 3.1. Then the game is called the *general redistribution game* and it is denoted as $\mathrm{GRG}(X, P, \alpha)$.

Given an m-player game $\mathrm{GRG}(X, P, \alpha)$, we construct the associated $(m+1)$-player zero-sum redistribution game $\widetilde{\mathcal{G}} = \mathrm{ZRG}_{m+1}(X, P, \alpha)$ by adding the *dummy player* with the following attributes:
(a) no exercising allowed,
(b) P_{m+1} is arbitrary, but, for simplicity, we set $P_{m+1} = -\sum_{i\in\mathcal{M}} P_i$ so that $\mathcal{G}'$ is zero-sum,
(c) the weights $w_{m+1}(\mathcal{E}) = \frac{\alpha_{m+1}}{1-\sum_{i\in\mathcal{E}}\alpha_i}$ where $\alpha_{m+1} = 1 - \sum_{i\in\mathcal{M}} \alpha_i > 0$,
(d) the payoff function satisfies

$$V_{m+1}(s) = P_{m+1} - w_{m+1}(\mathcal{E}(s)) \sum_{i\in\mathcal{E}(s)} (X_i - P_i).$$

The following properties are easy to check:
(i) the space of strategy profiles in $\widetilde{\mathcal{G}}$ is the space $\mathcal{S}$ in $\mathrm{GRG}(X, P, \alpha)$,
(ii) the payoffs of all players in $\mathcal{M}$ do not change in $\widetilde{\mathcal{G}}$,
(iii) any Nash/optimal equilibrium for the game $\widetilde{\mathcal{G}}$ is also a Nash/optimal equilibrium of $\mathrm{GRG}(X, P, \alpha)$ and vice versa,
(iv) the value of $\mathrm{GRG}(X, P, \alpha)$ is equal to the value of $\widetilde{\mathcal{G}}$, after restricting to the first m coordinates.

Since the game $\widetilde{\mathcal{G}}$ is zero-sum, $V_{m+1}(s) + \sum_{i\in\mathcal{M}} V_i(s) = 0$ for all $s \in \mathcal{S}$. Hence the space of payoff vectors for $\widetilde{\mathcal{G}}$ is the hyperplane

$$\widetilde{\mathbb{H}}(0) := \left\{ x \in \mathbb{R}^{m+1} : \sum_{i=1}^{m+1} x_i = 0 \right\},$$

which is endowed with the inner product $\langle \cdot, \cdot \rangle^0$ given by

$$\langle x, y \rangle^0 = \sum_{i=1}^{m} \left(\frac{x_i y_i}{\alpha_i} \right) + \frac{(-\sum_{i=1}^{m} x_i)(-\sum_{i=1}^{m} y_i)}{\alpha_{m+1}}. \tag{3.10}$$

Rewriting α_{m+1} in terms of $\alpha_1, \ldots, \alpha_m$, formula (3.10) motivates the following inner product

$$\langle x, y \rangle = \sum_{i=1}^{m} \left(\frac{x_i y_i}{\alpha_i} \right) + \frac{(\sum_{i=1}^{m} x_i)(\sum_{i=1}^{m} y_i)}{1 - \sum_{i=1}^{m} \alpha_i} \tag{3.11}$$

and the associated norm $\|\cdot\|$. For any subset $\mathcal{E} \subseteq \mathcal{M}$, we denote

$$\widetilde{\mathbb{H}}_{\mathcal{E}}(X) := \{x \in \mathbb{R}^{m+1} : x_i = X_i, \ \forall i \in \mathcal{E}\}$$

and

$$\widetilde{\mathbb{O}}(X) := \{x \in \mathbb{R}^{m+1} : x_i \geq X_i, \ \forall i \in \mathcal{M}\}.$$

Consider the isometry $\phi : \widetilde{\mathbb{H}}(0) \to \mathbb{R}^m$, which discards the $(m+1)$th coordinate. For any $\mathcal{E} \subseteq \mathcal{M}$, it maps $\widetilde{\mathbb{H}}_{\mathcal{E}}(X, 0) = \widetilde{\mathbb{H}}_{\mathcal{E}}(X) \cap \widetilde{\mathbb{H}}(0)$ and $\widetilde{\mathbb{S}}(X, 0) := \widetilde{\mathbb{O}}(X) \cap \widetilde{\mathbb{H}}(0)$ to $\mathbb{H}_{\mathcal{E}}(X)$ and $\mathbb{O}(X)$, respectively. Equipped with the new inner product $\langle \cdot, \cdot \rangle$, we can in fact discard the dummy player and return to the space $\mathbb{R}^m$ and the original game $\mathrm{GRG}(X, P, \alpha)$.

Lemma 3.14 and Theorem 3.15 are counterparts of Lemma 3.11 and Theorem 3.12, respectively. Recall that we consider here the case when $\sum_{i \in \mathcal{M}} \alpha_i < 1$.

Lemma 3.14. *In the game* $\mathrm{GRG}(X, P, \alpha)$, *for any strategy profile* $s \in \mathcal{S}$, *the payoff vector* $V(s)$ *equals*

$$V(s) = \pi_{\mathbb{H}_{\mathcal{E}(s)}(X)}(P)$$

where $\pi_{\mathbb{H}_{\mathcal{E}(s)}(X)} : \mathbb{R}^m \to \mathbb{H}_{\mathcal{E}(s)}(X)$ *is the projection under the norm* $\|\cdot\|$.

Proof. Let us denote $\mathbb{H}_{\mathcal{E}(s)} = \mathbb{H}_{\mathcal{E}(s)}(X)$. We observe that the vector $V(s)$ equals

$$V(s) = \big[V_i(s) = X_i, \, i \in \mathcal{E}(s), \, V_i(s) = P_i - w_i(\mathcal{E}(s))D(s), \, i \notin \mathcal{E}(s)\big]$$

and thus it certainly belongs to the hyperplane

$$\mathbb{H}_{\mathcal{E}(s)} = \{x \in \mathbb{R}^m : x_i = X_i, \ \forall i \in \mathcal{E}(s)\}.$$

Therefore, it suffices to check that the vector $v(s)$, which is given by

$$v(s) := P - V(s) = \big[v_i = P_i - X_i, \, i \in \mathcal{E}(s), \, v_i = w_i(\mathcal{E}(s))D(s), \, i \notin \mathcal{E}(s)\big],$$

is orthogonal to $\mathbb{H}_{\mathcal{E}(s)}$. Let $u = \big[u \in \mathbb{R}^m : u_i = 0, \, i \in \mathcal{E}(s)\big] = x - y$ for arbitrary vectors $x, y \in \mathbb{H}_{\mathcal{E}(s)}$.

Then

$$\begin{aligned}
\langle u, v(s)\rangle &= \sum_{i\notin\mathcal{E}(s)} \frac{u_i w_i(\mathcal{E}(s))D(s)}{\alpha_i} \\
&\quad + \frac{\left(\sum_{i\notin\mathcal{E}(s)} u_i\right)\left(\sum_{i\in\mathcal{E}(s)}(P_i - X_i) + \sum_{i\notin\mathcal{E}(s)} w_i(\mathcal{E}(s))D(s)\right)}{1-\sum_{i=1}^m \alpha_i} \\
&= \frac{\sum_{i\notin\mathcal{E}(s)} u_i D(s)}{1-\sum_{i\in\mathcal{E}(s)}\alpha_i} + \frac{\left(\sum_{i\notin\mathcal{E}(s)} u_i\right)\left(-D(s) + \frac{\sum_{i\notin\mathcal{E}(s)}\alpha_i}{1-\sum_{i\in\mathcal{E}(s)}\alpha_i} D(s)\right)}{1-\sum_{i=1}^m \alpha_i} \\
&= \frac{\sum_{i\notin\mathcal{E}(s)} u_i D(s)}{1-\sum_{i\in\mathcal{E}(s)}\alpha_i}\left(1 + \frac{-\left(1-\sum_{i\in\mathcal{E}(s)}\alpha_i\right) + \sum_{i\notin\mathcal{E}(s)}\alpha_i}{1-\sum_{i=1}^m \alpha_i}\right) \\
&= \frac{\sum_{i\notin\mathcal{E}(s)} u_i D(s)}{1-\sum_{i\in\mathcal{E}(s)}\alpha_i}(1-1) = 0,
\end{aligned}$$

as required. □

The existence of the value for the game $\mathcal{G}$ was first established in Theorem 3.6. Theorem 3.15 reaffirms that result by providing an explicit representation using a concise notation. The chosen optimal equilibrium s^* is determined by identifying the exercise set $\mathcal{E}(s^*)$ from the hyperplanes that are implicitly and uniquely specified by the equality of projections $\pi_{\mathbb{O}(X)}(P) = \pi_{\mathbb{H}_{\mathcal{E}(s^*)}(X)}(P)$.

Theorem 3.15. *The value of the game* $\mathrm{GRG}(X, P, \alpha)$ *is given by*

$$V^* = V(s^*) = [V_1(s^*), \ldots, V_m(s^*)] = \pi_{\mathbb{O}(X)}(P) = \pi_{\mathbb{H}_{\mathcal{E}(s^*)}(X)}(P)$$

where the projection π *is taken under the norm* $\|\cdot\|$ *and an optimal equilibrium* $s^* = [s_1^*, \ldots, s_m^*]$ *is given by*

$$s_i^* = 0 \iff \left[\pi_{\mathbb{O}(X)}(P)\right]_i = X_i.$$

In the proof Theorem 3.15, we will employ three lemmas. The first lemma is elementary and thus its proof is omitted.

Lemma 3.16. *If* $\mathbb{H} \subset \mathbb{R}^m$ *is a hyperplane, then the projection* π *is orthogonal, that is,* $\pi_{\mathbb{H}}(P)$ *is the unique vector in* $\mathbb{H}$ *such that*

$$\langle \pi_{\mathbb{H}}(P) - P, Q - \pi_{\mathbb{H}}(P)\rangle = 0, \quad \forall Q \in \mathbb{H}.$$

Furthermore, if $\mathbb{K}$ *is a convex subset of the hyperplane* $\mathbb{H}$, *then* $\pi_{\mathbb{K}}(P) = \pi_{\mathbb{K}}(\pi_{\mathbb{H}}(P))$.

Lemma 3.17. *Assume that* $P \in \mathbb{O}(X)$. *Then* $\pi_{\mathbb{H}_{\mathcal{E}}(X)}(P) \in \mathbb{O}(X)$ *for any subset* $\mathcal{E} \subseteq \mathcal{M}$.

Proof. We denote $\mathbb{H}_{\mathcal{E}} = \mathbb{H}_{\mathcal{E}}(X)$. In view of Lemma 3.14, the projection $\pi_{\mathbb{H}_{\mathcal{E}}}(P)$ corresponds to the payoff vector when $\mathcal{E}$ is the set of exercising players. Let s be the corresponding strategy profile. In particular, for any $i \in \mathcal{E}$

$$[\pi_{\mathbb{H}_{\mathcal{E}}}(P)]_i = X_i \leq P_i$$

and thus $D(s) = \sum_{i \in \mathcal{E}}(X_i - P_i) \leq 0$. Consequently, for any $j \in \mathcal{M} \setminus \mathcal{E}$,

$$[\pi_{\mathbb{H}_{\mathcal{E}}}(P)]_j = P_j - w_j(\mathcal{E}(s))(s)D(s) \geq P_j \geq X_j,$$

and thus $\pi_{\mathbb{H}_{\mathcal{E}}}(P) \in \mathbb{O}(X)$. □

Lemma 3.18. *For any* $k \in \mathcal{M}$, *if*

$$\pi_{\mathbb{O}(X)}(P) \notin \mathbb{H}_{\{k\}}(X) = \{x \in \mathbb{R}^m : x_k = X_k\}$$

then $P_k > X_k$. *Equivalently, if* $P_k \leq X_k$ *then* $\pi_{\mathbb{O}(X)}(P) \in \mathbb{H}_{\{k\}}(X)$.

Proof. For brevity, we write $\mathbb{O} = \mathbb{O}(X)$ and $\mathbb{H}_{\{k\}} = \mathbb{H}_{\{k\}}(X)$. We will argue by contradiction. Let us then suppose that $P_k \leq X_k$ and assume that $\pi_{\mathbb{O}}(P) \notin \mathbb{H}_{\{k\}}$. The projection $Q := \pi_{\mathbb{H}_{\{k\}}}(\pi_{\mathbb{O}}(P))$ is still in $\mathbb{O}$ (by Lemma 3.17) and it is distinct from $\pi_{\mathbb{O}}(P)$ (since, by assumption, $\pi_{\mathbb{O}}(P) \notin \mathbb{H}_{\{k\}}$). We will show that

$$\|P - Q\| < \|P - \pi_{\mathbb{O}}(P)\|, \tag{3.12}$$

which, since $Q \in \mathbb{O}$, contradicts the definition of the projection $\pi_{\mathbb{O}}(P)$.
Case 1: In the case of $P_k = X_k$, we have $P, Q \in \mathbb{H}_{\{k\}}$ and thus, by Lemma 3.16, the vector $\pi_{\mathbb{O}}(P) - Q$ is orthogonal to $P - Q \in \mathbb{H}_{\{k\}}$. Therefore,

$$\|P - Q\|^2 < \|P - Q\|^2 + \|\pi_{\mathbb{O}}(P) - Q\|^2 = \|P - \pi_{\mathbb{O}}(P)\|^2.$$

Case 2: To establish (3.12) when $P_k < X_k$, we introduce the auxiliary hyperplane $\widehat{\mathbb{H}}_{\{k\}}$ parallel to $\mathbb{H}_{\{k\}}$ by setting

$$\widehat{\mathbb{H}}_{\{k\}} := \{x \in \mathbb{R}^m : x_k = P_k\},$$

so that, manifestly, $P \in \widehat{\mathbb{H}}_{\{k\}}$. We observe that for all $x \in \mathbb{O}$

$$\pi_{\widehat{\mathbb{H}}_{\{k\}}}(\pi_{\mathbb{H}_{\{k\}}}(x)) = \pi_{\widehat{\mathbb{H}}_{\{k\}}}(x).$$

Hence for $R := \pi_{\widehat{\mathbb{H}}_{\{k\}}}(\pi_{\mathbb{O}}(P))$, we obtain

$$R = \pi_{\widehat{\mathbb{H}}_{\{k\}}}(\pi_{\mathbb{H}_{\{k\}}}(\pi_{\mathbb{O}}(P))) = \pi_{\widehat{\mathbb{H}}_{\{k\}}}(Q)$$

where the second equality follows from the definition of Q. Since $P_k < X_k$, the vectors $R \in \widehat{\mathbb{H}}_{\{k\}}$ and $\pi_{\mathbb{O}}(P) \in \mathbb{O} \setminus \mathbb{H}_{\{k\}}$ lie on opposite sides of the hyperplane $\mathbb{H}_{\{k\}}$. It is thus clear that

$$\|R - Q\| < \|R - Q\| + \|Q - \pi_{\mathbb{O}}(P)\| = \|R - \pi_{\mathbb{O}}(P)\| . \tag{3.13}$$

Finally, since $P - R \in \widehat{\mathbb{H}}_{\{k\}}$ is orthogonal to both $R - Q = \pi_{\widehat{\mathbb{H}}_{\{k\}}}(Q) - Q$ and $R - \pi_{\mathbb{O}}(P) = \pi_{\widehat{\mathbb{H}}_{\{k\}}}(\pi_{\mathbb{O}}(P)) - \pi_{\mathbb{O}}(P)$, we have

$$\|P - Q\|^2 = \|P - R\|^2 + \|R - Q\|^2$$

and

$$\|P - \pi_{\mathbb{O}}(P)\|^2 = \|P - R\|^2 + \|R - \pi_{\mathbb{O}}(P)\|^2 .$$

Therefore, (3.13) implies (3.12), which ends the proof of the lemma. □

Proof of Theorem 3.15. We begin by noting that in the subgame $\mathcal{G}_{\mathcal{M}'}$ defined in Lemma 3.7, the variables P'_k can be rewritten as

$$P'_k = P_k - w_k(\mathcal{E}) \sum_{i \in \mathcal{E}} (X_i - P_i) = \left[\pi_{\mathbb{H}_{\mathcal{E}}}(P)\right]_k$$

where we also used Lemma 3.14. The map $\phi : \mathbb{H}_{\mathcal{E}} \to \mathbb{R}^{m-|\mathcal{E}|}$, defined by discarding the coordinates with indices in $\mathcal{E}$, is an isometry to the space of $\mathcal{G}_{\mathcal{M}'}$ payoffs. Let us endow $\mathbb{R}^{m-|\mathcal{E}|}$ with the norm

$$\|x\|' = \left(\sum_{i \in \mathcal{M}'} \left(\frac{x_i^2}{\alpha'_i} \right) + \frac{\left(\sum_{i \in \mathcal{M}'} x_i\right)^2}{1 - \sum_{i \in \mathcal{M}'} \alpha'_i} \right)^{\frac{1}{2}}$$

and let π' be the corresponding projection map.

As before, we denote $\mathbb{O} = \mathbb{O}(X)$. To establish the assertion of the theorem, it is sufficient to show that the strategy profile s^* defined by

$$s_i^* = 0 \quad \Longleftrightarrow \quad \left[\pi_{\mathbb{O}}(P)\right]_i = X_i$$

is a Nash equilibrium (hence an optimal equilibrium, since the game GRG(X, P, α) is WUC). This goal will be achieved using the induction arguments used in the proof of Theorem 3.6, but with a few additions.

The base case of $m = 1$ can be easily checked. Let us this consider $m \geq 2$. If $P_i > X_i$ for all i, then P lies in the interior of $\mathbb{O}$. Therefore, $\pi_{\mathbb{O}}(P) = P$ and $s^* = 1$ is an equilibrium. If $P_k \leq X_k$ for some k, then we consider the $m-1$ player subgame $\mathcal{G}_{\{-k\}}$. By the induction hypothesis, the strategy profile $s' \in \mathcal{S}_{-k}$ defined by

$$s'_i = 0 \quad \Longleftrightarrow \quad \left[\pi'_{\mathbb{O}'}(P')\right]_i = X_i, \quad \forall i \in \mathcal{M} \setminus \{k\}$$

is an equilibrium of $\mathcal{G}_{\{-k\}}$. By applying the isometry ϕ^{-1} and using Lemma 3.16, we obtain

$$\left[\pi'_{\mathbb{O}'}(P')\right]_i = \left[\pi_{\mathbb{O}\cap\mathbb{H}_{\{k\}}}\left(\pi_{\mathbb{H}_{\{k\}}}(P)\right]_i = \left[\pi_{\mathbb{O}\cap\mathbb{H}_{\{k\}}}(P)\right]_i, \quad \forall\, i \in \mathcal{M}\setminus\{k\}.$$

By Lemma 3.18, the inequality $P_k \leq X_k$ implies that $\pi_{\mathbb{O}}(P) \in \mathbb{H}_{\{k\}}$. Hence $\pi_{\mathbb{O}\cap\mathbb{H}_{\{k\}}}(P) = \pi_{\mathbb{O}}(P)$ and s' can be rewritten as follows

$$s'_i = 0 \quad\Longleftrightarrow\quad \left[\pi_{\mathbb{O}}(P)\right]_i = X_i, \quad \forall\, i \in \mathcal{M}\setminus\{k\}.$$

Finally, the equality $\left[\pi_{\mathbb{O}}(P)\right]_k = X_k$ implies that $s^*_k = 0$ and thus $s^* = \left[s^*_k = 0,\ s^*_{-k} = s'\right]$. By the proof of Theorem 3.6, we conclude that s^* is an equilibrium of $\mathrm{GRG}(X, P, \alpha)$, as was required to show. □

3.5. *Extended Projection*

If the original game is already a zero-sum game $\mathrm{ZRG}(X, P, \alpha)$, then the introduction of the dummy player $m+1$ is problematic. Indeed, division by zero occurs because $\alpha_{m+1} = 1 - \sum_{i\in\mathcal{M}} \alpha_i = 0$ and thus the game $\widetilde{\mathcal{G}}$ and the inner product $\langle\cdot,\cdot\rangle^0$ from Subsection 3.4.1 are no longer well defined. We find it convenient to extend the concept of the projection map π so that we can henceforth cover simultaneously zero-sum and nonzero-sum games. This will be used in the next section for the multi-period game. We will argue that Lemma 3.14 and Theorem 3.15 can be applied to the zero-sum game if the following conventions are adopted. Let us fix $\epsilon > 0$ sufficiently small, and let us consider the $(m+1)$-player game $\mathcal{G}^\epsilon$, where $\alpha_1, \dots, \alpha_m$ and α_{m+1} are replaced by

$$\alpha^\epsilon_i = \alpha_i - \frac{\epsilon}{m} > 0,\ i \in \mathcal{M}, \quad \alpha^\epsilon_{m+1} = 1 - \sum_{i\in\mathcal{M}} \alpha^\epsilon_i = \epsilon.$$

We then take $\widetilde{\mathcal{G}}$ to be the limit of $\mathcal{G}^\epsilon$ as $\epsilon \downarrow 0$. Since for any fixed $k \in \mathcal{M}$ and any fixed $s \in \mathcal{S}$, the $\mathcal{G}^\epsilon$ payoff $V^\epsilon_k(s)$ is continuous (in fact, linear) in ϵ, the limit $\lim_{\epsilon\to 0} V^\epsilon_k(s)$ is indeed the desired payoff of $\widetilde{\mathcal{G}}$. Furthermore, the quantities

$$\underline{V}^\epsilon_k = \max_{s_k}\min_{s_{-k}} V^\epsilon_k\left(\left[s_k, s_{-k}\right]\right), \quad \overline{V}^\epsilon_k = \min_{s_{-k}}\max_{s_k} V^\epsilon_k\left(\left[s_k, s_{-k}\right]\right)$$

are also continuous in ϵ, so the value of $\mathcal{G}^\epsilon$ converges to the value of $\widetilde{\mathcal{G}}$ as well. Once again, the payoffs and the value of the m-player game $\mathrm{ZRG}(X, P, \alpha)$ are obtained from $\widetilde{\mathcal{G}}$ after discarding the $(m+1)$th coordinate. Although $\|\cdot\|$ is not well defined, Lemma 3.14 and Theorem 3.15 can still be recovered by redefining the projection π.

To this end, we formally adopt the following definition of the *extended projection* $\widehat{\pi}$ for the remainder of the paper. Obviously, this definition covers the case of an arbitrary game $\mathrm{RG}(X, P, \alpha)$.

Definition 3.19. When $\sum_{i \in \mathcal{M}} \alpha_i < 1$, then we define $\widehat{\pi}$ as the projection under the norm $\| \cdot \|$ given by (3.11). When $\sum_{i \in \mathcal{M}} \alpha_i = 1$, then we define the projection $\widehat{\pi}$ on $\mathbb{H}_{\mathcal{E}}$ where $\mathcal{E} \subseteq \mathcal{M}$ and $\mathbb{O}$ as follows, for all $P \in \mathbb{R}^m$,

$$\widehat{\pi}_{\mathbb{H}_{\mathcal{E}}}(P) = \lim_{\epsilon \downarrow 0} \pi^{\epsilon}_{\mathbb{H}_{\mathcal{E}}}(P), \quad \widehat{\pi}_{\mathbb{O}}(P) = \lim_{\epsilon \downarrow 0} \pi^{\epsilon}_{\mathbb{O}}(P),$$

where π^{ϵ} is the projection under the norm

$$\|x\|^{\epsilon} = \left(\sum_{i=1}^{m} \left(\frac{x_i^2}{\alpha_i - \frac{\epsilon}{m}} \right) + \frac{\left(\sum_{i=1}^{m} x_i\right)^2}{\epsilon} \right)^{\frac{1}{2}}. \tag{3.14}$$

4. Stopping Games with Redistribution of Payoffs

In this section, we study the multi-period stochastic extension of the single-period deterministic game. Let us stress that a multi-period stopping game is rarely WUC, even when its single-period building blocks happen to be. Nevertheless, we attempt to identify multi-period stochastic stopping games where optimal equilibria and value still exist.

All of the following definitions are taken under the probability space $(\Omega, \mathcal{F}, \mathbb{P})$ endowed with the filtration $\mathbb{F} = \{\mathcal{F}_t : t = 0, 1, \ldots, T\}$ representing the information flow observed by all the players. As before, the players are indexed by the set $\mathcal{M} = \{1, 2, \ldots, m\}$. In the present multi-period set-up, if a game starts at t, then each player has the right to exercise at any time in the interval $[t, T] := \{t, t+1, \ldots, T\}$ and the game stops as soon as anyone exercises. In addition, if no player exercises before time T, then everyone must exercise at time T. Further details about the specification of the multi-period redistribution game are given in the foregoing definition (see also Remarks 4.6 and 4.7 for additional comments).

Definition 4.1. For $t = 0, 1, \ldots, T$, the m-player *multi-period redistribution game* $\mathcal{G}_t = \mathrm{MRG}_t(X, \alpha)$ is defined on the time interval $[t, T]$ and specified by the inputs:
(a) the $\mathbb{F}$-adapted, $\mathbb{R}^m$-valued process $X_u = [X_u^1, \ldots, X_u^m]$ where $u = t, t+1, \ldots, T$,
(b) the weights $w_k(\mathcal{E}) = \frac{\alpha_k}{1 - \sum_{i \in \mathcal{E}} \alpha_i}$ for $k \notin \mathcal{E} \subset \mathcal{M}$ where $\alpha_i > 0$ is deterministic and $\sum_{i \in \mathcal{M}} \alpha_i \leq 1$,

and the rules of the game:
(i) the strategy s_t^k of player k is a stopping time from the space $\mathcal{S}_t^k$ of $\mathbb{F}$-stopping times with values in $[t,T]$; hence the strategy profile $s_t = [s_t^1,\ldots,s_t^m] \in \mathcal{S}_t$ is an m-tuple of stopping times; we denote by $\widehat{s}_t = s_t^1 \wedge \cdots \wedge s_t^m$ the minimal exercise time, also an $\mathbb{F}$-stopping time,
(ii) for each strategy profile $s_t \in \mathcal{S}_t$, the outcome of the game $\mathcal{G}_t$ is the expected payoff vector $V_t(s_t) = [V_t^1(s_t),\ldots,V_t^m(s_t)]$, which is given by

$$V_t^k(s_t) = \mathbb{E}_{\mathbb{P}}\big(X_{\widehat{s}_t}^k \mathbb{1}_{\{k\in\mathcal{E}(s_t)\}} + \widehat{X}_{\widehat{s}_t}^k \mathbb{1}_{\{k\notin\mathcal{E}(s_t)\}} \,\big|\, \mathcal{F}_t\big) \tag{4.1}$$

where the exercise set $\mathcal{E}(s_t) = \{i \in \mathcal{M} : s_t^i = \widehat{s}_t\}$ is the random set of earliest stopping players and

$$\widehat{X}_{\widehat{s}_t}^k = V_{\widehat{s}_t+1}^{*k} - w_k(\mathcal{E}(s_t)) \sum_{i\in\mathcal{E}(s_t)} \big(X_{\widehat{s}_t}^i - V_{\widehat{s}_t+1}^{*i}\big), \quad \widehat{s}_t < T, \tag{4.2}$$

where $V_u^* = (V_u^{*1},\ldots,V_u^{*m})$ is the value of the game $\mathcal{G}_u = \mathrm{MRG}_u(X,\alpha)$ for $u = t+1, t+2, \ldots, T$.

Remark 4.2. Since the game is stopped at time $\widehat{s}_t$, the indicator functions in (4.1) separate the exercising players from the others, specifically, $X_{\widehat{s}_t}^k$ is the payoff for an exercising player, whereas $\widehat{X}_{\widehat{s}_t}^k$ is the payoff for a non-exercising player. The game $\mathcal{G}_{\widehat{s}_t+1}$ can be considered as the continuation of the current game if it does not stop at time $\widehat{s}_t$. Note that in (4.2), the quantity $\widehat{X}_{\widehat{s}_t}^k$ is left unspecified for $\widehat{s}_t = T$. This is immaterial, however, because if the game is stopped at T, then every player must exercise and thus she receives the payoff X_T^k, rather than $\widehat{X}_T^k$.

The following lemma, which deals with the case of the single-period m-player stochastic stopping game, follows easily from results of preceding sections by taking expected values. For this reason, its proof is omitted. Note that the orthant $\mathbb{O}(X_0)$ may be random if the σ-field $\mathcal{F}_0$ is non-trivial.

Lemma 4.3. *The single-period m-player stochastic redistribution game $\mathcal{G}_0$ has the unique value given by $V(s^*) = \widehat{\pi}_{\mathbb{O}(X_0)}(\mathbb{E}_{\mathbb{P}}(X_1 \,|\, \mathcal{F}_0))$ where $\mathbb{O}(X_0)$ is the orthant defined by*

$$\mathbb{O}(X_0) = \big\{x \in \mathbb{R}^m : x_i \geq X_0^i,\ 1 \leq i \leq m\big\}$$

and the projection $\widehat{\pi}$ is given by Definition 3.19. A possible optimal equilibrium $s^* = [s_1^*,\ldots,s_m^*]$ is given by

$$s_i^* = 0 \iff \big[\widehat{\pi}_{\mathbb{O}(X_0)}\big(\mathbb{E}_{\mathbb{P}}(P \,|\, \mathcal{F}_0)\big)\big]_i = X_0^i.$$

We now proceed to the analysis of a multi-period stochastic redistribution game $\mathrm{MRG}(X, \alpha)$. The following lemma is an immediate consequence of Definitions 3.19 and 4.1 combined with Lemma 3.14.

Lemma 4.4. *The expected payoff* $V_t(s_t) = [V_t^1(s_t), \dots, V_t^m(s_t)]$ *can be represented using the projection*

$$V_t(s_t) = \mathbb{E}_{\mathbb{P}}\Big(\widehat{\pi}_{\mathbb{H}_{\mathcal{E}(s_t)}}(V^*_{\widehat{s}_t+1})\mathbb{1}_{\{\widehat{s}_t<T\}} + X_T \mathbb{1}_{\{\widehat{s}_t=T\}} \,\Big|\, \mathcal{F}_t\Big)$$

where $\mathbb{H}_{\mathcal{E}(s_t)}$ *is the* $\mathcal{F}_{\widehat{s}_t}$*-measurable hyperplane*

$$\mathbb{H}_{\mathcal{E}(s_t)} = \big\{x \in \mathbb{R}^m : x_i = X^i_{\widehat{s}_t},\ \forall\, i \in \mathcal{E}(s_t)\big\}.$$

Theorem 4.5. *Define recursively the* $\mathbb{F}$*-adapted process* $U = [U^1, \dots, U^m]$ *by setting* $U_T = X_T$ *and*

$$U_t = \widehat{\pi}_{\mathbb{O}(X_t)}\big(\mathbb{E}_{\mathbb{P}}(U_{t+1} \,|\, \mathcal{F}_t)\big), \quad \forall\, t = 0, 1, \dots, T-1, \tag{4.3}$$

where $\mathbb{O}(X_t)$ *is the* $\mathcal{F}_t$*-measurable orthant*

$$\mathbb{O}(X_t) = \big\{x \in \mathbb{R}^m : x_i \geq X_t^i,\ 1 \leq i \leq m\big\} \tag{4.4}$$

and the extended projection $\widehat{\pi}$ *is given by Definition 3.19. We define the collection* $\tau_t^* = (\tau_t^{*1}, \dots, \tau_t^{*m})$ *of* $\mathbb{F}$*-stopping times by setting*

$$\tau_t^{*i} = \min\big\{u \in [t, T] : U_u^i = X_u^i\big\}. \tag{4.5}$$

Then:
(i) the equality $U_t = V_t(\tau_t^*)$ *holds for every* $t = 0, 1, \dots, T$,
(ii) τ_t^* *is an optimal equilibrium of the* m*-player redistribution game* $\mathcal{G}_t = \mathrm{MRG}_t(X, \alpha)$ *and* $U_t = V_t^*$ *is the value of the game for every* $t = 0, 1, \dots, T$.

Proof. Both statements are proven by the backward induction. For $t = T$, we have $\tau_T^* = [T, \dots, T]$. The game $\mathcal{G}_T$ is always stopped at time T and thus the payoff vector $X_T = U_T = V_T(\tau^*) = V_T^*$ yields the value. Let us now assume that statements (i) and (ii) are valid for the game $\mathcal{G}_{t+1}$, so that its value is given by

$$V^*_{t+1} = U_{t+1} = V_{t+1}(\tau^*). \tag{4.6}$$

Note throughout the proof that if the game $\mathcal{G}_t$ is stopped at time t, then it is reduced to a single-period stochastic game with payoff vectors X_t and V^*_{t+1}. Let us denote this single-period game by $\mathcal{G}'$. Let us also write $\widehat{\tau}_t = \tau_t^{*1} \wedge \dots \wedge \tau_t^{*m}$. We first show that part (i) holds at time t.

Case 1: If $\widehat{\tau}_t = t$, then the game is stopped at time t. By Lemma 4.3 and equation (4.5), τ_t^* is an optimal equilibrium of the single-period game $\mathcal{G}'$, whose value is

$$U_t = \widehat{\pi}_{\mathbb{O}(X_t)}\big(\mathbb{E}_{\mathbb{P}}(U_{t+1}|\mathcal{F}_t)\big) = \widehat{\pi}_{\mathbb{O}(X_t)}\big(\mathbb{E}_{\mathbb{P}}(V_{t+1}^*|\mathcal{F}_t)\big) = \widehat{\pi}_{\mathbb{H}_{\mathcal{E}(\tau_t^*)}}\big(\mathbb{E}_{\mathbb{P}}(V_{t+1}^*|\mathcal{F}_t)\big).$$

The result then follows from Lemma 4.4, noting that $\mathbb{H}_{\mathcal{E}(\tau)}$ is $\mathcal{F}_t$-measurable, so that

$$U_t = \widehat{\pi}_{\mathbb{H}_{\mathcal{E}(\tau_t^*)}}\big(\mathbb{E}_{\mathbb{P}}(V_{t+1}^* \,|\, \mathcal{F}_t)\big) = \widehat{\pi}_{\mathbb{H}_{\mathcal{E}(\tau_t^*)}}\big(\mathbb{E}_{\mathbb{P}}(V_{\widehat{\tau}_t+1}^* \,|\, \mathcal{F}_t)\big)$$

Case 2: If $\widehat{\tau}_t \geq t+1$, then the game is not stopped at time t. By (4.3), (4.4) and (4.5), we obtain $U_t^i > X_t^i$ and thus the vector U_t lies in the interior of $\mathbb{O}(X_t)$ (a.s.). The induction hypothesis (4.6) yields

$$U_t = \widehat{\pi}_{\mathbb{O}(X_t)}\big(\mathbb{E}_{\mathbb{P}}(U_{t+1} \,|\, \mathcal{F}_t)\big) = \mathbb{E}_{\mathbb{P}}\big(U_{t+1} \,\big|\, \mathcal{F}_t\big) = \mathbb{E}_{\mathbb{P}}\big(V_{t+1}(\tau_t^*) \,\big|\, \mathcal{F}_t\big).$$

It is thus sufficient to show $V_t(\tau_t^*) = \mathbb{E}_{\mathbb{P}}\big(V_{t+1}(\tau_t^*) \,\big|\, \mathcal{F}_t\big)$. To establish this equality, we note that $\widehat{\tau}_t \geq t+1$ and we apply Lemma 4.4, to obtain

$$\begin{cases} V_t(\tau_t^*) = \mathbb{E}_{\mathbb{P}}\big(\mathbb{E}_{\mathbb{P}}\big(\widehat{\pi}_{\mathbb{H}_{\mathcal{E}(\tau_t^*)}}(V_{\widehat{\tau}_t+1}^*) \,\big|\, \mathcal{F}_{t+1}\big) \,\big|\, \mathcal{F}_t\big) = \mathbb{E}_{\mathbb{P}}\big(V_{t+1}(\tau_t^*) \,\big|\, \mathcal{F}_t\big) \text{ if } \widehat{\tau}_t < T \\ V_t(\tau_t^*) = \mathbb{E}_{\mathbb{P}}\big(\mathbb{E}_{\mathbb{P}}\big(X_T \,\big|\, \mathcal{F}_{t+1}\big) \,\big|\, \mathcal{F}_t\big) = \mathbb{E}_{\mathbb{P}}\big(V_{t+1}(\tau_t^*) \,\big|\, \mathcal{F}_t\big) \text{ if } \widehat{\tau}_t = T. \end{cases}$$

This completes the proof of part (i) for the game $\mathcal{G}_t$.

It remains to establish part (ii). By part (i), (4.3), and the induction hypothesis (4.6), we have

$$V_t(\tau_t^*) = U_t = \widehat{\pi}_{\mathbb{O}(X_t)}\big(\mathbb{E}_{\mathbb{P}}(U_{t+1} \,|\, \mathcal{F}_t)\big) = \widehat{\pi}_{\mathbb{O}(X_t)}\big(\mathbb{E}_{\mathbb{P}}(V_{t+1}^* \,|\, \mathcal{F}_t)\big).$$

To check that τ_t^* is an optimal equilibrium, we need to show that, for each $k \in \mathcal{M}$, the inequalities

$$V_t^k\left(\left[\tau_t^{*k}, s_{-k}\right]\right) \geq \left[\widehat{\pi}_{\mathbb{O}(X_t)}\big(\mathbb{E}_{\mathbb{P}}(V_{t+1}^* \,|\, \mathcal{F}_t)\big)\right]_k \geq V_t^k\left(\left[s_k, \tau_t^{*,-k}\right]\right)$$

hold for all $s_k \in \mathcal{S}_k$ and $s_{-k} \in \mathcal{S}_{-k}$. Let thus $s' = \left[\tau_t^{*k}, s_{-k}\right]$ and $s'' = \left[s_k, \tau_t^{*,-k}\right]$ be alternative strategy profiles with the respective minimal stopping times $\widehat{s}'$ and $\widehat{s}''$.

Case 1: Assume first that $\widehat{s}' = \widehat{s}'' = t$. Then both s' and s'' can be interpreted as strategy profiles of the single-period game $\mathcal{G}'$. Hence the result follows from Lemma 4.3, because the quantity

$$\left[\widehat{\pi}_{\mathbb{O}(X_t)}\big(\mathbb{E}_{\mathbb{P}}(V_{t+1}^* \,|\, \mathcal{F}_t)\big)\right]_k$$

is the value of the game $\mathcal{G}'$ for player k.

Case 2: Assume now that $\widehat{s}' \geq t+1$, so that s' is a valid strategy profile of $\mathcal{G}_{t+1}$. We also deduce that $\tau_t^{*k} \geq t+1$ is a maximin strategy, in view of the induction hypothesis and because it belongs to an optimal equilibrium of $\mathcal{G}_{t+1}$, by (4.5). From Lemma 4.4, we obtain

$$\begin{cases} V_t(s') = \mathbb{E}_{\mathbb{P}}\big(\mathbb{E}_{\mathbb{P}}\big(\widehat{\pi}_{\mathbb{H}_{\mathcal{E}(s')}}(V^*_{\widehat{s}'+1})\,\big|\,\mathcal{F}_{t+1}\big)\,\big|\,\mathcal{F}_t\big) = \mathbb{E}_{\mathbb{P}}\big(V_{t+1}(s')\,\big|\,\mathcal{F}_t\big) & \text{if } \widehat{s}' < T, \\ V_t(s') = \mathbb{E}_{\mathbb{P}}\big(\mathbb{E}_{\mathbb{P}}\big(X_T\,\big|\,\mathcal{F}_{t+1}\big)\,\big|\,\mathcal{F}_t\big) = \mathbb{E}_{\mathbb{P}}\big(V_{t+1}(s')\,\big|\,\mathcal{F}_t\big) & \text{if } \widehat{s}' = T. \end{cases}$$

Using the fact that τ_t^{*k} is a maximin strategy, we obtain

$$\begin{aligned} V_t^k(s') &= \mathbb{E}_{\mathbb{P}}\big(V_{t+1}^k(s')\,\big|\,\mathcal{F}_t\big) = \mathbb{E}_{\mathbb{P}}\big(V_t^k\left([\tau_t^{*k}, s_{-k}]\right)\,\big|\,\mathcal{F}_t\big) \\ &\geq \mathbb{E}_{\mathbb{P}}\big(V_{t+1}^{*k}\,\big|\,\mathcal{F}_t\big). \end{aligned} \tag{4.7}$$

Since $\tau_t^{*k} \geq t+1$, by (4.5), we must have

$$\big[\widehat{\pi}_{\mathbb{O}(X_t)}\big(\mathbb{E}_{\mathbb{P}}(V_{t+1}^*\,|\,\mathcal{F}_t)\big)\big]_k = \big[\widehat{\pi}_{\mathbb{O}(X_t)}\big(\mathbb{E}_{\mathbb{P}}(U_{t+1}\,|\,\mathcal{F}_t)\big)\big]_k > X_t^k. \tag{4.8}$$

By Lemma 4.3 and (4.8), player k does not exercise in the optimal equilibrium of the single-period game $\mathcal{G}'$. Interpreting $\mathbb{E}_{\mathbb{P}}\big(V_{t+1}^{*k}\,|\,\mathcal{F}_t\big)$ as the expected payoff of player k if no one exercises and using the definition of optimal equilibrium, we find that

$$\mathbb{E}_{\mathbb{P}}\big(V_{t+1}^{*k}\,\big|\,\mathcal{F}_t\big) \geq \big[\widehat{\pi}_{\mathbb{O}(X_t)}\big(\mathbb{E}_{\mathbb{P}}(V_{t+1}^*\,|\,\mathcal{F}_t)\big)\big]_k. \tag{4.9}$$

From (4.7) and (4.9), we obtain $V_t^k(s') \geq \big[\widehat{\pi}_{\mathbb{O}(X_t)}\big(\mathbb{E}_{\mathbb{P}}(V_{t+1}^*\,|\,\mathcal{F}_t)\big)\big]_k$, as required.

Case 3: For $\widehat{s}'' \geq t+1$, by arguments similar to the ones used for (4.7), we get

$$V_t^k(s'') = \mathbb{E}_{\mathbb{P}}\big(V_{t+1}^k(s'')\,\big|\,\mathcal{F}_t\big) \leq \mathbb{E}_{\mathbb{P}}\big(V_{t+1}^{*k}\,\big|\,\mathcal{F}_t\big). \tag{4.10}$$

For all $i \neq k$, since $\tau_t^{*i} \geq t+1$, by (4.5), we have

$$\big[\widehat{\pi}_{\mathbb{O}(X_t)}\big(\mathbb{E}_{\mathbb{P}}(V_{t+1}^*\,|\,\mathcal{F}_t)\big)\big]_i = \big[\widehat{\pi}_{\mathbb{O}(X_t)}\big(\mathbb{E}_{\mathbb{P}}(U_{t+1}\,|\,\mathcal{F}_t)\big)\big]_i > X_t^i. \tag{4.11}$$

By Lemma 4.3, if $i \neq k$ then player i does not exercise in the optimal equilibrium of the single-period game $\mathcal{G}'$. Again, interpreting $\mathbb{E}_{\mathbb{P}}\big(V_{t+1}^{*k}\,|\,\mathcal{F}_t\big)$ as the expected payoff of player k if no one exercises and using the definition of an optimal equilibrium, we conclude that

$$\mathbb{E}_{\mathbb{P}}\big(V_{t+1}^{*k}\,\big|\,\mathcal{F}_t\big) \leq \big[\widehat{\pi}_{\mathbb{O}(X_t)}\big(\mathbb{E}_{\mathbb{P}}(V_{t+1}^*\,|\,\mathcal{F}_t)\big)\big]_k. \tag{4.12}$$

Finally, formulae (4.10) and (4.12) imply that

$$V_t^k(s'') \leq \big[\widehat{\pi}_{\mathbb{O}(X_t)}\big(\mathbb{E}_{\mathbb{P}}(V_{t+1}^*\,|\,\mathcal{F}_t)\big)\big]_k$$

and thus the proof is completed. □

Remark 4.6. It is straightforward to further generalize the game by making the weights (hence the coefficients α_i) to be $\mathbb{F}$-adapted processes. The random weights at time $\widehat{s}$ are thus applied when the game is stopped. Theorem 4.5 will still hold with the projection $\widehat{\pi}$ also made time-dependent and randomized. For continuous-time counterparts of Definition 4.1 and Theorem 4.5, the interested reader is referred to Nie and Rutkowski [22] who solved the game problem through a judiciously chosen multi-dimensional BSDE with oblique reflection at the boundary of a stochastic orthant.

Remark 4.7. One could argue that the stopping game described by Definition 4.1 is perhaps not the most obvious generalization of the single-period game of Definition 3.1. A more straightforward, although not necessarily more practically appealing, extension would be for the non-exercising player k to receive the expected payoff based on the following expression

$$X^k_T - w_k(\mathcal{E}(s_t)) \sum_{i\in\mathcal{E}(s_t)} (X^i_{\widehat{s}_t} - X^k_T), \quad \widehat{s}_t < T,$$

when the game is stopped, that is, using X^k_T in the right-hand side of (4.2) instead of the value $V^{*k}_{\widehat{s}_t+1}$ of $\mathcal{G}_{t+1}$. However, even in the deterministic case, this convention does not always produce an optimal equilibrium in pure strategies. Let us consider, for example, the two-period 3-player stopping game with $X_0 = [-1,-1,0]$, $X_1 = [-2,-2,4]$, $X_2 = [0,0,0]$, and $\alpha_1 = \alpha_2 = \alpha_3 = 1/3$. Player 3 will always want to exercise at time 1, while there is a prisoner's dilemma between players 1 and 2 at time 0. This game has two Nash equilibria with different payoffs, but no optimal equilibrium in pure strategies exists.

5. Quitting Games with Redistribution of Payoffs

A *quitting game* is an alternative extension of a general redistribution game to the multi-period set-up. As opposed to a stopping game, a quitting game does not terminate when one of the players decides to exercise. Instead, the non-exercising players continue the game and may exercise at a later date. We focus on the deterministic case here, as the stochastic case does not always produce optimal equilibria (see Example 5.4). A strategy profile $s = [s_1, \dots, s_m]$ thus belongs to $[0,T]^m$.

Remark 5.1. It is worth stressing that the game in introduced below differs from the quitting game studied by Solan and Vieille [30] who dealt with an original variant of a multi-player stopping game.

Definition 5.2. A deterministic m-player *multi-player quitting game* $\mathfrak{G}$ on the interval $[0,T]$ is specified by the inputs:
(a) the vectors $X_t = [X_t^1, \ldots, X_t^m]$ where $X_t \in \mathbb{R}^m$ for $t = 0, 1, \ldots, T$,
(b) the weights $w_k(\mathcal{E}) = \frac{\alpha_k}{1-\sum_{i\in\mathcal{E}}\alpha_i}$ where $\alpha_i > 0$ and $\sum_{i\in\mathcal{M}} \alpha_i \le 1$,
and the following rules:
(i) if player k exercises at time $t \le T-1$ (i.e., $s_k = t \le T - t$), then she receives a payoff of X_t^k,
(ii) if player k does not exercise before time T (i.e., $s_k = T$), then she receives at time T the payoff

$$X_T^k - w_k(\mathcal{E}(s)) \sum_{i\in\mathcal{E}(s)} (X_t^i - X_T^i)$$

where $s = [s_1, \ldots, s_m]$ and $\mathcal{E}(s)$ is the set of players exercising before time T; in other words, the payoff vector $V(s) = [V^1(s), \ldots, V^m(s)]$ is given by

$$V^k(s) = \begin{cases} X_{s_k}^k, & \text{if } s_k < T, \\ X_T^k - w_k(\mathcal{E}(s)) \sum_{i\in\mathcal{E}(s)} (X_{s_i}^i - X_T^i), & \text{if } s_k = T. \end{cases}$$

5.1. *Value and Optimal Equilibrium*

Let us denote $M_t^i = \max_{0\le u\le t} X_u^i$ for $i = 1, 2, \ldots, m$. The following result corresponds to Theorem 4.5.

Theorem 5.3. *The quitting game $\mathfrak{G}$ has a unique value given by the formula $V^* = \widehat{\pi}_{\mathbb{O}(M_{T-1})}(X_T)$ where $\mathbb{O}(M_{T-1})$ is the orthant defined by*

$$\mathbb{O}(M_{T-1}) := \{x \in \mathbb{R}^m : x_i \ge M_{T-1}^i,\ i = 1, 2, \ldots, m\}$$

and the projection $\widehat{\pi}$ is given by Definition 3.19. A possible optimal equilibrium $s^* = [s_1^*, \ldots, s_m^*]$ is given by

$$s_i^* = \min\{t \in [0,T] : [\widehat{\pi}_{\mathbb{O}(M_{T-1})}(X_T)]_i \le X_t^i\}.$$

Proof. Consider a single-period redistribution game $\mathfrak{G}'$ with $X_k' = M_{T-1}^k$ and $P_k' = X_T^k$ for all $k \in \mathcal{M}$. By Theorem 3.15, the vector $V^* = \widehat{\pi}_{\mathbb{O}(M_{T-1})}(X_T)$ is the value of $\mathfrak{G}'$. If player k exercises, then $\widehat{\pi}_{\mathbb{O}(M_{T-1})}(X_T) = X_k' = M_{T-1}^k$. If player k does not exercise, then (recall that $D(s^*) \ge 0$ since s^* is an equilibrium; see Lemma 3.8)

$$V_k^* = X_T^k - w_k(\mathcal{E}(s^*))D(s^*) \le X_T^k.$$

Either way, the strategy s_k^* is well defined and the exercise decision corresponds to the optimal equilibrium of $\mathfrak{G}'$. It is thus easy to check that the equality $V^* = V(s^*)$ holds.

To prove that s^* is an optimal equilibrium of the quitting game $\mathfrak{G}$, we will first check that it is a Nash equilibrium, that is,

$$V_k(s_k^*, s_{-k}^*) \geq V_k(s_k, s_{-k}^*), \quad \forall\, s_k \in \mathcal{S}_k.$$

If $s_k^* < T$ and player k exercises, then she cannot improve her payoff by exercising at another time since $X^k_{s_k^*}$ is maximal. Moreover, she cannot improve by not exercising since s^* corresponds to a Nash equilibrium in $\mathfrak{G}'$.

If instead $s_k^* = T$ and player k does not exercise, it is sufficient to check that she cannot improve by exercising at any time, or $V_k^* \geq M^k_{T-1} = X'_k$. This is certainly true since, once again, s^* corresponds to a Nash equilibrium in $\mathfrak{G}'$. We conclude that s^* is a Nash equilibrium of $\mathfrak{G}$.

To complete the proof, it remains to check that

$$V_k(s_k^*, s_{-k}) \geq V_k(s_k^*, s_{-k}^*), \quad \forall\, s_{-k} \in \mathcal{S}_{-k}.$$

If $s_k^* = t < T$ so that player k exercises at time t, then her payoff is fixed and thus it cannot be decreased by the action of other players. Finally, suppose that $s_k^* = T$ so that k does not exercise. If we write $s = (s_k^*, s_{-k})$ for an arbitrary $s_{-k} \in \mathcal{S}_{-k}$, then we obtain

$$\begin{aligned} V_k(s) &= X_T^k - w_k(\mathcal{E}(s)) \sum_{i \in \mathcal{E}(s)} (X^i_{s_i} - X^i_T) \\ &\geq X_T^k - w_k(\mathcal{E}(s)) \sum_{i \in \mathcal{E}(s)} (M^i_{T-1} - X^i_T) \\ &= P'_k - w_k(\mathcal{E}(s)) \sum_{i \in \mathcal{E}(s)} (X'_i - P'_i) = V_k(s') \end{aligned} \tag{5.1}$$

where by s' we denote the strategy profile in which any exercising player under s chooses to exercise for the maximal payoff X'_i instead. But s' corresponds to a strategy profile in $\mathfrak{G}'$ with player k not exercising and thus, since s^* is an optimal equilibrium of $\mathfrak{G}'$, we have that $V_k(s') \geq V_k(s^*)$. When combined with formula (5.1), it implies that $V_k(s) \geq V_k(s^*)$. □

Example 5.4. In the stochastic case, unfortunately, a quitting game $\mathfrak{G}$ may fail to have an equilibrium in pure strategies. As a counter-example, one may consider the two-period quitting game with $\alpha_i = 1/3$, $i = 1, 2, 3$, the probability space $\Omega = \{\omega_1, \omega_2\}$ endowed with the filtration $\mathcal{F}_0 = \{\emptyset, \Omega\}$, $\mathcal{F}_1 = \mathcal{F}_2 = 2^\Omega$ and the probability $\mathbb{P}(\omega_1) = \mathbb{P}(\omega_2) = 1/2$, and the payoffs given by: $X_0 = [2.1, 3.5, -50]$, $X_2 = [0, 5, -5]$, and

$$X_1(\omega) = \begin{cases} [-50, -50, -5.05], & \omega = \omega_1, \\ [4, -50, -50], & \omega = \omega_2. \end{cases}$$

Since our ultimate goal is to design multi-period stochastic games with redistribution of payoffs for which optimal equilibria in pure strategies exist, we conclude that stochastic quitting gamesare unsuitable as potential models for multi-person financial game options.

Remark 5.5. Note that Theorem 5.3 does not specify the amount of information available to the players regarding the exercise decisions of others. Unlike a stopping game, the strategies in a quitting game can also depend on the observable actions of other players. However, if we denote the information flow available to player k by the filtration $\mathbb{A}^k = \{\mathcal{A}_t^k, t = 0, \ldots, T\}$, then the strategy s_k is an $\mathbb{A}^k$-stopping time. Theorem 5.3 shows that, in the quitting game, regardless of how much any player knows about the actions of others, the value is fixed and an optimal equilibrium attaining the value can be chosen independently of the information available to players.

5.2. *Subgame Perfect Optimal Equilibrium*

The following definition is classic.

Definition 5.6. A strategy profile $s^\dagger$ is said to be *subgame perfect optimal equilibrium* if it is an optimal equilibrium for any reachable subgame of $\mathfrak{G}$.

Let $\mathbb{A} = (\mathbb{A}^1, \mathbb{A}^2, \ldots, \mathbb{A}^m)$ be the information structure of the game. We will show that if $\mathfrak{G} = (\mathcal{G}, \mathbb{A})$ is a *perfect information* quitting game, that is, when for each k the filtration $\mathbb{A}^k$ is generated by the observations of actions of all other players, then a subgame perfect optimal equilibrium can be constructed. Denote the set of exercising player up to time $t-1$ by $\mathcal{E}_{t-1}(s)$ and the set of remaining players by $\mathcal{M}_t^\dagger := \mathcal{M} \setminus \mathcal{E}_{t-1}(s)$. By Lemma 3.7, the quitting subgame $(\mathfrak{G}_t^\dagger, \mathcal{M}_t^\dagger)$ on the interval $[t, T]$ is the quitting game amongst the remaining players $\mathcal{M}_t^\dagger$ with the following modifications:
(i) the exercise payoffs satisfy $X_u^{\dagger k} = X_u^k$ for all $t \le u \le T-1$ and $k \in \mathcal{M}_t^\dagger$,
(ii) the payoffs at T are given by,. for every $k \in \mathcal{M}_t^\dagger$,

$$X_T^{\dagger k} = X_T^k - w_k(\mathcal{E}_{t-1}(s)) \sum_{i \in \mathcal{E}_{t-1}(s)} (X_{s_i}^i - X_T^i),$$

(iii) the weights $w_k^\dagger(\mathcal{E})$ for every $\mathcal{E} \subset \mathcal{M}_t^\dagger$ are specified using the modified coefficients $\alpha_k^\dagger$s, for all $k \in \mathcal{M}_t^\dagger$,

$$\alpha_k^\dagger := w_k(\mathcal{E}_{t-1}(s)) = \frac{\alpha_k}{1 - \sum_{i \in \mathcal{E}_{t-1}(s)} \alpha_i},$$

(iv) a strategy profile $s_t = s_{\mathcal{M}_t^\dagger} \in \mathcal{S}_t := [t, T]^{m_t}$ where $m_t := |\mathcal{M}_t^\dagger|$.

Proposition 5.7. *In a perfect information quitting game $\mathfrak{G}$, a subgame perfect optimal equilibrium exists.*

Proof. Note that a subgame perfect optimal equilibrium $s^\dagger$ is also an optimal equilibrium and it attains the value of the game, that is, $V(s^\dagger) = V^*$. The optimal equilibrium s^* constructed in Theorem 5.3 is not necessarily subgame perfect, as the strategy s_k^* of player k does not take the actions of other players into account. For example, let us assume that another player, say i, deviates from s_i^* by exercising too early. The strategy s_k^* does not adjust to punish the mistake. Consequently, even though the value of V_k^* is still guaranteed, player k misses the chance to guarantee an even higher payoff, created by the sub-optimal deviation of player i.

The subgame perfect optimal equilibrium can be constructed in a recursive way, using the backward induction with respect to both time and remaining set of players. Specifically, for the quitting subgame $(\mathfrak{G}_t^\dagger, \mathcal{M}_t^\dagger)$, the subgame perfect optimal equilibrium $s_t^\dagger = \left[s_t^{\dagger k},\ k \in \mathcal{M}_t^\dagger\right]$ is given by: $s_T^{\dagger k} = T$ and, for all $t < T$,

$$s_t^{\dagger k} = \begin{cases} t, & \text{if } [\widehat{\pi}_{\mathbb{O}_t(M_{T-1})}(X_T^\dagger)]_k = X_t^k, \\ s_{t+1}^{\dagger k}, & \text{if } [\widehat{\pi}_{\mathbb{O}_t(M_{T-1})}(X_T^\dagger)]_k > X_t^k, \end{cases}$$

where $\mathbb{O}_t(M_{T-1}) = \left\{x_i \geq M_{T-1}^i,\ i \in \mathcal{M}_t^\dagger\right\}$ is an orthant in $\mathbb{R}^{m_t}$, the projection $\widehat{\pi}$ is given by Definition 3.19 with an obvious modification for $\mathcal{M}_t^\dagger$, and $s_{t+1}^{\dagger k}$ is the subgame perfect optimal equilibrium strategy for player k in the subgame $(\mathfrak{G}_{t+1}^\dagger, \mathcal{M}_t^\dagger)$. Then $s^\dagger := s_0^\dagger$ is the subgame perfect optimal equilibrium of the quitting game $\mathfrak{G}$ with the perfect information. □

We conclude by noting that in the quitting game with *imperfect information*, where the players' actions are partially or completely hidden to others, it is not always possible to determine the current subgame $(\mathfrak{G}_t^\dagger, \mathcal{M}_t^\dagger)$. Consequently, the subgame perfect optimal equilibria may not exist. However, by using the same idea as above, each player can construct a strategy that is the optimal equilibrium strategy in all observable subgames. And, as was already mentioned, the lack of subgame perfection does not change the value of the quitting game $\mathfrak{G}$.

Acknowledgement. The research of Ivan Guo and Marek Rutkowski was supported under Australian Research Council's Discovery Projects funding scheme (DP120100895).

References

[1] Cvitanić, J. and Karatzas, I. (1996). Backward stochastic differential equations with reflection and Dynkin games, *Ann. Probab.* **24**, pp. 2024–2056.

[2] Bielecki, T. R., Crépey, S., Jeanblanc, M. and Rutkowski, M. (2008). Arbitrage pricing of defaultable game options with applications to convertible bonds, *Quant. Finance* **8**, pp. 795–810.

[3] De Wolf, O. (1999). Optimal strategies in n-person unilaterally competitive games, working paper, Université Catholique de Louvain.

[4] Dolinsky, Y. (2010). Applications of weak convergence for hedging of game options, *Ann. Appl. Probab.* **20**, pp. 1891–1906.

[5] Dolinsky, Y. and Kifer, Y. (2007). Hedging with risk for game options in discrete time, *Stochastics* **79**, pp. 169–195.

[6] Dolinsky, Y. and Kifer, Y. (2008). Binomial approximations of shortfall risk for game options, *Ann. Appl. Probab.* **18**, pp. 1737–1770.

[7] Dynkin, E. B. (1969). Game variant of a problem on optimal stopping, *Soviet Math. Dokl.* **10**, pp. 270–274.

[8] Ekström, E. and Peskir, G. (2008). Optimal stopping games for Markov processes, *SIAM J. Control Optim.* **47**, pp. 684–702.

[9] Ferenstein, E. Z. (2007). Randomized stopping games and Markov market games. *Math. Methods Oper. Res.* **66**, pp. 531–544.

[10] Guo, I. (2013). Competitive multi-player stochastic games with applications to multi-person financial contracts, Ph.D. thesis, University of Sydney.

[11] Guo, I. and Rutkowski, M. (2012). A zero-sum competitive multi-player game, *Demonstratio Math.* **45**, pp. 415–433.

[12] Guo, I. and Rutkowski, M. (2013). Arbitrage pricing of multi-person game contingent claims, working paper, University of Sydney.

[13] Hamadène, S. and Hassani, M. (2011). The multi-player non zero-sum Dynkin game in continuous time, working paper, Université du Maine.

[14] Hamadène, S. and Hassani, M. (2012). The multi-player non zero-sum Dynkin game in discrete time, working paper, Université du Maine.

[15] Hamadène, S. and Zhang, J. (2010). The continuous time nonzero-sum Dynkin game problem and application in game options, *SIAM J. Control Optim.* **48**, pp. 3659–3669.

[16] Kallsen, J. and Kühn, C. (2004). Pricing derivatives of American and game type in incomplete markets, *Finance Stoch.* **8**, pp. 261–284.

[17] Kats, A. and Thisse, J. F. (1992). Unilaterally competitive games, *Internat. J. Game Theory* **21**, pp. 291–299.

[18] Kifer, Y. (2000). Game options, *Finance Stoch.* **4**, pp. 443–463.

[19] Kifer, Y. (2013). Dynkin games and Israeli options, *ISRN Probability and Statistics,* ID 856458, 17 pages.

[20] Laraki, R. and Solan, E. (2005). The value of zero-sum stopping games in continuous time, *SIAM J. Control Optim.* **43**, pp. 1913–1922.

[21] Laraki, R. and Solan, E. (2012). Equilibrium in two-player non-zero-sum Dynkin games in continuous time, *Stochastics*, pp. 1–18, DOI:10.1080/17442508.2012.726222.

[22] Nie, T. and Rutkowski, M. (2013). Multi-player stopping games with redistribution of payoffs and multi-dimensional BSDEs with oblique reflection, working paper, University of Sydney.

[23] Ohtsubo, Y. (1986). Optimal stopping in sequential games with or without a constraint of always terminating, *Math. Oper. Res.* **11**, pp. 591–607.

[24] Ohtsubo, Y. (1987). A nonzero-sum extension of Dynkin's stopping problem, *Math. Oper. Res.* **12**, pp. 277–296.

[25] Peskir, G. (2008). Optimal stopping games and Nash equilibrium, *Theory Probab. Appl.* **53**, pp. 623–638.

[26] Pruzhansky, V. (2011). Some interesting properties of maximin strategies, *Internat. J. Game Theory* **40**, pp. 351–365.

[27] Ramasubramanian, S. (2000). A subsidy-surplus model and the Skorokhod problem in an orthant, *Math. Oper. Res.* **25**, pp. 509–538.

[28] Rosenberg, D., Solan, E. and Vieille, N. (2001). Stopping games with randomized strategies, *Probab. Theory Rel. Fields* **119**, pp. 433–451.

[29] Shmaya, E. and Solan, E. (2004). Two-player nonzero-sum stopping games in discrete time, *Ann. Probab.* **32**, pp. 2733–2764.

[30] Solan, E. and Vieille, N. (2001). Quitting games, *Math. Oper. Res.* **26**, pp. 265–285.

[31] Solan, E. and Vieille, N. (2003). Deterministic multi-player Dynkin games, *J. Math. Econ.* **39**, pp. 911–929.

[32] Touzi, N. and Vieille, N. (2002). Continuous-time Dynkin games with mixed strategies, *SIAM J. Control Optim.* **41**, pp. 1073–1088.

[33] Yasuda, M. (1985). On a randomized strategy in Neveu's stopping problem, *Stochastic Process. Appl.* **21**, pp. 159–166.

6. Appendix

The proof of Proposition 3.5 hinges on the following two lemmas.

Lemma 6.1. *Given a subset $\mathcal{E} \subset \mathcal{M}$ with $0 \leq |\mathcal{E}(s)| \leq m-2$ and $i, j \notin \mathcal{E}(s)$, we set $\mathcal{E}' = \mathcal{E} \cup \{j\}$. If s and s' are strategy profiles with $\mathcal{E}(s) = \mathcal{E}$ and $\mathcal{E}(s') = \mathcal{E}'$, then*

$$\begin{aligned} V_i(s) - V_i(s') + w_i(\mathcal{E}')(V_j(s) - V_j(s')) \\ = \big(w_i(\mathcal{E}')[1 - w_j(\mathcal{E})] - w_i(\mathcal{E})\big)D(s). \end{aligned} \tag{6.1}$$

Proof. Note that player j exercises in s', but not in s. We have that $D(s') = D(s) + X_j - P_j$. Furthermore,

$$\begin{aligned} V_i(s) - V_i(s') &= w_i(\mathcal{E}')D(s') - w_i(\mathcal{E})D(s) \\ &= (w_i(\mathcal{E}') - w_i(\mathcal{E}))D(s) + w_i(\mathcal{E}')(X_j - P_j) \end{aligned}$$

and

$$\begin{aligned} w_i(\mathcal{E}')(V_j(s) - V_j(s')) &= w_i(\mathcal{E}')(P_j - w_j(\mathcal{E})D(s) - X_j) \\ &= -w_i(\mathcal{E}')w_j(\mathcal{E})D(s) - w_i(\mathcal{E}')(X_j - P_j). \end{aligned}$$

Adding these expressions yields the desired result. □

Lemma 6.2. *The game $\mathcal{G}$ is WUC for all $X, P \in \mathbb{R}^m$ if and only if the following conditions hold:*
(i) for any $\mathcal{E} \subset \mathcal{M}$ such that $1 \le |\mathcal{E}| \le m-2$ and $i, j \notin \mathcal{E}$, we have

$$w_i(\mathcal{E} \cup \{j\})(1 - w_j(\mathcal{E})) = w_i(\mathcal{E}),$$

(ii) for any $\mathcal{E}' \subset \mathcal{M}$ such that $1 \le |\mathcal{E}'| \le m-1$ and $i \notin \mathcal{E}'$, we have $w_i(\mathcal{E}') > 0$.

Proof. We prove the statement in three steps.
Step 1: [WUC $\Rightarrow$ (i)] Take s and s' to be strategy profiles with $\mathcal{E}(s) = \mathcal{E}$ and $\mathcal{E}(s') = \mathcal{E}' = \mathcal{E} \cup \{j\}$. If the game $\mathcal{G}$ is WUC for all $X, P \in \mathbb{R}^m$, then

$$V_j(s) = V_j(s') \implies V_i(s) = V_i(s').$$

Using Lemma 6.1, we thus obtain

$$\big(w_i(\mathcal{E}')(1 - w_j(\mathcal{E})) - w_i(\mathcal{E})\big)D(s) = 0.$$

When $|\mathcal{E}| \ge 1$, we can choose $X, P \in \mathbb{R}^m$ so that $D(s) \ne 0$. Hence $w_i(\mathcal{E}')(1 - w_j(\mathcal{E})) = w_i(\mathcal{E})$, and thus condition (i) is proven.
Step 2: [WUC and (i) $\Rightarrow$ (ii)] Now assume condition (i) holds, hence $w_i(\mathcal{E}')(1 - w_j(\mathcal{E})) - w_i(\mathcal{E}) = 0$ if $|\mathcal{E}| \ge 1$. Note that if $|\mathcal{E}| = 0$, then $D(s) = 0$. In either case, (6.1) always simplifies to

$$V_i(s) - V_i(s') = w_i(\mathcal{E}')(V_j(s') - V_j(s)).$$

But $\mathcal{G}$ being WUC requires that the inequality $V_j(s') - V_j(s) > 0$ implies that $V_i(s) - V_i(s') \ge 0$. Since the weights are required to be non-zero, we have $w_i(\mathcal{E}') > 0$ for all $1 \le |\mathcal{E}'| \le m-1$ and thus condition (ii) holds as well.
Step 3: [(i) and (ii) $\Rightarrow$ WUC] As before, condition (i) yields

$$V_i(s) - V_i(s') = w_i(\mathcal{E}')(V_j(s') - V_j(s)).$$

When condition (ii) also holds, it is easy to check that the game has indeed the WUC property. □

Proof of Proposition 3.5. Recall that we consider here the case where $m \geq 4$. It is sufficient to completely solve the system under conditions (i)–(ii) of Lemma 6.2. By condition (ii), we have $0 < w_i(\{j\}) < 1$ for all $i \neq j$. By condition (i),

$$\frac{w_i(\{j\})}{1 - w_k(\{j\})} = w_i(\{j,k\}) = \frac{w_i(\{k\})}{1 - w_j(\{k\})} \iff \frac{w_i(\{j\})}{w_i(\{k\})} = \frac{1 - w_k(\{j\})}{1 - w_j(\{k\})}.$$

Since only the left-hand side depend on i, we have, for i, j, k, l all distinct,

$$\frac{w_i(\{j\})}{w_i(\{k\})} = \frac{w_l(\{j\})}{w_l(\{k\})}.$$

This system of equations is known to admit the parametric solution $w_i(\{j\}) = \alpha_i \beta_j$ with $\alpha_i, \beta_i \neq 0$. Substituting back, we obtain

$$\frac{\alpha_i \beta_j}{1 - \alpha_k \beta_j} = \frac{\alpha_i \beta_k}{1 - \alpha_j \beta_k} \iff \alpha_j + \frac{1}{\beta_j} = \alpha_k + \frac{1}{\beta_k} \iff \alpha_i + \frac{1}{\beta_i} = c$$

where c is a constant for all i. Solving for β_j yields $w_i(\{j\}) = \alpha_i \beta_j = \frac{\alpha_i}{c - \alpha_j}$. After scaling all α_is by a factor of $1/c$, we get

$$w_i(\{j\}) = \frac{\alpha_i}{1 - \alpha_j}.$$

Substituting into condition (i), while recursively incrementing the size of $\mathcal{E}$, we obtain

$$w_k(\mathcal{E}) = \frac{\alpha_k}{1 - \sum_{i \in \mathcal{E}} \alpha_i}.$$

Condition (ii) adds the following restrictions: $\alpha_k > 0$ and $\sum_{i \neq k} \alpha_i < 1$. This solution can easily be checked to always satisfy conditions (i) and (ii) of Lemma 6.2. □

Remark 6.3. When $m = 2$ or $m = 3$, the same arguments as in the proof above show that if the weights are defined as in Proposition 3.5, then the game is always WUC. The converse is not necessarily true, however. Specifically, there exist other choices of weights, which also lead to WUC games. They may be found by solving the system described by conditions (i) and (ii) of Lemma 6.2.

Let us first examine the case $m = 2$. Then condition (i) of Lemma 6.2 is not applicable and thus the required weights are precisely given by condition (ii), that is, $w_1(\{2\}) > 0$ and $w_2(\{1\}) > 0$. It is possible to parameterize them by $w_1(\{2\}) = \frac{\alpha_1}{1-\alpha_2}$ and $w_2(\{1\}) = \frac{\alpha_2}{1-\alpha_1}$, but in this case there is no reason for α_1 and α_2 to lie between 0 and 1.

It remains to study the case when $m = 3$. After writing $w_{ij} = w_i(\{j\})$, conditions (i) and (ii) of Lemma 6.2 can be reduced to: $0 < w_{ij} < 1$ and

$$w_{12}(1 - w_{23}) = w_{13}(1 - w_{32}),$$
$$w_{23}(1 - w_{31}) = w_{21}(1 - w_{13}),$$
$$w_{31}(1 - w_{12}) = w_{32}(1 - w_{21}).$$

We may parameterize the left-hand side variables by $w_{12} = \frac{\alpha_1}{1-\alpha_2}$, $w_{23} = \frac{\alpha_2}{1-\alpha_3}$ and $w_{31} = \frac{\alpha_3}{1-\alpha_1}$. The constraint $0 < w_{ij} < 1$ translates to $\alpha_1, \alpha_2, \alpha_3 > 0$ and $\alpha_1 + \alpha_2 < 1$, $\alpha_2 + \alpha_3 < 1$, $\alpha_1 + \alpha_3 < 1$. Now, the required system of equations is given by: $0 < w_{ij} < 1$ and

$$\frac{\alpha_1(1 - \alpha_2 - \alpha_3)}{(1 - \alpha_2)(1 - \alpha_3)} = w_{13}(1 - w_{32}),$$
$$\frac{\alpha_2(1 - \alpha_1 - \alpha_3)}{(1 - \alpha_1)(1 - \alpha_3)} = w_{21}(1 - w_{13}),$$
$$\frac{\alpha_3(1 - \alpha_1 - \alpha_2)}{(1 - \alpha_1)(1 - \alpha_2)} = w_{32}(1 - w_{21}).$$

It is clear that if w_{13} is known in terms of α_1, α_2 and α_3, then the other variables, w_{32} and w_{21}, are uniquely determined. Eliminating w_{32} and w_{21}, we arrive at a quadratic equation in w_{13}. The exact quadratic is omitted for brevity, but its two solutions can be easily verified to be:

$$(w_{13}, w_{21}, w_{32}) = \left(\frac{\alpha_1}{1 - \alpha_3}, \frac{\alpha_2}{1 - \alpha_1}, \frac{\alpha_3}{1 - \alpha_2}\right)$$

or

$$\left(\frac{1 - \alpha_2 - \alpha_3}{1 - \alpha_3}, \frac{1 - \alpha_1 - \alpha_3}{1 - \alpha_1}, \frac{1 - \alpha_1 - \alpha_2}{1 - \alpha_2}\right).$$

The first solution coincides with the conclusion in Proposition 3.5. The second solution leads to the following set of conditions on weights: $w_i(\{j\}) > 0$ and

$$w_1(\{3\}) + w_2(\{3\}) = w_1(\{2\}) + w_3(\{2\}) = w_2(\{1\}) + w_3(\{1\}) = 1,$$
$$w_1(\{2, 3\}) = w_2(\{1, 3\}) = w_3(\{1, 2\}) = 1.$$

This describes a class of zero-sum games, which strictly contains the class of zero-sum redistribution games (when $m = 3$), which are discussed in Section 3.4.1 (see also [11]).

Proposition 6.4. *Let the weights be defined by* (3.1). *If* $\sum_{i \in \mathcal{M}} \alpha_i > 1$, *then there exist* $X, P \in \mathbb{R}^m$ *such that no equilibrium exists in pure strategies. In particular, it suffices to take* $X = [0, 0, \ldots, 0]$ *and* $P = [1, -1, 0, \ldots, 0]$.

Proof. We first note that, for $i \neq j$,

$$w_i(\mathcal{M} \setminus \{i\}) = \frac{\alpha_i}{1 - \sum_{j \neq i} \alpha_j} > 1.$$

Moreover, as before, $w_i(\mathcal{E}) < 1$ if $|\mathcal{E}| \leq m - 2$. We will simply check all the possibilities. If $\{3, \ldots, m\} \subseteq \mathcal{E}$ or if there are only two players, then we need to examine the following cases:

- If both players 1 and 2 exercise, then $V_1 = V_2 = 0$, but player 2 can receive $w_2(\mathcal{M} \setminus \{2\}) - 1 > 0$ if she does not exercise.
- If only player 1 exercises, then $V_1 = 0$, but player 1 can receive $P_1 = 1$ if she does not exercise.
- If only player 2 exercises, then $V_1 = 1 - w_1(\mathcal{M} \setminus \{1\}) < 0$, but player 1 can receive $X_1 = 0$ if she also exercises.
- If neither player 1 nor 2 exercises, then $V_2 = -1$, but player 2 can receive $X_2 = 0$ if she also exercises.

If some player $k \neq 1, 2$ does not exercise, so that $k \notin \mathcal{E}$, then we have the following cases:

- If both players 1 and 2 exercise, then $V_1 = V_2 = 0$, but $|\mathcal{E} \setminus \{1\}| \leq m - 2$ and player 1 can receive $1 - w_1(\mathcal{E} \setminus \{1\}) > 0$ if she does not exercise.
- If only player 1 exercises, then $|\mathcal{E}| \leq m - 2$ and $V_2 = w_1(\mathcal{E}) - 1 < 0$, but player 2 can receive $X_2 = 0$ if she exercises.
- If only player 2 exercises, then $V_k = 0 - w_k(\mathcal{E}) < 0$, but player k can receive $X_k = 0$ if she also exercises.
- If neither player 1 nor 2 exercises, then $V_2 = -1$, but player 2 can receive $X_2 = 0$ if she also exercises.

We conclude that no equilibrium exists in pure strategy profiles. □

A Note on BSDEs with Singular Driver Coefficients

Monique Jeanblanc*, Anthony Réveillac†

Abstract. In this note we study a class of BSDEs which admits a particular singularity in the driver. More precisely, we assume that the driver is not integrable and degenerates when approaching to the terminal time of the equation.

1. Introduction

Since the seminal works of Bismut [2] and of Pardoux and Peng [13], a lot of attention has been given to the study of Backward Stochastic Differential Equations (BSDEs) as this object naturally arises in stochastic control problems and was found to be an ad hoc tool for many financial applications as illustrated in the famous guideline paper [9]. Recall that a BSDE takes the following form:

$$Y_t = Y_T - \int_t^T f(s, Y_s, Z_s)ds - \int_t^T Z_s dW_s, \quad t \in [0, T],$$

where W is a multi-dimensional Brownian motion. The historical natural assumption for providing existence and uniqueness (in the appropriate spaces) is to assume the driver f to be Lipschitz plus some integrability conditions on the terminal condition. However, in applications one may deal with drivers which are not Lipschitz continuous, and which exhibit e.g. a quadratic growth in z (in the context of incomplete markets in Finance), or only some monotonicity in the y-variable. One way of relaxing the Lipschitz growth condition in y is the so-called stochastic Lipschitz assumption which basically consists in replacing the usual Lipschitz constant by a stochastic process satisfying appropriate integrability conditions. As noted in Section

*Université d'Evry Val d'Essonne, Laboratoire Analyse et Probabilités,
Email: monique.jeanblanc@univ-evry.fr
†CEREMADE UMR CNRS 7534, Université Paris Dauphine,
Email: anthony.reveillac@ceremade.dauphine.fr

2.1.2 "Pathology" in [8], even in the stochastic linear framework, one has to be very careful when relaxing the integrability conditions on the driver of the equation. As an illustration, consider the following example presented in [8] (cf. [8, (2.9)]):

$$Y_t = 0 + \int_t^T [rY_s + \sigma Z_s + \gamma Y_s (e^{\gamma(T-s)} - 1)^{-1}] ds + \int_t^T Z_s dW_s, \quad t \in [0,T], \tag{1.1}$$

where W is a one-dimensional Brownian motion, $r, \sigma, \gamma > 0$ and T is a fixed positive real number. It is proved in [8] that the BSDE (1.1) has an infinite number of solutions. Note that here the driver is not Lipschitz continuous in y due to the exploding term $(e^{\gamma(T-t)} - 1)^{-1}$ as t goes to T, and completely escapes the existing results of the literature.

The aim of this note is to elaborate on the pathology mentioned in [8] and to try to understand better what kind of behavior can appear as soon as the usual integrability conditions are relaxed. In light of Example (1.1), multiple solutions is one of the behaviour which can be observed. However, is it the only type of problem that can occur ? For instance is it clear that existence is guaranteed? This note is an attempt in this direction and is motivated by the work in preparation [11] where equations with this specific pathology appear naturally in the financial application under interest in [11].

We proceed as follows. First we make precise the context of our study and we explain what is the notion of solution we use for dealing with non-integrable drivers. Then we deal with the particular case of affine equations in Section 3. These equations already allow us to present several type of pathologic behaviour. We then study in Section 4 a class of non-linear drivers which will be of interest for a specific financial application presented in [11]. In particular, in our main result Theorem 4.4 we provide an existence and uniqueness result under a monotonicity assumption on the mapping f in (2.1) defined below.

2. Preliminaries

In this note T denotes a fixed positive real number and d a given positive integer. We set $(W_t)_{t\in[0,T]} := (W_t^1, \ldots, W_t^d)_{t\in[0,T]}$ a d-dimensional standard Brownian motion defined on a filtered probability space $(\Omega, \mathcal{F}, \mathbb{F} := (\mathcal{F}_t)_{t\in[0,T]}, \mathbb{P})$ where $\mathbb{F}$ denotes the natural filtration of W (completed and right-continuous) and $\mathcal{F} = \mathcal{F}_T$. Throughout this paper "$\mathbb{F}$-predictable" (rep. $\mathbb{F}$-adapted) processes will be referred to predictable (resp. adapted) processes. For later use we set for $p \geq 1$:

$$\mathbb{S}^p := \Big\{(Y_t)_{t\in[0,T]} \text{ continuous adapted one dimensional process },\\ \mathbb{E}\left[\sup_{t\in[0,T]} |Y_t|^p\right] < +\infty\Big\},$$

$$\mathbb{H}^p(\mathbb{R}^m) := \Big\{(Z_t)_{t\in[0,T]} \text{ predictable } m\text{-dimensional process },\\ \mathbb{E}\left[\left(\int_0^T \|Z_t\|^2 dt\right)^{p/2}\right] < +\infty\Big\},$$

where $\|\cdot\|$ denotes the Euclidian norm on $\mathbb{R}^m$ ($m \geq 1$). For any element Z of $\mathbb{H}^1(\mathbb{R}^d)$, we set $\int_0^\cdot Z_s dW_s := \sum_{i=1}^d \int_0^\cdot Z_s^i dW_s^i$. We also set $L^p := L^p(\Omega, \mathcal{F}_T, \mathbb{P})$.

Let $\lambda : (\lambda_t)_{t\in[0,T]}$ be a one-dimensional non-negative predictable process. For convenience we set $\Lambda_t := \int_0^t \lambda_s ds$, $t \in [0,T]$. We make the following

Standing assumption on λ:

$$\Lambda_t < +\infty, \ \forall t < T, \ \text{ and } \Lambda_T = +\infty, \ \mathbb{P}-a.s.$$

The typical example we have in mind is a coefficient λ of the form $\lambda_t := (e^{\gamma(T-t)} - 1)^{-1}$ as in the introducing example (1.1), or when λ is the intensity process related to a prescribed random time τ in the context of enlargement of filtration as presented in [11]. In this note, we aim in studying BSDEs of the form:

$$Y_t = A - \int_t^T [\varphi_s + \lambda_s f(Y_s)]ds - \int_t^T Z_s dW_s, \quad t \in [0,T], \tag{2.1}$$

where A is a regular enough $\mathcal{F}_T$-measurable random variable, $f : \mathbb{R} \to \mathbb{R}$ is a deterministic map and φ is a predictable processes with some integrability conditions to be specified. Before going further, we would like to stress that in contradistinction to the classical case where λ is bounded (and A and φ are square-integrable), the space $\mathbb{S}^2 \times \mathbb{H}^2(\mathbb{R}^d)$ is no more the natural space for solutions of our BSDEs. For instance if $f(x) := x$, the fact that (Y,Z) belongs to $\mathbb{S}^2 \times \mathbb{H}^2(\mathbb{R}^d)$ does not guarantee that

$$\mathbb{E}\left[\int_0^T |\lambda_s Y_s|^p ds\right] < +\infty$$

for some $p \geq 1$ (which would be immediately satisfied with $p = 2$ if λ were bounded) leading to a possible definition problem for the term $\int_0^t \lambda_s Y_s ds$ in equation (2.1). For this reason we make very precise the notion of solution in our context.

Definition 2.1 (Solution). *Let A be an element of L^1 and $f : \Omega \times [0,T] \times \mathbb{R} \times \mathbb{R}^d \to \mathbb{R}$ such that for any (y,z) in $\mathbb{R} \times \mathbb{R}^d$ the stochastic process $(t,\omega) \mapsto f(t,y,z)$ (where as usual we omit the ω-variable in the expression of f) is progressively measurable. We say that a pair of predictable processes (Y,Z) with values in $\mathbb{R} \times \mathbb{R}^d$ is a solution to the BSDE*

$$Y_t = A - \int_t^T f(s, Y_s, Z_s) ds - \int_t^T Z_s dW_s, \quad t \in [0,T], \tag{2.2}$$

if

$$\mathbb{E}\left[\int_0^T |f(t, Y_t, Z_t)| dt + \left(\int_0^T \|Z_t\|^2 dt\right)^{1/2}\right] < +\infty, \tag{2.3}$$

and Relation (2.2) is satisfied for any t in $[0,T]$, $\mathbb{P}$-a.s.

Remark 2.2. This notion of solution is related to the theory of L^1-solution (see e.g. [3, Definition 2.1] or [4, 5]) where in Relation (2.3) the expectation is replaced by a $\mathbb{P}$-a.s. criterion. The fact that Z is an element of $\mathbb{H}^1(\mathbb{R}^d)$ implies that the martingale $\int_0^\cdot Z_s dW_s$ is uniformly integrable. Combining this property with the $(\Omega \times [0,T], \mathbb{P} \otimes dt)$-integrability of $f(\cdot, Y, Z)$, it immediately follows that the solution process Y is of class (D) (which then finds similarities with the notion of solution used in [3]).

Remark 2.3. We would like to stress that even in the case where the terminal condition A is in L^2 we do not require Y to be an element of $\mathbb{S}^2$. This fact bears some similarities with the papers [4, 5] and with [3, Section 6].

Remark 2.4 (Classical L^2 setting). *If f is uniformly (in time) Lipschitz in (y,z) and if $\mathbb{E}\left[|A|^2 + \int_0^T |f(s,0,0)|^2 ds\right] < +\infty$, then the fact that there exists (Y,Z) in $\mathbb{S}^2 \times \mathbb{H}^2(\mathbb{R}^d)$ satisfying (2.2) implies that the process $f(\cdot, Y_\cdot, Z_\cdot)$ is in $\mathbb{H}^2(\mathbb{R}^d)$ and thus Relation (2.3) is satisfied.*

Another important issue in our context is uniqueness. The uniqueness for the Z component will be understood in the $\mathbb{H}^1(\mathbb{R}^d)$ sense. Concerning the Y component, since we do not impose Y to belong to $\mathbb{S}^1$ we will say that $Y^1 = Y^2$ if the processes are indistinguishable (by definition of a solution, both processes are continuous, and hence uniqueness boils down to require Y^1 to be a modification of Y^2). This definition for uniqueness in our very special setting coincides with the notion of uniqueness with respect to a particular norm. More precisely, according to Remark 2.2 a solution process Y is of class (D). This space can be naturally equipped with the norm $\|\cdot\|_{(D)}$ defined as[‡]:

$$\|X\|_{(D)} := \sup_{\tau \in \mathcal{T}} \mathbb{E}[|X_\tau|], \quad X \text{ of class } (D),$$

where $\mathcal{T}$ denotes the set of stopping time smaller or equal to T. By [6, Theorem IV.86], uniqueness with respect to the norm $\|\cdot\|_{(D)}$ is equivalent to indistinguishability.

From now on, by solution to a BSDE we mean a solution in the sense of Definition 2.1. For any pair of ($\mathcal{F}_T$-measurable) random variables (A, B), we write $A \not\equiv B$ if $\mathbb{P}[A \neq B] > 0$. Similarly, $A = B$, $\mathbb{P}$-a.s. will be denoted as $A \equiv B$. Throughout this paper C will denote a generic constant which can differ from line to line.

3. Affine Equations with Exploding Coefficients

As the reader will figure out later, it seems pretty complicated to define a general theory since many situations (non-existence, non-uniqueness) can be found under our assumption on λ for BSDEs of the form (2.1). These very different behaviours can be clearly illustrated by studying affine equations, that is when f in (2.1) stands for the identity (or minus the identity). In some sense, our results find immediate counterparts in the deterministic realm while considering the corresponding ODEs when all the coefficients of the equation are deterministic. However, for this latter case, techniques of time-reversion can be employed to provide immediate results which unfortunately can not be applied in the stochastic framework due to the measurability feature of the solution to a BSDE calling for different techniques.

In this section, we consider stochastic affine BSDEs of one of the following forms:

$$Y_t = A - \int_t^T (\varphi_s - \lambda_s Y_s) ds - \int_t^T Z_s dW_s; \quad t \in [0, T], \tag{3.1}$$

[‡]This norm is referred as $\|\cdot\|_1$ in [7, Definition VI.20], we do not use this notation here to avoid any confusion.

$$Y_t = A - \int_t^T (\varphi_s + \lambda_s Y_s)ds - \int_t^T Z_s dW_s; \quad t \in [0,T]. \tag{3.2}$$

We start with Equation (3.1).

Proposition 3.1. *Let A be in L^1 and $\varphi := (\varphi_t)_{t\in[0,T]}$ be an element of $\mathbb{H}^1(\mathbb{R})$. The Brownian BSDE*

$$dY_t = (\varphi_t - \lambda_t Y_t)dt + Z_t dW_t; \quad Y_T = A \tag{3.3}$$

admits no solution if $A \not\equiv 0$. If $A \equiv 0$, the BSDE (3.3) may admit infinitely many solutions.

Proof. Step 1: non-existence of solution if $A \not\equiv 0$

Let (Y, Z) be a solution to (3.3). Assume there exists a set $\mathcal{A}$ in $\mathcal{F}_T$ such that $A > 0$ on $\mathcal{A}$. By definition of a solution, it holds that

$$\int_0^T |\lambda_s Y_s| ds < \infty, \quad \mathbb{P}\text{-a.s.} \tag{3.4}$$

since

$$\int_0^T |\lambda_s Y_s| ds \le \int_0^T |\varphi_s - \lambda_s Y_s| ds + \int_0^T |\varphi_s| ds.$$

For ω in $\mathcal{A}$, let $t_0(\omega) := \sup\{t \in [0,T],\ Y_t(\omega) < A/2\}$. By continuity of Y and the fact that $Y_T = A$, for $\mathbb{P}$-almost all ω in $\mathcal{A}$, $t_0(\omega) < T$ and $Y_t(\omega)\mathbf{1}_{[t_0(\omega),T]}(t) \ge A/2$. Note that t_0 is not a stopping time but only a $\mathcal{F}_T$-measurable random variable. As a consequence, on $\mathcal{A}$, it holds that

$$\int_0^T |\lambda_s Y_s| ds \ge \int_{t_0}^T \lambda_s Y_s ds \ge A/2 \underbrace{\int_{t_0}^T \lambda_s ds}_{=+\infty},$$

which contradicts (3.4). As a consequence, $A \le 0$, $\mathbb{P}$-a.s.. Similarly, one proves that $A \ge 0$, $\mathbb{P}$-a.s..

Step 2: Multiplicity of solutions if $A \equiv 0$

If $A \equiv 0$, we will provide examples of non-uniqueness of solution. Remark that if $(\mathcal{Y}, \mathcal{Z})$ is a particular solution to the BSDE and that (Y, Z) is a solution to the (fundamental) BSDE:

$$dY_t = -\lambda_t Y_t dt + Z_t dW_t, \quad Y_T = 0, \tag{3.5}$$

then as for ODEs, the sum of any of these fundamental solutions and $\mathcal{Y}$ is a solution to (3.3) (together with the sum of the associated Z processes). In addition, Equation (3.5) admits an infinite number of solutions (like $Y_t = Y_0 e^{-\Lambda_t}$ and $Z \equiv 0$ which is an adapted continuous solution to the BSDE satisfying $\mathbb{E}\left[\int_0^T \lambda_s |Y_s| ds\right] = |Y_0|$ for any chosen real number Y_0). An example of particular solution can be given by the process $\mathcal{Y}_t := -\mathbb{E}\left[\int_t^T e^{\int_t^s \lambda_u du} \varphi_s ds | \mathcal{F}_t\right]$ if it is well-defined, such that Relation (2.3) is satisfied. In that case, the existence of $\mathbb{E}\left[\int_t^T \varphi_s e^{\Lambda_s} ds | \mathcal{F}_t\right]$ entails that it converges to 0 as t goes to T, and hence that $\mathcal{Y}_T = 0$. One can check that $\mathcal{Y}$ together with the process $\mathcal{Z} := \tilde{Z} e^{-\Lambda}$ is solution to (3.3), where $\tilde{Z}$ is such that $\mathbb{E}\left[\int_0^T \varphi_s e^{\Lambda_s} ds | \mathcal{F}_t\right] = \mathbb{E}\left[\int_0^T \varphi_s e^{\Lambda_s} ds\right] - \int_0^t \tilde{Z}_s dW_s$ $(t \in [0,T])$. We conclude the proof with an example: set $\varphi_t := e^{-\Lambda_t}$. With this choice, the process $\mathcal{Y}$ satisfies all the requirements above providing infinitely many solutions to (3.3). □

Note that in the previous proof the non-existence when $A \not\equiv 0$ relies on the assumption that $\int_0^T |\varphi_s| ds < +\infty$, $\mathbb{P}-a.s.$ If the latter is not satisfied, one may find existence of solutions for $A \not\equiv 0$ as the following proposition illustrates in the deterministic case.

Proposition 3.2. *Let A be a given constant and $\varphi := (\varphi_t)_{t\in[0,T]}$ be a deterministic map. We assume that λ is a deterministic function such that $\Lambda_t = \int_0^t \lambda_s ds < +\infty$, for $t < T$, and $\int_0^T \lambda_s ds = +\infty$. Then*

(i) If $e^{-\Lambda_t} \int_0^t e^{\Lambda_s} \varphi_s ds$ converges to C when t goes to T, then the ODE

$$dY_t = (\varphi_t - \lambda_t Y_t)dt; \quad Y_T = A \tag{3.6}$$

admits no solution if $A \neq C$. If $A = C$, the ODE (3.6) admits infinitely many solutions given by $Y_t = e^{-\Lambda_t}\left(Y_0 + \int_0^t e^{\Lambda_s} \varphi_s ds\right)$ provided that $\int_0^T |\varphi_t - \lambda_t Y_t| dt < \infty$.

(ii) If $e^{-\Lambda_t} \int_0^t e^{\Lambda_s} \varphi_s ds$ does not converge, the ODE (3.6) has no solution.

Remark 3.3. Note that the assumption in (i) of Proposition 3.2 when $C = 0$, can be met only if $\int_0^T |\varphi_s| ds = +\infty$. Indeed, assume that $\int_0^T |\varphi_s| ds < \infty$.

Let $\varepsilon > 0$ and $t < T$. We have:

$$e^{-\Lambda_t}\left|\int_0^t e^{\Lambda_s}\varphi_s ds\right| \leq e^{-\Lambda_t}\int_0^{t-\varepsilon} e^{\Lambda_s}|\varphi_s|ds + e^{-\Lambda_t}\int_{t-\varepsilon}^t e^{\Lambda_s}|\varphi_s|ds$$
$$\leq e^{-\Lambda_t}e^{\Lambda_{t-\varepsilon}}\int_0^T |\varphi_s|ds + \int_{t-\varepsilon}^t |\varphi_s|ds.$$

Hence as t goes to T, we have that $\lim_{t\to T} e^{-\Lambda_t}\left|\int_0^t e^{\Lambda_s}\varphi_s ds\right| \leq \int_{T-\varepsilon}^T |\varphi_s|ds$, and hence

$$\lim_{t\to T} e^{-\Lambda_t}\left|\int_0^t e^{\Lambda_s}\varphi_s ds\right| = 0,$$

which contradicts the assumption of (i).

Remark 3.4. Since λ is unbounded, assuming A, λ and φ to be deterministic in Equation (3.1) does not lead to deterministic solutions (and so differs from the ODE framework of Proposition 3.2) as the following example illustrates. Assume $A \equiv 0$, $\varphi \equiv 0$ and λ is a deterministic mapping. Then for any element $\beta := (\beta_t)_{t\in[0,T]}$ in $\mathbb{H}^1(\mathbb{R}^d)$, the pair of adapted processes (Y, Z) defined as:

$$Y_t := Y_0 e^{-\Lambda_t} + e^{-\Lambda_t}\int_0^t \beta_s dW_s, \quad Z_t := e^{-\Lambda_t}\beta_t, \quad t \in [0,T]$$

is a solution to (3.1). This provides in turn a generalization of the fundamental solution to Equation (3.5).

We continue with the BSDE:

$$dY_t = (\varphi_t + \lambda_t Y_t)dt + Z_t dW_t; \quad Y_T = A.$$

Proposition 3.5. *Let A be in L^1 and $\varphi := (\varphi_t)_{t\in[0,T]}$ be a bounded predictable process. The Brownian BSDE*

$$dY_t = (\varphi_t + \lambda_t Y_t)dt + Z_t dW_t; \quad Y_T = A \tag{3.7}$$

admits no solution unless $A \equiv 0$. If $A \equiv 0$, then the BSDE admits a unique solution.

Proof. Let (Y, Z) be a solution and set $\tilde{Y} := Ye^{-\Lambda} - \int_0^\cdot e^{-\Lambda_s}\varphi_s ds$. We have that

$$d\tilde{Y}_t = e^{-\Lambda_t}Z_t dW_t, \text{ and } \tilde{Y}_T = -\int_0^T e^{-\Lambda_s}\varphi_s ds.$$

Hence $\tilde{Y}$ is a L^1-martingale and

$$\tilde{Y}_t = -\mathbb{E}\left[\int_0^T e^{-\Lambda_s}\varphi_s ds|\mathcal{F}_t\right], \quad t \in [0,T],$$

leading to

$$Y_t = -\mathbb{E}\left[\int_t^T e^{-\int_t^s \lambda_u du}\varphi_s ds|\mathcal{F}_t\right], \quad t \in [0,T]. \tag{3.8}$$

In particular, $Y_T = 0$. Indeed, since φ is bounded

$$e^{\Lambda_t}\left|\int_t^T e^{-\Lambda_s}\varphi_s ds\right| \le e^{\Lambda_t}e^{-\Lambda_t}\int_t^T |\varphi_s|ds \le \|\varphi\|_\infty(T-t).$$

This proves that there is no solution to the equation unless $A \equiv 0$. We now assume that $A \equiv 0$. In that case, we prove that the process given by (3.8) together with a suitable process Z is a solution to the BSDE. We begin with the integrability condition

$$\mathbb{E}\left[\int_0^T |\lambda_s Y_s|ds\right] < +\infty.$$

We have

$$\mathbb{E}\left[\int_0^T |\lambda_s Y_s|ds\right] = \mathbb{E}\left[\int_0^T \left|\lambda_s \mathbb{E}\left[\int_s^T e^{-\int_s^u \lambda_r dr}\varphi_u du|\mathcal{F}_s\right]\right| ds\right]$$

$$\le \mathbb{E}\left[\int_0^T \lambda_s e^{\Lambda_s}\int_s^T e^{-\Lambda_u}|\varphi_u|duds\right]$$

$$= \mathbb{E}\left[\lim_{s\to T, s<T}[e^{\Lambda_s}\int_s^T e^{-\Lambda_u}|\varphi_u|du] - \int_0^T e^{-\Lambda_u}|\varphi_u|du + \int_0^T e^{\Lambda_s}e^{-\Lambda_s}|\varphi_s|ds\right]$$

$$< +\infty,$$

where we have used the estimate $e^{-\Lambda_u} \le e^{-\Lambda_s}$ for $u \ge s$. We now turn to the definition of the Z process in the equation. Consider the L^2 martingale $\hat{L}$ defined as:

$$\hat{L}_t := \mathbb{E}\left[\int_0^T e^{-\Lambda_s}\varphi_s ds|\mathcal{F}_t\right], \quad t \in [0,T].$$

By the martingale representation theorem, there exists a process $\hat{Z}$ in $\mathbb{H}^2(\mathbb{R}^d)$ such that $\hat{L}_t = \hat{L}_0 + \int_0^t \hat{Z}_s dW_s$. Now let $Z_t := -e^{\Lambda_t}\hat{Z}_t$ and

$L_t := \int_0^t Z_s dW_s$ which is a local martingale. With this definition, it is clear that the pair (Y, Z) has the dynamics:

$$dY_t = (\varphi_t + \lambda_t Y_t)dt + Z_t dW_t, \quad t \in [0, T].$$

Note that a priori $\int_0^\cdot Z_s dW_s$ is only a local martingale. From the equation, there exists a constant $C > 0$ such that

$$\mathbb{E}\left[\sup_{t\in[0,T]}\left|\int_0^t Z_s dW_s\right|\right] \leq C\left(2\mathbb{E}[\sup_{t\in[0,T]}|Y_t|] + T\|\varphi\|_\infty + \mathbb{E}\left[\int_0^T |\lambda_s Y_s| ds\right]\right)$$

is finite since by definition Y is bounded. Hence Z is an element of $\mathbb{H}^1(\mathbb{R}^d)$ by Burkholder-Davis-Gundy's inequality. Finally note that this argument provides uniqueness of the solution since we have characterized any solution via the process $\tilde{Y}$. □

Remark 3.6. Up to a Girsanov transformation, the previous result can be generalized to equations of the form:

$$Y_t = A - \int_t^T (\varphi_s + \sigma_t Z_t - \lambda_s Y_s)ds - \int_t^T Z_s dW_s; \quad t \in [0, T],$$

$$Y_t = A - \int_t^T (\varphi_s + \sigma_t Z_t + \lambda_s Y_s)ds - \int_t^T Z_s dW_s; \quad t \in [0, T],$$

where $\sigma := (\sigma_t)_{t\in[0,T]}$ is any bounded predictable process. In particular, our results contain the motivating example (2.1) from [8].

4. A Class of Non-linear Equations

From the results of Section 3 it appears clearly that there is no hope to provide a general theory for equations of the form (2.1) with a non-integrable coefficient λ. However, motivated by financial applications, we need to prove that the particular equation (2.1) with $f(x) := \alpha^{-1}(1 - e^{-\alpha x})$ admits a unique solution if and only if $Y_T = 0$. In addition, in order to provide a complete answer to the financial problem associated to this equation, we need to prove that the process Y is bounded and that the martingale $\int_0^\cdot Z_s dW_s$ is a BMO-martingale (whose definition will be recalled below). This section is devoted to the study of a class of equations which generalizes this particular case. We start with a generalization of Proposition 3.1.

Proposition 4.1. *Let φ be an element of $\mathbb{H}^1(\mathbb{R})$ and A in L^1. Let $f : \mathbb{R} \to \mathbb{R}$ be an increasing (respectively decreasing) map with $f(0) = 0$. The BSDE*

$$Y_t = A - \int_t^T [\varphi_s + \lambda_s f(Y_s)]ds - \int_t^T Z_s dW_s, \quad t \in [0,T] \tag{4.1}$$

admits no solution if $A \not\equiv 0$.

Proof. The proof follows the lines of the one of Proposition 3.1 and of Proposition 3.5. □

Remark 4.2. Note that the previous result does not contradict the conclusion of Proposition 3.2 in the deterministic setting, since according to Remark 3.3 the assumption of (i) in Proposition 3.2 on λ is not compatible with the $\mathbb{H}^1(\mathbb{R})$-requirement of Proposition 4.1.

The following lemma will be of interest for proving the main result of this section.

Lemma 4.3. *Let $f : \mathbb{R} \to \mathbb{R}$ satisfying $f(0) = 0$, f is non-decreasing and $f(x) - x \leq 0$, $\forall x \in \mathbb{R}$. Then the equation*

$$Y_t = 0 - \int_t^T \lambda_s f(Y_s)ds - \int_t^T Z_s dW_s, \quad t \in [0,T] \tag{4.2}$$

admits $(0,0)$ as unique solution.

Proof. It is clear that $(0,0)$ solves (4.2). Let (Y,Z) be any solution and $\tilde{Y} := e^{-\Lambda}Y$. It holds that $\tilde{Y}_T = 0$ and that

$$d\tilde{Y}_t = \lambda_t e^{-\Lambda_t}(-Y_t + f(Y_t))dt + e^{-\Lambda_t} Z_t dW_t.$$

Since $f(x) - x \leq 0$, for all $x \in \mathbb{R}$, $\tilde{Y}_t \geq 0$. Hence by definition, $Y_t \geq 0$, $\forall t \in [0,T]$, $\mathbb{P}$-a.s. From Equation (4.2), since $Y \geq 0$ we have that $f(Y_t) \geq 0$ which implies that

$$Y_t = 0 - \mathbb{E}\left[\int_t^T \lambda_s f(Y_s)ds | \mathcal{F}_t\right] \leq 0, \quad \forall t \in [0,T], \ \mathbb{P} - a.s.$$

As a consequence $Y_t = 0$ for all t, $\mathbb{P}$-a.s. which in turn gives $Z = 0$ (in $\mathbb{H}^1(\mathbb{R}^d)$), which concludes the proof. □

We now consider a class of nonlinear BSDEs.

Theorem 4.4. *Let φ be a non-negative bounded predictable process and $f : \mathbb{R} \to \mathbb{R}$ a continuously differentiable map satisfying: $f(0) = 0$, f is non-decreasing, there exists $\delta > 0$ such that*

$$f(x) - x \leq 0, \ \forall x \in \mathbb{R} \text{ and } f'(x) \geq \delta, \ \forall x \leq 0.$$

Assume in addition that $\mathbb{E}[\Lambda_t] < +\infty$, $\forall t < T$. Then the BSDE

$$Y_t = A - \int_t^T (\varphi_s + \lambda_s f(Y_s))ds - \int_t^T Z_s dW_s, \quad t \in [0,T] \tag{4.3}$$

admits a solution if and only if $A \equiv 0$. In that case, the solution is unique, Y is bounded and $\int_0^\cdot Z_s dW_s$ is a BMO-martingale, that is:

$$\text{esssup}_{\tau \in \mathcal{T}} \mathbb{E}\left[\int_\tau^T \|Z_s\|^2 ds | \mathcal{F}_\tau\right] < \infty,$$

where we recall that $\mathcal{T}$ denotes the set of stopping time smaller or equal to T.

Proof. We have seen in Proposition 4.1 that the only possible value for A to admit a solution is 0. From now on, we assume that $A \equiv 0$.

Step 1: some estimates
We start with some estimates on the (possible) solution to the BSDE. Assume that there exists a solution (Y, Z) to Equation (4.3). Since φ is non-negative, (Y, Z) is a sub-solution to the BSDE:

$$\mathcal{Y}_t = 0 - \int_t^T \lambda_s f(\mathcal{Y}_s)ds - \int_t^T \mathcal{Z}_s dW_s, \quad t \in [0,T],$$

which admits $(0, 0)$ as unique solution by Lemma 4.3. Indeed, this sub-solution property is classical for (L^2) Lipschitz BSDE and follows from the comparison theorem. However, here the BSDE (4.2) is not Lipschitz due to the unboundedness of λ. In our context the result can be proved explicitly. Since $f(0) = 0$ the BSDE (4.3) can be written as[§]

$$Y_t = 0 - \int_t^T \tilde{\lambda}_s Y_s ds - \int_t^T Z_s dW_s - \int_t^T \varphi_s ds, \quad t \in [0,T], \tag{4.4}$$

with $\tilde{\lambda}_t := \lambda_t \int_0^1 f'(\theta Y_t) d\theta$ which is non-negative. Following the lines of Proposition 3.5 with λ replaced by $\tilde{\lambda}$, and using the non-negativity of φ,

[§]by: $f(x) - f(y) = (x - y) \int_0^1 f'(y + \theta(x - y))d\theta$, $(x, y) \in \mathbb{R}^2$

we get that

$$Y_t = -\mathbb{E}\left[\int_t^T e^{-\int_t^s \tilde{\lambda}_u du}\varphi_s ds|\mathcal{F}_t\right] \leq 0.$$

From the non-positivity of Y, we can deduce that $\tilde{\lambda} \geq \delta\lambda$ from which we get that

$$Y_t \geq -\mathbb{E}\left[\int_t^T e^{-\delta\int_t^s \lambda_u du}\varphi_s ds|\mathcal{F}_t\right], \quad t \in [0,T].$$

To summarize, we have proven that

$$-(T-t)\|\varphi\|_\infty \leq -\mathbb{E}\left[\int_t^T e^{-\delta\int_t^s \lambda_u du}\varphi_s ds|\mathcal{F}_t\right] \leq Y_t \leq 0, \quad \forall t \in [0,T],\ \mathbb{P}\text{–}a.s.. \tag{4.5}$$

We now prove that the process $\int_0^\cdot Z_s dW_s$ is a BMO-martingale. Let τ be any stopping time such that $\tau \leq T$. By Itô's formula, we have that

$$|Y_\tau|^2 = 0-2\int_\tau^T \varphi_s Y_s ds-2\int_\tau^T Y_s Z_s dW_s-\int_\tau^T \|Z_s\|^2 ds-2\int_\tau^T Y_s f(Y_s)\lambda_s ds.$$

Since Y is bounded and Z is an element of $\mathbb{H}^1(\mathbb{R}^d)$, the stochastic integral process is a true martingale, and since Y is non-positive, the last term of the previous expression is non-positive. As a consequence, it holds that

$$\mathbb{E}\left[\int_\tau^T \|Z_s\|^2 ds|\mathcal{F}_\tau\right] \leq -2\mathbb{E}\left[\int_\tau^T \varphi_s Y_s ds|\mathcal{F}_\tau\right] \leq 2T^2\|\varphi\|_\infty^2.$$

So the claim is proved.

Step 2: existence

Now, we prove the existence of a solution for the BSDE (4.3). For any positive integer n, we set $\lambda_\cdot^n := \lambda_\cdot \wedge n$, $\tilde{f}(x) := f(x)\mathbf{1}_{\{[-T\|\varphi\|_\infty,0]\}}(x) + f(-T\|\varphi\|_\infty)\mathbf{1}_{\{(-\infty,-T\|\varphi\|_\infty]\}}(x)$, and (Y^n, Z^n) the unique (classical) solution in $\mathbb{S}^2 \times \mathbb{H}^2(\mathbb{R}^d)$ to the BSDE

$$Y_t^n = 0 - \int_t^T (\varphi_s + \tilde{f}(Y_s^n)\lambda_s^n)ds - \int_t^T Z_s^n dW_s, \quad t \in [0,T]. \tag{4.6}$$

It is clear that this equation admits a unique solution since $\tilde{f}$ is Lipschitz continuous and λ^n is bounded. In addition, by definition, $\tilde{f}(Y_s^n) \leq 0$, and so $Y_t^n \geq -\|\varphi\|_\infty(T-t)$. Thus (Y^n, Z^n) solves the same equation with $\tilde{f}$ replaced by $\hat{f}(x) := f(x)\mathbf{1}_{\{x\geq 0\}}$. Note that $\hat{f}(x) \leq x$ for any x in $\mathbb{R}$. Since

φ is non-negative, Y^n is a classical sub-solution to the BSDE (4.2) with f replaced by $\hat{f}$, and so by Lemma 4.3 we deduce that $Y_t^n \leq 0$. Thus

$$|Y_t^n| \leq (T-t)\|\varphi\|_\infty, \quad \forall t \in [0,T],\ \mathbb{P}-a.s.. \tag{4.7}$$

Hence we can re-write Equation (4.6) as:

$$Y_t^n = 0 - \int_t^T (\varphi_s + f(Y_s^n)\lambda_s^n)ds - \int_t^T Z_s^n dW_s, \quad t \in [0,T]. \tag{4.8}$$

Repeating the same argument used in the previous step we can prove that

$$\sup_n \mathbb{E}\left[\int_0^T \|Z_t^n\|^2 dt\right] < \infty. \tag{4.9}$$

By comparison theorem for Lipschitz BSDEs the sequence $(Y^n)_n$ is non-decreasing. Hence it converges pointwise to some element $Y := \limsup_{n\to\infty} Y^n$. We would like to point out at this stage that by construction Y takes values in $[-T\|\varphi\|_\infty, 0]$. In view of Dini's theorem, to obtain convergence uniformly in time, we need to prove that Y is continuous. This is done in two steps. Fix $0 < t_0 < T$, $n \geq 1$ and $p,q \geq n$. We show that the sequence $(Y_n \mathbf{1}_{[0,t_0]})_n$ is a Cauchy sequence in $\mathbb{S}^2$. Let $\delta Y := Y^p - Y^q$, $\delta Z := Z^p - Z^q$. Itô's formula gives for every $t \in [0,t_0]$ that

$$|\delta Y_t|^2 + \int_t^{t_0} \|Z_s\|^2 ds \leq |\delta Y_{t_0}|^2 - 2\int_t^{t_0} \delta Y_s f(Y_s^q)(\lambda_s^p - \lambda_s^q)ds - 2\int_t^{t_0} \delta Y_s \delta Z_s dW_s, \tag{4.10}$$

where we have used the fact that $\delta Y_s(f(Y_s^p) - f(Y_s^q)) \geq 0$ since f is non-decreasing. From this relation we deduce in particular for $t = 0$ that

$$\mathbb{E}\left[\int_0^{t_0} \|Z_s\|^2 ds\right] \leq C\mathbb{E}\left[|\delta Y_{t_0}|^2 + \int_0^{t_0} |\lambda_s^p - \lambda_s^q| ds\right], \tag{4.11}$$

since Y^p and Y^q are uniformly (in p,q) bounded. Taking the supremum over $[0,t_0]$ in Relation (4.10) leads to

$$\begin{aligned}&\mathbb{E}[\sup_{t\in[0,t_0]} |\delta Y_t|^2] \\ &\leq C\left(\mathbb{E}[|\delta Y_{t_0}|^2] + \mathbb{E}\left[\int_0^{t_0} |\delta Y_s f(Y_s^q)||\lambda_s^p - \lambda_s^q| ds + \sup_{t\in[0,t_0]}\left|\int_t^{t_0} \delta Y_s \delta Z_s dW_s\right|\right]\right)\end{aligned}$$

$$\leq C\left(\mathbb{E}[|\delta Y_{t_0}|^2] + \mathbb{E}\left[\int_0^{t_0} |\lambda_s^p - \lambda_s^q| ds\right] + \mathbb{E}\left[\left(\int_0^{t_0} |\delta Y_s|^2 \|\delta Z_s\|^2 ds\right)^{1/2}\right]\right)$$

$$\leq C\left(\mathbb{E}[|\delta Y_{t_0}|^2] + \mathbb{E}\left[\int_0^{t_0} |\lambda_s^p - \lambda_s^q| ds\right]\right) + \frac{1}{2}\mathbb{E}\left[\sup_{t\in[0,t_0]} |\delta Y_t|^2\right]$$

$$+ \frac{C^2}{2}\mathbb{E}\left[\int_0^{t_0} \|\delta Z_s\|^2 ds\right],$$

where we have used the fact that $|\delta Y_s f(Y_s^q)|$ is bounded uniformly in p, q, the Burkholder inequality and the inequality $Cab \leq \frac{1}{2}a^2 + \frac{C^2 b^2}{2}$. Combining the previous estimate with Estimate (4.11) proves that

$$\mathbb{E}[\sup_{t\in[0,t_0]} |\delta Y_t|^2] \leq C\left(\mathbb{E}[|\delta Y_{t_0}|^2] + \mathbb{E}\left[\int_0^{t_0} |\lambda_s^p - \lambda_s^q| ds\right]\right),$$

where C does not depend on p, q. Recalling the definition of $\delta Y = Y^p - Y^q$ it follows that

$$\lim_{n\to\infty} \sup_{p,q\geq n} \mathbb{E}[\sup_{t\in[0,t_0]} |\delta Y_t|^2] \leq C \lim_{n\to\infty}\left(\mathbb{E}[|Y_{t_0}^n - Y_{t_0}|^2] + \mathbb{E}\left[\int_0^{t_0} |\lambda_s^n - \lambda_s| ds\right]\right) = 0,$$

by Lebesgue's dominated convergence Theorem (since $\mathbb{E}[\Lambda_{t_0}] < \infty$)¶. Hence $(Y^n \mathbf{1}_{[0,t_0]})_n$ is a Cauchy sequence in $\mathbb{S}^2$ which thus converges to $Y\mathbf{1}_{[0,t_0]}$. So Y is continuous on $[0, t_0]$ for any $t_0 < T$. It remains to prove that Y is continuous at T. Let $\varepsilon > 0$. By Inequality (4.7) it holds that

$$|Y_{T-\varepsilon}| = \lim_{n\to\infty} |Y_{T-\varepsilon}^n| \leq \varepsilon \|\varphi\|_\infty,$$

proving that Y is continuous at T. Hence, $(Y^n)_n$ is a non-decreasing sequence of continuous bounded processes converging to a continuous process Y, thus by Dini's Theorem, $(Y_n)_n$ converges in $\mathbb{S}^2$ to Y.

We now prove that Y together with a suitable process Z solves the BSDE (4.3). To this end, we aim at applying [1, Theorem 1]. We have obtained already that $\lim_{n\to\infty} \mathbb{E}[\sup_{t\in[0,T]} |Y_t^n - Y_t|] = 0$. To satisfy the assumptions of [1, Theorem 1], we have to prove that for every n

$$\sup_n \mathbb{E}\left[\left(\int_0^T \|Z_s^n\|^2 ds\right)^{1/2}\right] \leq C \qquad (4.12)$$

¶Here we did not use the classical a priori estimates for Lipschitz BSDEs since they would lead to an estimate of the form $\mathbb{E}\left[\int_0^{t_0} |\lambda_s^p - \lambda_s^q|^2 ds\right]$ which is not compatible with our L^1 assumption: $\mathbb{E}[\Lambda_{t_0}] < \infty$.

(which by Burkholder's inequality implies that $\mathbb{E}\left[\sup_{t\in[0,T]}\left|\int_0^t Z_s^n dW_s\right|\right] \leq C$ for every n) and that

$$\sup_n \mathbb{E}\left[\int_0^T |\lambda_s^n f(Y_s^n)|\, ds\right] \leq C, \quad \forall n \geq 1, \tag{4.13}$$

since the process $\int_0^\cdot \lambda_s^n f(Y_s^n)ds$ is non-increasing (recall that $Y^n \leq 0$ and the assumptions on f). Relation (4.12) is a direct consequence of (4.9). With this estimate at hand we can deduce Relation (4.13). Indeed, using Equation (4.8) and the uniform estimates on the Y^n obtained above we deduce that

$$\begin{aligned}\mathbb{E}\left[\int_0^T |\lambda_s^n f(Y_s^n)|ds\right] &= \mathbb{E}\left[\left|\int_0^T \lambda_s^n f(Y_s^n)ds\right|\right] \\ &= \mathbb{E}\left[\left|Y_0^n + \int_0^T \varphi_s ds + \int_0^T Z_s^n dW_s\right|\right] \leq C, \quad n \geq 1,\end{aligned}$$

where C depends only on T and $\|\varphi\|_\infty$ (and not on n). Thus, by [1, Theorem 1], Y writes down as $Y_t = A_t + \int_0^t \varphi_s ds + \int_0^t Z_s dW_s$, with $Z \in \mathbb{H}^1(\mathbb{R}^d)$, and

$$\lim_{n\to\infty} \mathbb{E}[\sup_{t\in[0,T]} |A_t - \int_0^t \lambda_s^n f(Y_s^n)ds|] = 0. \tag{4.14}$$

We now identify the process A. We proceed in two steps: first we prove that $A_t = \int_0^t f(Y_s)\lambda_s ds$ for $t < T$ and then we prove the relation for $t = T$. Fix $t < T$. We have that

$$\begin{aligned}&\left|\int_0^t f(Y_s^n)(\lambda_s^n - \lambda_s)ds\right| \\ &\leq C\int_0^t |\lambda_s^n - \lambda_s|ds \to_{n\to\infty} 0, \quad \mathbb{P}-a.s.\end{aligned}$$

by the monotone convergence theorem, since the Y^n are uniformly bounded and $\Lambda_t < \infty$, $\mathbb{P}$-a.s. Hence up to a subsequence,

$$\lim_{n\to\infty}\left|A_t - \int_0^t f(Y_s^n)\lambda_s ds\right| = 0.$$

Recalling that $Y^n \leq Y$, we have that

$$\begin{aligned}&\left|\int_0^t (f(Y_s^n) - f(Y_s))\lambda_s ds\right| \\ &\leq C\int_0^t |Y_s^n - Y_s|\lambda_s ds \to_{n\to\infty} 0\end{aligned}$$

where once again we have used monotone convergence Theorem. This leads to

$$A_t = \int_0^t f(Y_s)\lambda_s ds, \quad \mathbb{P}-a.s.$$

for any $t < T$. The relation for $t = T$ follows from the continuity of A by (4.14). Finally according to Definition 2.1 it remains to prove Relation (2.3). This is done as follows by combining the monotone convergence theorem together with (4.13) and (4.14):

$$\mathbb{E}\left[\int_0^T |f(Y_s)|\lambda_s ds\right] = \lim_{t\to T}\mathbb{E}\left[\int_0^t |f(Y_s)|\lambda_s ds\right] = \lim_{t\to T}\mathbb{E}[|A_t|] < \infty.$$

Step 3: uniqueness

Assume there exist two solutions (Y^1, Z^1) and (Y^2, Z^2) to the BSDE (4.3). Then, the difference processes $(\delta Y := Y^1 - Y^2, \delta Z := Z^1 - Z^2)$ satisfies

$$\delta Y_t = 0 - \int_t^T \lambda_s(f(Y_s^1) - f(Y_s^2))ds - \int_t^T \delta Z_s dW_s, \quad t \in [0, T].$$

From the existence part we know that both processes Y^1 and Y^2 are uniformly bounded. As a consequence the mapping f restricted to the set $[-T\|\varphi\|_\infty, 0]$ has a non-negative derivative. Hence the equation re-writes as:

$$\delta Y_t = 0 - \int_t^T \tilde{\lambda}_s \delta Y_s ds - \int_t^T \delta Z_s dW_s, \quad t \in [0, T],$$

where $\tilde{\lambda}_t := \lambda_t \int_0^1 f'(Y_t^2 + \theta(Y_t^1 - Y_t^2))d\theta$ is a non-negative process which satisfies $\int_0^t \tilde{\lambda}_s ds < \infty$ for $t < T$, $\mathbb{P}$-a.s. and $\int_0^T \tilde{\lambda}_s ds = \infty$, $\mathbb{P}$-a.s. Similarly to Proposition 3.5 with λ replaced with $\tilde{\lambda}$, we deduce that $(\delta Y, \delta Z) = (0, 0)$ is the unique solution. □

Remark 4.5. Note that our previous result is not contained in the theory of monotonic drivers for BSDEs (see e.g. [3, 12] or [10]) where conditions of the form [3, (H5) and (H1")] are not satisfied in our setting due to the non-integrability at T of Λ.

Acknowledgments

The authors are very grateful to Nicole El Karoui and Jean Jacod and for helpful comments and discussions. This research benefited from the support of the "Chair Markets in Transition" under the aegis of Louis Bachelier laboratory, a joint initiative of École polytechnique, Université d'Évry Val d'Essonne and Fédération Bancaire Française.

References

[1] M.T. Barlow and P. Protter. On convergence of semimartingales. In *Séminaire de Probabilités, XXIV, 1988/89*, volume 1426 of *Lecture Notes in Math.*, pages 188–193. Springer, Berlin, 1990.

[2] J.-M. Bismut. Contrôle des systèmes linéaires quadratiques: applications de l'intégrale stochastique. In *Séminaire de Probabilités, XII (Univ. Strasbourg, Strasbourg, 1976/1977)*, volume 649 of *Lecture Notes in Math.*, pages 180–264. Springer, Berlin, 1978.

[3] Ph. Briand, B. Delyon, Y. Hu, E. Pardoux, and L. Stoica. L^p solutions of backward stochastic differential equations. *Stochastic Process. Appl.*, 108(1):109–129, 2003.

[4] F. Confortola, M. Fuhrman, and J. Jacod. BSDEs and point processes: the multi-jump case. In preparation.

[5] F. Confortola, M. Fuhrman, and J. Jacod. BSDEs and point processes: the one jump case. In preparation.

[6] C. Dellacherie and P.-A. Meyer. *Probabilities and potential*, volume 29 of *North-Holland Mathematics Studies*. North-Holland Publishing Co., Amsterdam, 1978.

[7] C. Dellacherie and P.-A. Meyer. *Probabilities and potential. B*, volume 72 of *North-Holland Mathematics Studies*. North-Holland Publishing Co., Amsterdam, 1982. Theory of martingales, Translated from the French by J. P. Wilson.

[8] N. El Karoui. Backward stochastic differential equations: a general introduction. In *Backward stochastic differential equations (Paris, 1995–1996)*, volume 364 of *Pitman Res. Notes Math. Ser.*, pages 7–26. Longman, 1997.

[9] N. El Karoui, S. Peng, and M. C. Quenez. Backward stochastic differential equations in finance. *Math. Finance*, 7(1):1–71, 1997.

[10] S.-J. Fan and L. Jiang. Uniqueness result for the BSDE whose generator is monotonic in y and uniformly continuous in z. *C. R. Math. Acad. Sci. Paris*, 348(1-2):89–92, 2010.

[11] M. Jeanblanc, N. Nguyen Hai, and A. Réveillac. Utility maximization with random horizon: a BSDE approach. In preparation.

[12] E. Pardoux. BSDEs, weak convergence and homogenization of semilinear PDEs. In *Nonlinear analysis, differential equations and control (Montreal, QC, 1998)*, volume 528 of *NATO Sci. Ser. C Math. Phys. Sci.*, pages 503–549. Kluwer Acad. Publ., Dordrecht, 1999.

[13] E. Pardoux and S. Peng. Adapted solution of a backward stochastic differential equation. *Systems Control Lett.*, 14(1):55–61, 1990.

A Portfolio Optimization Problem with Two Prices Generated by Two Information Flows

Caroline Hillairet *

Abstract. We study a model in which the presence of different prices for the same asset or contingent claim is a consequence of different information settings. We consider a specific case involving defaults : here the two different levels of information concern the knowledge of the occurence of a random time. Under no-short-sale restriction, we give conditions on the price dynamics in order to make the market arbitrage free despite the presence of price discrimination, and we construct the arbitrage portfolio if any. We then compute the optimal portfolio of an investor whose risk aversion is characterized by a logarithmic utility.

1. Introduction

In most financial models, an information set is given from the start and supposed to be common to all investors. Except models specifically dealing with the topic of insider trading (Grorud and Pontier [6], Imkeller et al. [1], Hillairet [7], Elliott and Jeanblanc [5]) or equilibrium models of the type of Kyle and Back (Cho [3], Kyle [12], Back [2]), few models studied the impact of the information on the financial prices.
Nevertheless, the amount and quality of the information affect the risk perception of investors when quantifying the financial risks or elaborating hedging strategies. For example, during the recent financial crisis, it was observed that defaults of important companies can have an important effect upon companies that might have seemed uncorrelated with the defaulted company and therefore, the default event represents itself a piece of information that needs to be explicitly taken into account in models.

This paper proposes a toy financial model where the information upon which investors rely is explicitly modeled, as a key element. The information is modeled by a filtration, that is, an increasing sequence of sigma

*CMAP, Ecole Polytechnique, email: caroline.hillairet@polytechnique.edu

fields, reflecting the fact that one keeps track of the history. We first characterize the *informational inefficiency* of the market. In our broad definition of market, we can distinguish among two different market sectors (or two differently informed type of agents). The agents of the two market sectors have access to a different amount of information. Therefore the market sectors involve a different information settings, i.e., two filtrations $\mathbb{F}$ and $\mathbb{G}$. Furthermore, we assume that the agents of one market sector are better informed that the others, thus the following condition holds

$$\forall t,\, \mathcal{F}_t \subset \mathcal{G}_t.$$

The main mathematical tool used is the theory of enlargements of filtrations. In each market sector, investors price the assets according to their information, and their pricing measure. We *assume* here that the pricing probability is the same for the whole market, i.e. the same for the two sectors. Even under a unique pricing measure $\mathbb{P}$, asymmetry of information gives rise to different *pricing rules* for the same product. We consider a contingent claim $\zeta \in \mathcal{F}_T$ which may be not perfectly hedgeable (think of a weather derivative for example). The usual approach is to define the price of the claim as the discounted expected value of future cash flows, conditional to the information available to the agent. In the two market sectors, the time-t price of the claim is calculated according to different information flows (assuming null interest rate, without loss of generality):

$$\begin{cases} \mathbb{E}[\zeta|\mathcal{F}_t] =: x_t \\ \mathbb{E}[\zeta|\mathcal{G}_t] =: X_t \end{cases} \tag{1.1}$$

Note that the information on ζ is fully known for both markets at time T so that $X_T = x_T$, and that, if $\mathcal{F}_0 = \mathcal{G}_0$ are trivial (which is questionable, since, even at time 0, the market has information on the past) the prices of ζ are equal at time 0.

In general, the presence of multiple prices for the same asset is an indicator of an arbitrage opportunity, as a result of market inefficiency. In the case of a liquid market, arbitrage theory predicts that the price of the product must be unique. If it is not the case, arbitrageurs can make profit just by buying the asset from the market with a low price and in the same time short-selling it in the other market with a high price. However, analysis of real markets shows that discrepancies in prices for the same asset still arise. The presence of selling restrictions can prevent investors from exploiting arbitrages. Thus we introduce a no-short-sale restriction in our market; anyway we prove that this might be not sufficient to prevent market from arbitrage.

This paper studies the particular case where the additional information is synthesized in one event: the occurrence of a random time (which can be interpreted as a default time, or the time of switching to a new regime of the economy). The distribution of this time depends on the information available on a financial market, but its realization is not observable. The knowledge about its occurrence supposes some knowledge of the future evolution of the market prices, or of the states of the economy. Thus two financial entities or market sectors, having two different information sets (one entity being better informed than the other), will propose different bid prices (as for instance in the over the counter markets, where the prices are not transparent). In this framework, we study dynamically as information evolves the different prices evolution and their dependencies and we identify non arbitrage conditions. In the situation where the two prices converge to the same terminal value at some maturity date, there are trivial arbitrages in the market, which consist in short-selling the more expensive asset and buying the cheaper, then holding the portfolio until the maturity. We show that even with short-selling constraints, some non trivial arbitrages may exist, for instance based on the sign of the volatility (but these arbitrages may not be possible to realize). Since short-selling interdictions are natural on these markets where several prices may coexist, we provide a systematic study of the implications of these constraints in order to characterize the no-arbitrage conditions. Using results by Jouini and Kallal [11], and by Pulido [13], we are able to provide necessary and sufficient conditions for the parameters of the model such that the financial market is arbitrage free in absence of short-sales possibilities.

The rest of the paper is organized as follow: Section 2 introduces the informational framework. The two information flows are modeled through a Brownian filtration and its enlargement with a random time. This leads to two different price processes whose dynamics are specified. Attempts to compare those prices are given. Section 3 studies arbitrage opportunities in this frictionless market and under short-selling constraints, and presents the computation of the optimal portfolio for a logarithmic utility function.

2. The informational framework and the prices dynamics

2.1. *The default-free filtration and its progressive enlargement with a random time*

We consider a default-free (sub-)market, whose information flow is conveyed by a filtration $\mathbb{F}$, where $\mathbb{F}$ is the natural filtration of a Brownian motion B. As the price $(x_t)_{0\leq t\leq T}$ of the contingent claim ζ, relatively to this information, is an $\mathbb{F}$-martingale, the Predictable Representation Property (PRP) in $\mathbb{F}$ implies the following dynamics

$$dx_t = \nu_t dB_t, \quad 0 \leq t \leq T$$

where $(\nu_t)_{0\leq t\leq T}$ is an $\mathbb{F}$-predictable process. Besides, other investors or an other submarket is aware of the eventuality of a default that may affect the valuation of the contingent claim ζ. Here the information flow is characterized by the filtration $\mathbb{G}$, that is the filtration $\mathbb{F}$ progressively enlarged with a random time τ (a strictly positive random variable),

$$\mathcal{G}_t = \cap_{s>t}\{\mathcal{F}_s \vee \sigma(\tau \wedge s)\}.$$

For example the random time τ represents the default time. In order to make explicit the dynamics of the contingent claim price $(X_t)_{0\leq t\leq T}$ in this sub-market, we assume the following hypotheses.

Assumption 2.1. We assume that $(\mathcal{H}')$ hypothesis is satisfied, that is any $\mathbb{F}$-martingale is a $\mathbb{G}$-semi-martingale. More precisely, we assume that there exists an integrable $\mathbb{G}$-adapted process μ such that $dB_t = dW_t + \mu_t dt$ where W is a $\mathbb{G}$-Brownian Motion[a].

We also assume that there exists a $\mathbb{G}$-adapted non-negative process $\lambda^{\mathbb{G}}$ such that $M_t = \mathbf{1}_{\tau\leq t} - \int_0^t \lambda_s^{\mathbb{G}} ds$ is a $\mathbb{G}$-martingale. Note that $\lambda^{\mathbb{G}}$ vanishes after τ and that there exists an $\mathbb{F}$-adapted process $\lambda^{\mathbb{F}}$, such that $\lambda_t^{\mathbb{G}} = \mathbf{1}_{t<\tau}\lambda_t^{\mathbb{F}}$.
Under mild conditions, any $\mathbb{G}$-martingale Y can be represented as a sum of two stochastic integrals (Predictable Representation Property (PRP) in $\mathbb{G}$)

$$Y_t = y + \int_0^t \widehat{y}_s dW_s + \int_0^t \widetilde{y}_s dM_s$$

where $\widehat{y}$ and $\widetilde{y}$ are $\mathbb{G}$-predictable processes (see, e.g., Jeanblanc and Le Cam [8], or Song [14]).

[a] It can be noted that the reverse does not hold, that is the existence of μ does not imply $(\mathcal{H}')$ hypothesis, due to integrability conditions (see Jeulin and Yor [10])

The above conditions are satisfied in the case where the random time τ admits a conditional density with respect to $\mathbb{F}$, that is if

$$\mathbb{P}(\tau > u|\mathcal{F}_t) = \int_u^\infty p_t(x)dx$$

where, for any x, $p(x)$ is a positive $\mathbb{F}$-martingale. Due to the PRP (in the filtration $\mathbb{F}$) of the Brownian motion B, the martingale $p(x)$ admits a representation of the form

$$dp_t(x) = p_t(x)\sigma_t(x)dB_t \quad t \geq 0$$

where for any x, $\sigma(x)$ is an $\mathbb{F}$-predictable process. Note that, since, by hypothesis, $\mathbb{P}(\tau > 0) = 1$, one has $\int_0^\infty p_t(x)dx = 1$, for any $t \geq 0$. Under some smoothness conditions allowing to differentiate this equality under the integral sign (with respect to the running time t), this implies that $\int_0^\infty d_t p_t(x)dx = 0 = \left(\int_0^\infty p_t(x)\sigma_t(x)dx\right)dB_t$, that is $\int_0^\infty p_t(x)\sigma_t(x)dx = 0$. In that case, using Itô-Ventcell formula, the conditional survival probability $G_t = \mathbb{P}(\tau > t|\mathcal{F}_t)$ admits the Doob-Meyer decomposition

$$G_t = 1 + \int_0^t \xi_s dB_s - \int_0^t p_s(s)ds$$

with $\xi_t = \int_t^\infty p_t(x)\sigma_t(x)dx$.

2.2. *Prices dynamics*

Under Hypothesis 2.1, the two prices dynamics $x_t = \mathbb{E}(\zeta|\mathcal{F}_t)$ and $X_t = \mathbb{E}(\zeta|\mathcal{G}_t)$ of the contingent claim ζ, with respect to the information flow $\mathbb{F}$ and $\mathbb{G}$, are the following : there exist processes ν, α, β such that ν is $\mathbb{F}$-predictable, α and β are $\mathbb{G}$-predictable, and for $t \in [0, T]$

$$dx_t = \nu_t dB_t = \nu_t(dW_t + \mu_t dt) \tag{2.1}$$

$$dX_t = \alpha_t dW_t + \beta_t dM_t\,. \tag{2.2}$$

We recall from Jeulin [9] Lemma 4.4 and Song [14], the decomposition of any $\mathbb{G}$-predictable or optional process. Let φ be a $\mathbb{G}$-predictable (resp. optional) process. Then there exist an $\mathbb{F}$-predictable (resp. optional) process $\varphi^{\mathbb{F}}$, and a family of processes $\{\tilde{\varphi}_t(\theta), \theta \leq t \leq T, \theta \in [0, T]\}$ where $\tilde{\varphi}_t(\theta)$ is $\mathcal{P}(\mathbb{F}) \otimes \mathcal{B}(\mathbb{R}^+)$-measurable (resp. $\mathcal{O}(\mathbb{F}) \otimes \mathcal{B}(\mathbb{R}^+)$-measurable), such that

$$\varphi_t = \varphi_t^{\mathbb{F}} \mathbf{1}_{t \leq \tau} + \tilde{\varphi}_t(\tau)\mathbf{1}_{\tau > t}, \; 0 \leq t \leq T$$

$$(\text{resp. } \; \varphi_t = \varphi_t^{\mathbb{F}} \mathbf{1}_{t < \tau} + \tilde{\varphi}_t(\tau)\mathbf{1}_{\tau \geq t}, \; 0 \leq t \leq T).$$

The question of comparison of this two prices naturally follows. Unfortunately, it is impossible to find a model where the information drift (μ_s) has a constant sign, as proved in the Lemma 2.2:

Lemma 2.2. *If B is an $\mathbb{F}$-Brownian motion and a $\mathbb{G}$-semimartingale such that there exists a $\mathbb{G}$-Brownian motion W and a $\mathbb{G}$-adapted process (μ_s) satisfying $dB_t = dW_t + \mu_t dt$, then for any s, $\mathbb{E}(\mu_s|\mathcal{F}_s) = 0$.*

Proof. First recall that $\mathbb{E}(\int_0^t \mu_s ds|\mathcal{F}_t) - \int_0^t \mathbb{E}(\mu_s|\mathcal{F}_s)ds$ is an $\mathbb{F}$-martingale. Then taking expectation of $B_t = W_t + \int_0^t \mu_s ds$ implies that

$$B_t = \mathbb{E}(W_t|\mathcal{F}_t) + mart_t^{\mathbb{F}} + \int_0^t \mathbb{E}(\mu_s|\mathcal{F}_s)ds$$

which implies that $\int_0^t \mathbb{E}(\mu_s|\mathcal{F}_s)ds = 0$, hence $\mathbb{E}(\mu_s|\mathcal{F}_s) = 0$ for any s. Indeed, the finite variation process $\int_0^t \mathbb{E}(\mu_s|\mathcal{F}_s)ds$ is a continuous $\mathbb{F}$-martingale, as sum of different martingales in the same Brownian filtration $\mathbb{F}$, hence it is constant. Being null at $t = 0$, this process is null, and its derivative (which exists) is null too. □

2.3. *Comparison of prices*

One can ask the following question: can we compare the price of the contingent claim ζ computed in the filtration $\mathbb{F}$ and the price of ζ computed in the filtration $\mathbb{G}$?
It is known that, $\mathbb{G}$ being the progressive enlargement filtration of $\mathbb{F}$,

$$\mathbb{E}(\zeta|\mathcal{G}_t) = \mathbf{1}_{t<\tau}\frac{\mathbb{E}(\zeta\mathbf{1}_{t<\tau}|\mathcal{F}_t)}{G_t} + \mathbf{1}_{\tau\le t}\mathbb{E}(\zeta|\mathcal{G}_t), \text{ where } G_t = \mathbb{P}(t < \tau|\mathcal{F}_t).$$

Thus before τ, this prices comparison reduces to the comparison between $\mathbb{E}(\zeta|\mathcal{F}_t)$ and $\frac{\mathbb{E}(\zeta\mathbf{1}_{t<\tau}|\mathcal{F}_t)}{G_t}$, or, equivalently to the comparison between $G_t\mathbb{E}(\zeta|\mathcal{F}_t)$ and $\mathbb{E}(\zeta\mathbf{1}_{t<\tau}|\mathcal{F}_t)$.

This result is related with the sign of the conditional covariance

$$\mathbb{E}(\zeta\mathbf{1}_{t<\tau}|\mathcal{F}_t) - G_t\mathbb{E}(\zeta|\mathcal{F}_t) = \mathbb{E}(\zeta\mathbf{1}_{t<\tau}|\mathcal{F}_t) - \mathbb{E}(\mathbf{1}_{t<\tau}|\mathcal{F}_t)\mathbb{E}(\zeta|\mathcal{F}_t) = Cov_t(\zeta, \mathbf{1}_{t<\tau}).$$

We give an example. Recall that B is a Brownian motion with natural filtration $\mathbb{F}$. Assume that $\mathbb{P}(\tau > \theta|\mathcal{F}_t) = G_t(\theta) := \Phi\Big(\frac{m_t - h(\theta)}{\sigma(t)}\Big)$ where $m_t = \int_0^t f(u)dB_u$, f is a deterministic function such that $\int_0^\infty f^2(s)ds = 1$ and Φ is the survival function for a standard Gaussian law. The positive function σ is defined as $\sigma^2(t) = \int_t^\infty f^2(s)ds$ and h is a deterministic increasing function such that $h(0) = -\infty$ (see El Karoui et al. [4] for more details on the construction of τ). Then, for a fixed θ, $(G_t(\theta))_{t\geq 0}$ is an $\mathbb{F}$-martingale and

$$dG_t(\theta) = -G_t(\theta)\frac{1}{\Phi\left(\frac{m_t-h(\theta)}{\sigma(t)}\right)}\varphi\left(\frac{m_t - h(\theta)}{\sigma(t)}\right)\frac{f(t)}{\sigma(t)}dB_t = -G_t(\theta)\Sigma_t(\theta)dB_t\,,$$

where $\Sigma_t(\theta) = \left(\Phi\left(\frac{m_t-h(\theta)}{\sigma(t)}\right)\right)^{-1}\varphi\left(\frac{m_t-h(\theta)}{\sigma(t)}\right)\frac{f(t)}{\sigma(t)}$. Assume now that the price $x_t = \mathbb{E}(\zeta|\mathcal{F}_t)$, where ζ is positive, has a positive volatility, i.e.

$$dx_t = x_t\sigma_t dB_t$$

with $\sigma_t > 0$ (this would be the case for a European call). If $\zeta \in \mathcal{F}_T$, $\mathbb{E}(\zeta\mathbf{1}_{t<\tau}|\mathcal{F}_t) = \mathbb{E}(x_T G_T(t)|\mathcal{F}_t)$. Then, from integration by parts

$$x_T G_T(t) = x_t G_t(t) + \int_t^T G_s(t)dx_s + \int_t^T x_s d_s G_s(t) - \int_t^T x_s\sigma_s G_s(t)\Sigma_s(t)ds\,,$$

where the notation $d_sG_s(t)$ makes precise that the stochastic differential is w.r.t. the running time s. From our hypothesis on the sign of the volatility of x, and assuming that f is non negative we obtain (since, obviously $\Sigma_s(t) > 0$)

$$\mathbb{E}(x_T G_T(t)|\mathcal{F}_t) = x_t G_t(t) - \int_t^T \mathbb{E}(x_s\sigma_s G_s(t)\Sigma_s(t)|\mathcal{F}_t)ds < x_t G_t(t) = x_t G_t$$

therefore, on the set $t < \tau$, one has $X_t = \frac{1}{G_t}\mathbb{E}(x_T G_T(t)|\mathcal{F}_t) < x_t$.

2.4. *Link between the price dynamics coefficients*

Proposition 2.3. *Let x and X be two processes with dynamics given by (2.1) and (2.2). Then*

$$\nu_t = \mathbb{E}(\alpha_t + \mu_t X_t|\mathcal{F}_t)\,.$$

Proof. The equality $x_t = \mathbb{E}(X_t|\mathcal{F}_t)$ (and $x_0 = X_0$) implies $\int_0^t \nu_s dB_s = \mathbb{E}(\int_0^t \alpha_s dW_s + \int_0^t \beta_s dM_s|\mathcal{F}_t)$ which is equivalent to

$$\mathbb{E}(\int_0^t \nu_s dB_s \int_0^t z_s dB_s) = \mathbb{E}((\int_0^t \alpha_s dW_s + \int_0^t \beta_s dM_s)\int_0^t z_s dB_s)$$

for any $\mathbb{F}$-adapted bounded process z. By integration by parts, the left-hand side is $E(\int_0^t \nu_s z_s ds)$.

We now transform the right-hand side. On the one hand, from the relation between W and B, using again integration by parts,

$$\mathbb{E}(\int_0^t \alpha_s dW_s \int_0^t z_s dB_s) = \mathbb{E}(\int_0^t \alpha_s z_s ds + \int_0^t \alpha_s dW_s \int_0^t z_s \mu_s ds)\,.$$

Using integration by parts, one obtains

$$\begin{aligned}\mathbb{E}(\int_0^t \alpha_s dW_s \int_0^t z_s \mu_s ds) &= \mathbb{E}\left(\int_0^t z_s \mu_s \left(\int_0^s \alpha_u dW_u\right) ds\right) \\ &= \mathbb{E}\left(\int_0^t z_s \mathbb{E}\left(\mu_s \int_0^s \alpha_u dW_u | \mathcal{F}_s\right) ds\right)\,.\end{aligned}$$

On the other hand, using the fact that the martingales W and M are orthogonal,

$$\begin{aligned}\mathbb{E}(\int_0^t \beta_s dM_s \int_0^t z_s dB_s) &= \mathbb{E}(\int_0^t \beta_s dM_s \int_0^t z_s \mu_s ds) + \mathbb{E}(\int_0^t \beta_s dM_s \int_0^t z_s dW_s) \\ &= \mathbb{E}(\int_0^t \beta_s dM_s \int_0^t z_s \mu_s ds)\end{aligned}$$

and by integration by parts once again

$$\mathbb{E}(\int_0^t \beta_s dM_s \int_0^t z_s \mu_s ds) = \mathbb{E}(\int_0^t z_s \mathbb{E}(\mu_s \int_0^s \beta_u dM_u) | \mathcal{F}_s) ds)\,.$$

To summarize, for any z

$$\mathbb{E}(\int_0^t \nu_s z_s ds) = \mathbb{E}(\int_0^t z_s \mathbb{E}(\alpha_s + \mu_s \int_0^s (\alpha_u dW_u + \beta_u dM_u) | \mathcal{F}_s))$$

which implies, since ν is $\mathbb{F}$-adapted,

$$\nu_s = \mathbb{E}(\alpha_s + \mu_s \int_0^s (\alpha_u dW_u + \beta_u dM_u) | \mathcal{F}_s) = \mathbb{E}(\alpha_s + \mu_s (X_s - X_0) | \mathcal{F}_s)\,.$$

Using the fact that $\mathbb{E}(\mu_s | \mathcal{F}_s) = 0$, we get

$$\nu_s = \mathbb{E}(\alpha_s + \mu_s X_s | \mathcal{F}_s).$$

□

3. Portfolio Optimization problem

We now study the problem of maximizing the utility from terminal wealth of an investor, in a market where the two price processes $(x_t)_{0 \le t \le T}$ and $(X_t)_{0 \le t \le T}$ coexist. More precisely, we consider a market where the two risky assets correspond to the two price processes $(x_t)_{0 \le t \le T}$ and $(X_t)_{0 \le t \le T}$;

this market consists of :
• a risk free asset with $r = 0$,
• a first risky asset with price dynamics $dx_s = \nu_s(dW_s + \mu_s ds)$,
• a second risky asset with price dynamics $dX_s = \alpha_s dW_s + \beta_s dM_s$.

Are there reasonable constraints (such that the short-selling constraints) that allow the existence of these two different prices for a single contingent claim, in an arbitrage free market? Even though it seems difficult to draw a general comparison between the two price processes (x_t) and (X_t), adding short selling constraints seems necessary to limit arbitrage opportunities in this market, as mentioned in the introduction. Indeed, for example, suppose that at time t, prices are such that $A_t = \{x_t > X_t\}$ on a set of positive probability. An arbitrage strategy is the following: on the set A_t, buy ζ at price X_t, short sale ζ at price x_t and invest the difference $x_t - X_t$ in the riskless asset. On the complementary set A_t^c, do nothing. At maturity T, the asset ζ is received and immediately delivered; the amount invested in the riskless asset represents the gain for the arbitrageur.

3.1. *No Arbitrage conditions*

We prove that short-sale restriction might be sufficient to prevent market from arbitrages, if the market parameters satisfy the following inequalities.

Proposition 3.1. *If there exist two $\mathbb{G}$-predictable processes ψ and γ such that, $dt \otimes d\mathbb{P}$ almost surely,*

$$\begin{cases} \nu_t(\mu_t + \psi_t) \leq 0 \\ \psi_t \alpha_t + \lambda_t \beta_t \gamma_t \leq 0 \\ \gamma_t \geq -1 \end{cases} \tag{3.1}$$

then excluding short-selling of the assets x and X (defined in (2.1) and (2.2)) precludes arbitrage opportunities, in the market with zero interest rate ($r = 0$).

Proof. The proof is based on Jouini and Kallal [11] and Pulido [13] result. The authors have established the equivalence between the existence of a supermartingale probability measure and the absence of arbitrage opportunities while excluding short-selling. A supermartingale probability measure $\mathbb{Q}$ is a probability measure, equivalent to the historical one $\mathbb{P}$, under which discounted prices are super-martingales. We note L the Radon Nikodym density of $\mathbb{Q}$ w.r.t. $\mathbb{P}$,

$$dL_t = L_{t-}(\psi_t dW_t + \gamma_t dM_t) \quad \text{with } \gamma > -1.$$

Applying Itô's formula, it is straightforward to see that xL and XL satisfy

$$\begin{aligned} qd(x_t L_t) &= L_t \nu_t(\mu_t + \psi_t)dt + x_t L_{t-}(\psi_t dW_t + \gamma_t dM_t) + L_t \nu_t dW_t \\ d(X_t L_t) &= L_t(\psi_t \alpha_t + \lambda_t \beta_t \gamma_t)dt + X_{t-}L_{t-}(\psi_t dW_t + \gamma_t dM_t) \\ &\quad + L_{t-}(\alpha_t dW_t + \beta_t dM_t) \end{aligned}$$

hence the result holds. □

Let us study the implications of the non arbitrage equations (3.1). It is easy to see that
(a) On the set $\{(t,\omega)$ such that $\alpha_t(\omega)\mu_t(\omega) > 0\}$, defining $\psi_t(\omega) = -\mu_t(\omega)$ and $\gamma_t(\omega) = 0$ satisfy (3.1).
(b) On the set $\{(t,\omega)$ such that $\mu_t(\omega)\nu_t(\omega) < 0\}$, defining $\psi_t(\omega) = \gamma_t(\omega) = 0$ satisfy (3.1).
(c) On the remaining set $\{(t,\omega)$ such that $\mu_t(\omega) < 0$ and $\mu_t(\omega)\nu_t(\omega) > 0\} = \{(t,\omega)$ such that $\alpha_t(\omega)\nu_t(\omega) < 0$ and $\mu_t(\omega)\nu_t(\omega) > 0\}$, the non arbitrage equations (3.1) are studied as follow:

(1) On the set $\{(t,\omega)$ such that $\nu_t(\omega) < 0 < \alpha_t(\omega)$ and $\mu_t(\omega)\nu_t(\omega) > 0\}$ then necessarily $\mu_t(\omega) < 0$. The first condition of (3.1) is $\psi_t(\omega)\nu_t(\omega) \leq -\nu_t(\omega)\mu_t(\omega)$. Hence $\psi_t(\omega) \geq -\mu_t(\omega) > 0$.
The non arbitrage condition, after τ is $\psi_t(\omega)\alpha_t(\omega) \leq 0$, which is impossible. Before τ, the NA condition is $\psi_t(\omega)\alpha_t(\omega) + \lambda_t(\omega)\beta_t(\omega)\gamma_t(\omega) \leq 0$.
   - On the subset $\{(t,\omega), \beta_t(\omega) < 0\}$, the condition is satisfied by taking $\gamma_t(\omega) = \frac{-\mu_t(\omega)\alpha_t(\omega)}{\lambda_t(\omega)\beta_t(\omega)}$.
   - On the subset $\{(t,\omega), \beta_t(\omega) > 0\}$ the NA condition requires that $-\mu_t(\omega)\alpha_t(\omega) + \lambda_t(\omega)\beta_t(\omega)\gamma_t(\omega) \leq 0$, for a $\gamma_t(\omega) > -1$ hence there are arbitrages on the subset $\{(t,\omega), \mu_t(\omega)\alpha_t(\omega) + \lambda_t(\omega)\beta_t(\omega) < 0$ and $\beta_t(\omega) > 0\}$.

(2) On the set $\{(t,\omega)$ such that $\alpha_t(\omega) < 0 < \nu_t(\omega)$ and $\mu_t(\omega)\nu_t(\omega) > 0\}$ then necessarily $\mu_t(\omega) > 0$. The first condition of (3.1) is $\psi_t(\omega) \leq -\mu_t(\omega) < 0$.
The non arbitrage condition, after τ is $\psi_t(\omega)\alpha_t(\omega) \leq 0$, which is impossible. Before τ, the NA condition is $\psi_t(\omega)\alpha_t(\omega) + \lambda_t(\omega)\beta_t(\omega)\gamma_t(\omega) \leq 0$.

- On the subset $\{(t,\omega), \beta_t(\omega) < 0\}$, the condition is satisfied by taking $\gamma_t(\omega) = \frac{-\mu_t(\omega)\alpha_t(\omega)}{\lambda_t(\omega)\beta_t(\omega)}$.
- On the subset $\{(t,\omega), \beta_t(\omega) > 0\}$ the NA condition requires that $-\mu_t(\omega)\alpha_t(\omega) + \lambda_t(\omega)\beta_t(\omega)\gamma_t(\omega) \leq 0$, for a $\gamma_t(\omega) > -1$. Hence there are arbitrages on the subset $\{(t,\omega), \mu_t(\omega)\alpha_t(\omega) + \lambda_t(\omega)\beta_t(\omega) < 0 \text{ and } \beta_t(\omega) > 0\}$.

Proposition 3.2. *Under the constraint of no short selling, the market is arbitrage free*
- *after τ iff, $dt \otimes d\mathbb{P}$ almost surely, $\alpha\mu > 0$ or $\alpha\nu < 0$,*
- *before τ iff, $dt \otimes d\mathbb{P}$ almost surely, $\alpha\mu > 0$ or $\alpha\nu < 0$ or $\beta < 0$ or $\mu\alpha + \lambda\beta > 0$.*

In other words, arbitrages are possible (after τ) if and only if $\alpha > 0 > \nu, \mu < 0$ or $\alpha < 0 < \nu, \mu > 0$ on a $dt \otimes d\mathbb{P}$ non null set.
Arbitrages are possible (before τ) if and only if $\alpha > 0 > \nu, \mu < 0, \mu\alpha + \lambda\beta < 0, \beta > 0$ or $\alpha < 0 < \nu, \mu > 0, \mu\alpha + \lambda\beta < 0, \beta > 0$ on a $dt \otimes d\mathbb{P}$ non null set.

One can easily explicit arbitrage strategies in each situation where constraints (3.1) are not satisfied. On the event $\Omega_t(\omega) := \{(t,\omega), \nu(t)(\omega) < 0 < \alpha_t(\omega)\}$ the portfolio that consists of investing $\pi_t^1(\omega) = \alpha_t(\omega)\mathbf{1}_{\Omega_t}(\omega)$ in the asset x and $\pi_t(\omega) = -\nu_t(\omega)\mathbf{1}_{\Omega_t}(\omega)$ in the asset X is admissible. The value of this portfolio is Y_t such that, for $t > \tau$, $dY_t = \alpha_t\mu_t\nu_t\, dt$. This is a risk-free asset, with return greater than the one of the savings account. For $t < \tau$, $dY_t = \nu_t(\alpha_t\mu_t + \beta_t\lambda_t)dt$ with $\nu_t(\alpha_t\mu_t + \beta_t\lambda_t) > 0$ if the conditions (3.1) are violated.

Similarly on the event $\bar{\Omega}_t(\omega) := \{(t,\omega), \alpha_t(\omega) < 0 < \nu_t(\omega)\}$ the portfolio that consists of investing $\pi_t^1(\omega) = -\alpha_t(\omega)\mathbf{1}_{\bar{\Omega}_t}(\omega)$ in the asset x and $\pi_t(\omega) = \nu_t(\omega)\mathbf{1}_{\bar{\Omega}_t}(\omega)$ in the asset X is admissible. The value of this portfolio is Y_t such that, for $t > \tau$, $dY_t = -\alpha_t\mu_t\nu_t\, dt$. This is a risk-free asset, with return greater than the one of the savings account. For $t < \tau$, $dY_t = -\nu_t(\alpha_t\mu_t + \beta_t\lambda_t)dt$ with $-\nu_t(\alpha_t\mu_t + \beta_t\lambda_t) > 0$ if the conditions (3.1) are violated.

Remark: We have found conditions under which there does not exist arbitrage before AND after τ. The arbitrages we exhibit (if these conditions are not satisfied) are based on the knowledge of the sign of the volatility, and that sign is not adapted to the filtration of the prices. Indeed, if the diffusion term can vanish, $\mathbb{F}^x$ is strictly smaller that $\mathbb{F}^B$ and does not contain the sign of ν. Note that the quantity $\alpha\nu$ is the covariation of x and X, hence is observable. Since these arbitrages can not be realized by investors having only information on prices, we are in a well known situation in Economic: investors know that something exists, but they can not realize it.

3.2. *Optimal portfolio for a logarithmic utility*

We solve here the optimization problem for a logarithmic utility, as all computations can be done explicitly. Let (δ^1, δ) be a self financing strategy, with (δ^1, δ) the number of shares invested in the assets (x, X) respectively. The corresponding wealth $Y^{\delta^1,\delta}$ satisfies the following dynamics (we recall that the interest rate is zero)

$$dY_t^{\delta^1,\delta} = \delta_t^1 \nu_t \mu_t dt + (\delta_t^1 \nu_t + \delta_t \alpha_t) dW_t + \delta_t \beta_t dM_t.$$

The short selling strategies are excluded (otherwise, arbitrage strategies always exist): $\delta^1 \geq 0$ and $\delta \geq 0$. The set of admissible strategies is restricted to the no-short selling strategies leading to a positive wealth process. We will see that the no arbitrage constraints (3.1) naturally appear to exclude infinite value functions.

For a logarithmic utility $U(x) = \ln(x)$, the optimization of $\mathbb{E}(\ln(Y_T^{\delta^1,\delta}))$ can be computed explicitly. We normalize the portfolio process (δ^1, δ) by the corresponding wealth , by considering the proportion of wealth (π^1, π) invested in each asset: for $t \in [0, T]$, $(\pi_t^1, \pi_t) = \frac{1}{Y_{t-}^{\delta^1,\delta}}(\delta_t^1, \delta_t)$, leading to the exponential form of the wealth :
$\pi^1 \geq 0$, $\pi \geq 0$ and

$$\begin{aligned}
dY_t^{\pi^1,\pi} &= Y_{t-}^{\pi^1,\pi}(\pi_t^1 \nu_t \mu_t dt + (\pi_t^1 \nu_t + \pi_t \alpha_t) dW_t + \pi_t \beta_t dM_t). \\
Y_T^{\pi^1,\pi} &= Y_0 \exp\left(\int_0^T (\pi_t^1 \nu_t + \pi_t \alpha_t) dW_t - \frac{1}{2}(\pi_t^1 \nu_t + \pi_t \alpha_t)^2 dt + \pi_t^1 \nu_t \mu_t dt\right) \\
&\quad + \exp\left(\int_0^T (-\pi_t) \beta_t \lambda_t^{\mathbb{G}} dt + \ln(1 + \pi_t \beta_t)(dM_t + \lambda_t^{\mathbb{G}} dt)\right) \\
\mathbb{E}(\ln(Y_T^{\pi^1,\pi}) &= \ln(Y_0) \\
&\quad + \mathbb{E}\left(\int_0^T \left(\pi_t^1 \nu_t \mu_t - \frac{1}{2}(\pi_t^1 \nu_t + \pi_t \alpha_t)^2 - \pi_t \beta_t \lambda_t^{\mathbb{G}} + \lambda_t^{\mathbb{G}} \ln(1 + \pi_t \beta_t)\right) dt\right).
\end{aligned}$$

Thus the optimization problem relies on maximizing for every $t \in [0, T]$

$$\pi_t^1 \nu_t \mu_t - \frac{1}{2}(\pi_t^1 \nu_t + \pi_t \alpha_t)^2 - \pi_t \beta_t \lambda_t^{\mathbb{G}} + \lambda_t^{\mathbb{G}} \ln(1 + \pi_t \beta_t), \tag{3.2}$$

under the constraint $\pi_t^1 \geq 0, \pi_t \geq 0$.

We first exclude the cases for which $\mathbb{E}(\ln(Y_T^{\pi^1,\pi}))$ is not bounded by above. This can be achieved only for π^1 or π going to $+\infty$ AND conditions on the market parameters such that the corresponding value tends to $+\infty$. Expliciting those conditions (in particular, we have to exclude the possibility of having $\pi_t^1\nu_t + \pi_t\alpha_t = 0$ with the rest of the term in dt being positive), we again find the conditions of no arbitrage (3.1) and the arbitrage strategies. Those cases being excluded, the Lagrangian of this optimization problem under constraints is

$$\mathcal{L}(\pi_t^1, \pi_t) = \pi_t^1\nu_t\mu_t - \frac{1}{2}(\pi_t^1\nu_t + \pi_t\alpha_t)^2 - \pi_t\beta_t\lambda_t^{\mathbb{G}} + \lambda_t^{\mathbb{G}}\ln(1+\pi_t\beta_t) + k_t^1\pi_t^1 + k_t\pi_t$$

$$\text{with } k_t^1\pi_t^1 = 0, k_t\pi_t = 0, k_t^1 \geq 0, k_t \geq 0.$$

Due to the convexity of (3.2), the Kuhn Tucker first order conditions are sufficient (under no arbitrage conditions (3.1)):

$$\begin{aligned}
&-(\pi_t^1\nu_t + \pi_t\alpha_t)\alpha_t - \beta_t\lambda_t^{\mathbb{G}} + \frac{\lambda_t^{\mathbb{G}}\beta_t}{1+\pi_t\beta_t} + k_t = 0\\
&\nu_t\mu_t - (\pi_t^1\nu_t + \pi_t\alpha_t)\nu_t + k_t^1 = 0\\
&k_t^1\pi_t^1 = 0, k_t\pi_t = 0, k_t^1 \geq 0, k_t \geq 0.
\end{aligned}$$

Using the condition $\pi_t\beta_t > -1$ for $t \leq \tau$ (so that the wealth remains positive after τ), a systematic (but tedious) study of the Kuhn-Tucker conditions before τ and after τ (recall that $\lambda^{\mathbb{G}} = 0$ after τ) gives

- after τ, the optimal strategy is $\pi_t = 0$ and $\pi_t^1 = \frac{\mu_t}{\nu_t}$ if ν_t,μ_t and α_t have the same signs, and $\pi_t = \pi_t^1 = 0$ otherwise
- before τ, the optimal strategy is $\pi_t = \frac{-\mu_t\alpha_t}{\beta_t(\lambda_t^{\mathbb{G}}\beta_t+\mu_t\alpha_t)}$ and $\pi_t^1 = \frac{\mu_t}{\nu_t}(1 + \frac{\alpha_t^2}{\beta_t(\lambda_t^{\mathbb{G}}\beta_t+\mu_t\alpha_t)})$ if $\nu_t\mu_t > 0$, $\mu_t\alpha_t < 0$ and $\beta_t(\lambda_t^{\mathbb{G}}\beta_t + \mu_t\alpha_t) > 0$. Otherwise, $\pi_t = \pi_t^1 = 0$.

Remark that the optimal portfolio often stands at the limit of the constrained domain, that is $\pi^1 = 0$ or $\pi = 0$.

Remark: For a general utility function U, solving

$$\sup_{\delta^1\geq 0,\delta\geq 0} \mathbb{E}(U(Y_T^{\delta^1,\delta})),$$

leads to the problem of the replication with no short selling portfolios. Classical technics (duality or BSDE approaches) of portfolio optimization in incomplete market do not apply here. The particularity of our framework is that there does not exist any probability measure under which the prices are local martingales. In the literature in incomplete markets, there always exists at least one martingale measure and the existence of such measure is crucial in the replication methodology.

This problem of replication in an incomplete market with no martingale measure is a challenging problem that exceeds the aim of this paper.

Acknowledgements

We thank the financial support of European Institute of Finance for its financial support of the project "Role of Information" and the others members of this project for stimulating discussions: D. Coculescu, M. Jeanblanc, Y. Jiao, T. Lim, L. Stizia.

References

[1] J. Amendinger, P. Imkeller, and M. Schweizer. Additional logarithmic utility of an insider. *Stochastic Processes and Their Applications*, 75:263–286, 1998.

[2] K. Back. Continuous trading with asymmetric information and imperfect competition. *IMA, Mathematical Finance; M.H.A. Davis and D. Duffie and W.H. Fleming and S.E. Shreve*, 1994.

[3] K-H. Cho and N. El Karoui. Insider trading and nonlinear equilibria: Single auction case. *Annales d'Economie et de Statistique*, 60, 2000.

[4] N. El Karoui, M. Jeanblanc, Y. Jiao, and B. Zargari. Conditional default probability and density. *To appear in Musiela festschrift, Springer*, 2011.

[5] R.J. Elliott and M. Jeanblanc. Incomplete markets and informed agents. *Mathematical Method of Operations Research*, 50:475–492, 1998.

[6] A. Grorud and M. Pontier. Insider trading in a continuous time market model. *International Journal of Theoretical and Applied Finance*, 1:331–347, 1998.

[7] C. Hillairet. Comparison of insiders' optimal strategies depending on the type of sideinformation. *Stochastic Processes and Their Applications*, 115:1603–1627, 2005.

[8] M. Jeanblanc and Y. Le Cam. Progressive enlargement of filtration with initial times.*Stochastic Processes and their Applications*. 119:2523–2543, 2009.

[9] Th. Jeulin. Semi-martingales et grossissement de filtration, *volume 833 of Lecture Notes in Maths. Springer-Verlag*, 1980.

[10] Th. Jeulin and M. Yor. Inégalité de Hardy, semimartingales et faux-amis. *In P-A. Meyer, editor, Séminaire de Probabilités XIII, volume 721 of Lecture Notes in Maths.*, pages 332–359. Springer-Verlag, 1979.

[11] E. Jouini and H. Kallal. Arbitrage in securities markets with short sales constraints. *Mathematical Finance, 5(3)*:197–232, 1995.

[12] A.S. Kyle. Continuous auctions and insider trading. *Econometrica*, 53, 1985.

[13] S. Pulido. The fundamental theorem of asset pricing, the hedging problem

and maximal claims in financial markets with short sales prohibitions. Forthcoming in *Annals of Applied Probability*, 2011.

[14] S. Song. Optional splitting formula in a progressively enlarged filtration. *Preprint* on http://arxiv.org/abs/1208.4149, 2013.

Option Pricing under Stochastic Volatility, Jumps and Cost of Information

Sana Mahfoudh, Monique Pontier *

Abstract. In this paper we analyze and value an extension of the Black Scholes and Merton's model, in the context of Merton's (1987) "Simple Model of Capital Market Equilibrium with Incomplete Information". We show the derivation of partial differential equations for option prices in the presence of 'shadow costs' of incomplete information, stochastic volatility and jumps. We use two types of hedging strategies: the first one hedges the diffusive movement applying a Black-Scholes type strategy. The second one hedges both diffusion and jumps according to a mean variance minimizing strategy. These two strategies are compared in two simple cases (stochastic volatility case without jumps, jump case without stochastic volatility) with respect to their cost process values. Finally, the option price functions are defined as viscosity solutions of partial integro-differential equations.

1. Introduction

Option pricing based on a log-normal stock price model was first derived by Black and Scholes [8]. Various extensions include option pricing with stochastic volatility, and interest rates and dividend yield (e.g., Geske [17]; Hull and White [21, 22]; Heston [20]). Cox, Ross and Rubinstein [13] showed that the log-normal model is a limiting case of a discretized binomial model. The abitrage free framework is based on martingale theory and the fundamental theorem of asset pricing (Harrison and Pliska [18, 21]). Karatzas [24] suggests that the arbitrage free price (or fair price) of any simple European option must be the conditional expectation of the discounted payoff at maturity, where the expectation is taken under some appropriate risk-neutral measure. In the case of the jump-diffusion model for stock prices introduced by Merton [25], the cumulative returns of the stock is modeled by two components. The first component is a scaled Brownian

*Lab:BESTMOD, Inst. Math. de Toulouse (IMT),
emails: sana.mahfoudh@isg.rnu.tn, pontier@math.univ-toulouse.fr

motion with drift, as in the Black-Scholes model. The second component is a compound Poisson process with normally distributed jumps. Such a model is inherently incomplete because we have one assets and more than one random factor driving the dynamics of the return process. Furthermore, the density of the return distributions at any fixed time does not belong to an exponential family. It belongs instead to a class of mixtures of exponential families. On another hand, the Capital Asset Pricing Model with incomplete information developed simultaneously and independently by Merton [27] is perhaps the most influential object in modern finance in terms of imperfection of the markets. It provides the theoretical foundation for relating risks linearly with expected return of an assets. Bellalah and Jacquillat [4] combine the Black Scholes equation and the Merton's 'shadow costs' to explain the impact of the information costs through a partial integro-differential equation satisfied by the option price function. Our aim is to extend this option pricing equation to the case of stochastic volatility, jumps and 'shadow costs'. To deal with this idea, we choose a mixed jump diffusion process for the underlying assets and a continuous diffusion process for the volatility.

Naik and Lee [31] proved that Merton's model [25, 26] is not a complete overview of Harrison and Pliska [18, 19]. The models with stochastic volatility and jump lead to an incomplete market; for these models, the no arbitrage argument must be used carefully for the valuation of contingent claims since the problem is now to give the option price *and* to construct a portfolio strategy to replicate the option. Föllmer-Sondermann [16], Föllmer-Schweizer [15], Schweizer [41] proposed a solution to this problem. On the other hand, the stochastic volatility models have been examined by several authors, for example Engle [14], Hull and White [21], Romano and Touzi [36], Johnson and Shanno [23], Stein and Stein [44], Heston [20].

The paper is structured as follows. In Section 2, we review some standard results. In Section 3 we define the model with jumps and stochastic volatility. Section 4 proposes a first method for the pricing of options under stochastic volatility, jumps and information costs using strategy that hedges the diffusive part of the assets process. In Section 5 we apply the risk variance minimizing strategy method in presence of information costs, stochastic volatility and jumps. Section 6, in two simple examples, compare these two strategies, Section 7 proves that the portofolio value functions are viscosity solutions of partial integro-differential equations and Section 8 concludes the paper.

2. Survey

One of the main problems in modern Finance is the option valuation. In 1973, Black-Scholes and Merton proposed an option valuation equation in an arbitrage free model. This result is used in practice to give the price of financial products. Merton [25] generalized Black-Scholes formula to the case of an underlying assets whose the price process is discontinuous. Merton's jump diffusion (below denoted as MJD) model is an exponential Lévy model of the form:

$$S_t = S_0 e^{L_t}$$

where the stock price process $(S_t)_{0 \leq t \leq T}$ is modeled as an exponential of a Lévy process $(L_t)_{0 \leq t \leq T}$. Merton's choice of Lévy process is a Brownian motion with drift (continuous diffusion process) plus a compound Poisson process as following:

$$L_t = (\mu - \frac{\sigma^2}{2} - \alpha k)t + \sigma W_t + \sum_{i=1}^{N_t} Y_i$$

where $(W_t)_{0 \leq t \leq T}$ is a standard Brownian motion process. The only difference between Black-Scholes model and the MJD model is the addition of the term $\sum_{i=1}^{N_t} Y_i$. This compound Poisson process $\sum_{i=1}^{N_t} Y_i$ contains two sources of randomness. The first one is the Poisson process $(N_t)_{0 \leq t \leq T}$ with intensity α (so $(\tilde{N}_t = N_t - \alpha t)$ is a martingale) which causes the assets price to jump randomly. Once the assets price process jumps, jump size is also a random variable. Merton assumes that log stock price jump size follows normal distribution, $Y_i \sim \mathrm{N}(\mu, \delta^2)$. Thus three extra parameters μ, α, δ are added to the original Black-Scholes model. From definition, the law of N_t is Poisson law with parameter αt, and if T_i denotes the ith jump, the inter jump times $T_{i+1} - T_i$ are independent, with exponential parameter α law. Standard assumption is that $(W_t)_{0 \leq t \leq T}$, $(N_t)_{0 \leq t \leq T}$ and (Y_i) are independent.

Concerning hedging in case of incomplete market, the way is twofold: One way is "locally risk minimizing strategy" and yields to the notion of "minimal martingale measure" (first introduced by Föllmer and Sondermann [16], Schweizer [38, 39]); the other way concerns the "mean variance hedging" (Schweizer [41] among others). The references on this topic are in a great amount: see some review papers and their bibliography (Moller [28, 29], Pham [33], Schweizer [43]). More recent papers are published and certainly a new review would be useful. Anyway, we quote the ones concerned by models with jumps and/or stochastic volatility: Poulsen et al. [35]. In Menouken-Panen and Moneya [30] all the diffusion coefficients depend on a non observed Markov chain. In Vandaele and Vanmaele [45] the stock prices processes are exponential Lévy processes. Biagini et al. [6, 7] deal with jumps model but without stochastic volatility. Our model is very close to the one in Colwell-Elliott [9]: in our paper, we stress the role of information costs, the so-called 'shadow costs' introduced by Merton [27] and we add stochastic volatility.

3. The Model

Consider an option pricing model as in Merton [25, 26] where the pricing of derivative securities in the presence of a stochastic volatility needs the use of two processes: one for the underlying assets and one for the volatility. Exactly, given an underlying filtered probability space $(\Omega, \mathcal{A}, \mathbf{F} = (\mathcal{F}_t)_{0\leq t\leq T}), \mathbb{P})$, consider the following dynamics for the underlying assets S

$$S_t = S_0 + \int_0^t \mu_s S_s ds + \int_0^t \sigma_s S_s dW_s^1 + \int_0^t \gamma_s S_{s-} d\tilde{N}_s, \ t \in [0,T], \quad (3.1)$$

where $S_0 \in R^+$ is given and deterministic, W^1 is an $(\mathbf{F}, \mathbb{P})$-standard Brownian motion, $d\tilde{N}_t = dN_t - \int_0^t \alpha_s ds$ is the compensated martingale of an inhomogeneous $(\mathbf{F}, \mathbb{P})$-Poisson process with jump times T_i. We suppose (W^1, N) are adapted with respect to the filtration $\mathbf{F}$, with deterministic and bounded intensity $\alpha_t \geq 0$; the drift μ_t, the variance σ_t^2 and the jump size $\gamma_t S_{t-}$ of the process S are assumed to be $\mathbf{F}$- predictable processes satisfying $\gamma_t > -1$ and integrability conditions: $\mathbb{P}$ almost surely,

$$\int_0^T | \mu_s | ds < \infty; \int_0^T \sigma_s^2 ds < \infty; \int_0^T \gamma_s^2 \alpha_s ds < \infty; \int_0^T | \gamma_s | \alpha_s ds < \infty. \quad (3.2)$$

The process $(S_t)_{0\leq t\leq T}$ is a special semi-martingale, its continuous martingale part is $\int_0^t \sigma_s S_s dW_s^1$. The jump times of the process $(S_t)_{0\leq t\leq T}$ are those of $(N_t)_{0\leq t\leq T}$, the jump at time t is $S_t - S_{t-} = \gamma_t S_{t-}$. The predictable bracket of S is $\langle S\rangle_t = \int_0^t \sigma_s^2 S_s^2 ds + \int_0^t \alpha_s \gamma_s^2 S_s^2 ds$, the quadratic variation process is $[S]_t = \int_0^t \sigma_s^2 S_s^2 ds + \int_0^t \gamma_s^2 S_s^2 dN_s$.

We consider the following process modeling the stochastic volatility:

$$d\sigma_t = \nu(t, S_t, \sigma_t)dt + a(t, S_t, \sigma_t)dW_t^2, \ \sigma_0 \in \mathbb{R}. \quad (3.3)$$

Processes W^1 and W^2 are $\mathbf{F}$-Brownian motions with correlation coefficient ρ, $\rho^2 < 1$, the pair (W^1, W^2) is independent from N. The functions ν and

a on $[0,T] \times \mathbb{R}^+ \times \mathbb{R}$ are specified under appropriate hypotheses such that the system (3.1)(3.3) admits a unique strong solution. Below, for short, we note $\nu(t, S_t, \sigma_t)$ as ν_t and $a(t, S_t, \sigma_t)$ as a_t.

According to Merton [27], for any assets X (that may be underlying assets S or any other claims that could be introduced below) we define a 'shadow cost' λ^X such that the drift of X, $\mu_t^X = r_t + \lambda_t^X$. Taking into account these 'shadow costs', in the same time

$$e^{-\int_0^t (r_s + \lambda_s^X) ds} X_t$$

are local $(\mathbf{F}, \mathbb{P})$-martingales, thus

$$(r_t + \lambda_t^X) X_t dt = \mathbb{P} - \text{ finite variation part of } (dX_t) \tag{3.4}$$

and can be understood as Merton did: "the effect of incomplete information on equilibrium price is similar to applying an additional discount rate" (cf. [27] p. 493). Actually, these coefficients λ represent the marginal cost that the buyer agrees to pay to cover his lack of knowledge on the market.

Finally, we suppose there exists a riskless assets, namely $S_t^0 = e^{\int_0^t r_s ds}$, (r_s) being a deterministic discounting rate process. Anyway, from now on, we suppose that all the assets or contingent claims are discounted, meaning S_t^0 identical to 1, $r_t = 0$.

Remark 3.1. This modelling is inspired by Merton [27]. In case of discrete time, he proposes a Capital Asset Pricing Model in the presence of *'shadow costs'* of incomplete information. He argues that financial models based on complete information do not present the real complexity of rationality in action. A lot of factors and constraints, like entry into the business system, are not costless and have an influence on the security prices. The treatment of information and its associated costs play a role in capital markets. If an investor doesn't know about a trading opportunity, he cannot choose the appropriate benefitting strategy.

From Merton's model in discrete time (only two times, that is times $t = 0, 1$), it appears that taking into account the effect of incomplete information on the price of an assets is similar to applying an additional discount rate to this assets' future return, modelled as

$$R_k = E(R_k) + \sigma_k Y + \rho_k \varepsilon_k, \tag{3.5}$$

with Y a common factor, $E(Y) = 0, Var(Y) = 1$, ε_k a specific noise. He introduces the 'shadow cost' in (10) page 491 [27] relative to the riskless security return R:

$$\lambda_k := E(R_k) - R. \tag{3.6}$$

Similarly in continuous case, $E(R_k)$ is the drift, σ_k the volatility and Y is the randomness (W, N). Here we do not have any specific noise since there exists only one risky assets on the market.
But we must emphasize that we draw upon Merton's framework only for the introduction of information costs. Our approach is very different from his own: he is concerned with CAPM, that is an equilibrium problem between several agents; we are dealing with the evaluation of an option, using a non arbitrage approach, and its best hedging strategy with respect to a minimum quadratic cost.

4. A first strategy that hedges the volatility

In our model, the market consists in two assets, a risky assets S which follows the dynamics (3.1) and the risk-free assets S_t^0 (constant and equal to 1 so it is completely known and $\lambda_t^{S_0} = 0$). The aim is to hedge the contingent claim $g(S_T)$ where g is a Borel function, for instance $(S_T - K)^+$, and we look for a regular function V on $[0,T] \times \mathbb{R}_*^+ \times \mathbb{R}$ so that $V(T, S_T, \sigma_T) = g(S_T)$. Many authors work at once under a risk neutral probability measure $\widehat{\mathbb{P}}$ (e.g. Föllmer-Sondermann [16]) so the price process is $V(t, S_t, \sigma_t) = E_{\widehat{\mathbb{P}}}[g(S_T)/\mathcal{F}_t]$. This process is commonly called "intrinsic value process".

In our case, we don't know if some risk neutral probability measure exists or not, or are infinitely numerous because of stochastic volatility and jump part in the process S. To insure there exists at least one risk neutral probability measure, we assume that

$$\text{the Doléans exponential } G_. := \mathcal{E}(-\int_0^. \frac{\lambda_t^S}{\sigma_t}.dW_t^1) \text{ is a } \mathbb{P}\text{-martingale}$$

and we denote $\mathbb{P}^\lambda := G_T.\mathbb{P}$ and E^λ the expectation with respect to $\mathbb{P}^\lambda$.

We now work under this equivalent probability measure depending on the shadow cost λ^S. Thus Merton's characterization has to be expressed under this new probability measure:

$$(r_t + \lambda_t^X)X_t dt = \mathbb{P}^\lambda - \text{ finite variation part of } (dX_t) \tag{4.1}$$

Proposition 4.1. *The claim price function V, supposed to be a smooth function, is solution to the following partial integro-differential equation*

(PIDE for short), $t \geq 0,\ x > 0, \sigma \in \mathbb{R}$:

$$\begin{aligned}
&\frac{\partial V}{\partial t} + \frac{1}{2}\sigma^2 x^2 \frac{\partial^2 V}{\partial S^2} + \rho\sigma a(t,x,\sigma) x \frac{\partial^2 V}{\partial S \partial \sigma} + \frac{1}{2} a^2(t,x,\sigma) \frac{\partial^2 V}{\partial \sigma^2} \\
&+\nu(t,x,\sigma)\frac{\partial V}{\partial \sigma} - \lambda_t^V V(t,x,\sigma) \qquad (4.2)\\
&+ \alpha_t[V(t, x(1+\gamma_t), \sigma) - V(t,x,\sigma) - \frac{\partial V}{\partial S}\gamma_t x] = 0, \\
&V(T,x,\sigma) = g(x),\ (t,x,\sigma) \in [0,T] \times \mathbb{R}^+ \times \mathbb{R}.
\end{aligned}$$

In Section 7 we shall prove that there exists a unique viscosity solution of this PIDE. We notice that the V-information costs or 'shadow costs' λ^V appears naturally in this equation. The S-information cost is taken into account through the equivalent probability measure $\mathbb{P}^\lambda$.

Proof. According to (4.1) we have

$$V_\cdot - \int_0^\cdot \lambda_t^V V_t dt$$

is a $(\mathbf{F}, \mathbb{P}^\lambda)$-local martingale. Applying Itô formula to the semi-martingale V_t that is supposed to be a smooth function V of (t, S, σ) :

$$\begin{aligned}
dV_t - [&\frac{\partial V}{\partial t} + \frac{1}{2} S_t^2 \sigma_t^2 \frac{\partial^2 V}{\partial^2 S} + \rho a_t S_t \frac{\partial^2 V}{\partial S \partial \sigma} + \frac{1}{2} a_t^2 \frac{\partial^2 V}{\partial \sigma^2} + \nu_t \frac{\partial V}{\partial \sigma} \\
&+ \alpha_t (V(t, S_t, \sigma_t) - V(S_{t-}, \sigma_t) - \gamma_t S_t \frac{\partial V}{\partial S}) - \lambda_t^V V_t] dt
\end{aligned}$$

is the stochastic differential of a $(\mathbf{F}, \mathbb{P}^\lambda)$-local martingale. Using $(V_t)_{0 \leq t \leq T}$ decomposition uniqueness, by identification and since the support of the diffusion $(S_t, \sigma_t)_{0 \leq t \leq T}$ is the whole of $\mathbb{R}_*^+ \times \mathbb{R}$, we get the result. □

Remark 4.2. This equation is a generalization form of Black-Scholes, Merton and Bellalah-Jacquillat's equations:

- In other hand, if we set that there is no stochastic volatility but a constant volatility, no jumps and incomplete information Equation (4.2) is reduced to Bellalah-Jaquillat's equation [4, 5].
- Whether we set that there is no stochastic volatility but a constant volatility, jumps and complete information we find Merton's equation.
- If there is stochastic volatility but no jump, Equation (4.2) is reduced to:

$$\frac{\partial V}{\partial t} + \frac{1}{2}\sigma^2 S^2 \frac{\partial^2 V}{\partial S^2} + \rho\sigma a_t S \frac{\partial^2 V}{\partial S \partial \sigma} + \frac{1}{2} a_t^2 \frac{\partial^2 V}{\partial \sigma^2} + \frac{\partial V}{\partial \sigma}\nu_t - \lambda_t^V V = 0$$

- If $\sigma_t = 0$ (no volatility but jumps) Equation (4.2) becomes:

$$\frac{\partial V}{\partial t} - \lambda_t^V V + \alpha_t[V(t, S(1+\gamma_t)) - V(t,S) - \gamma S \partial_S V] = 0.$$

This equation is Merton's one [25] but including 'shadow costs'.

We introduce some notations to describe the agents trading in the market (S^0, S). We define a self-financing strategy (x, ϕ), x is the initial wealth and ϕ is an $\mathbf{F}$- predictable process who describes the quantity of assets S in the portfolio such that:

(1) $\phi \in L^2(d\mathbb{P} \otimes d\langle S\rangle)$,
(2) $\phi_t^0 = x + \int_0^t \phi_s dS_s - \phi_t S_t$ is the quantity on the risk-free assets.

Since the volatility is not a traded assets, a problem arises because this new source of randomness can not be easily hedged away. We introduce the "cost process"

$$C_t(\phi) := V(t, S_t, \sigma_t) - x - \sum_{j\geq 1} \int_0^t \phi_s^j dS_s^j$$

which could be considered as the hedging error of the contingent claim $V(T, S_T, \sigma_T) = g(S_T)$. Remark that $C_0(\phi) = 0$ thus $x = V(0, S_0, \sigma_0)$.

Below, we propose a first hedging strategy.

Hedging strategy using a fictitious assets

Here we consider that there exists another $C^{1,2}$ function V^1 such that $V^1(t, S_t, \sigma_t)$ is the process price of a fictitious assets supposed to hedge the stochastic volatility. We look for ϕ_t units of the underlying assets S_t, and ϕ_t^1 units of $V^1(t, S_t, \sigma_t)$.

Under the condition of self financing strategy, the cost process $C_t(\phi)$ satisfies

$$dC_t(\phi) = dV_t - \phi_t dS_t - \phi_t^1 dV_t^1. \tag{4.3}$$

Proposition 4.3. *With the choice of fictitious assets $V^1 : (t, S, \sigma) \mapsto \sigma$, an hedging strategy is $\phi_t^1 = \frac{\partial V}{\partial \sigma}(t, S_t, \sigma_t)$, $\phi_t = \frac{\partial V}{\partial S}(t, S_t, \sigma_t)$, the cost process satisfies*

$$dC_t(\phi) = dV(t, S_t, \sigma_t) - \partial_S V(t, S_{t-}, \sigma_t) dS_t - \partial_\sigma V(t, S_{t-}, \sigma_t) d\sigma_t, \tag{4.4}$$

$$\begin{aligned} dC_t(\phi) &= (\lambda_t^V V_t - \nu_t \partial_\sigma V_t) dt \\ &\quad + [V(S_{t-}(1+\gamma_t), \sigma_t, t) - V(S_{t-}, \sigma_t, t) - \frac{\partial V}{\partial S} \gamma_t S_{t-}] d\tilde{N}_t. \end{aligned} \tag{4.5}$$

Proof. We apply Itô formula to processes $(V_t)_{0\leq t\leq T}$ and $(V_t^1)_{0\leq t\leq T}$ since V and V^1 are supposed to be regular functions of (t, S_t, σ_t).

$$dC_t = \left(\frac{\partial V}{\partial t} + \frac{1}{2}\sigma_t^2 S_t^2 \frac{\partial^2 V}{\partial S^2} + \rho\sigma_t a_t S_t \frac{\partial^2 V}{\partial S \partial \sigma} + \frac{1}{2}a_t^2 \frac{\partial^2 V}{\partial \sigma_t^2}\right) dt \tag{4.6}$$

$$-\phi_t^1 \left(\frac{\partial V^1}{\partial t} + \frac{1}{2}\sigma_t^2 S_t^2 \frac{\partial^2 V^1}{\partial S^2} + \rho\sigma_t a_t S_t \frac{\partial^2 V^1}{\partial S \partial \sigma} + \frac{1}{2}a_t^2 \frac{\partial^2 V^1}{\partial \sigma^2}\right) dt$$

$$+\left(\frac{\partial V}{\partial S} - \phi_t^1 \frac{\partial V^1}{\partial S} - \phi_t\right) dS_t + \left(\frac{\partial V}{\partial \sigma} - \phi_t^1 \frac{\partial V^1}{\partial \sigma}\right) d\sigma_t$$

$$+[V(t, S_{t^-}(1+\gamma_t), \sigma_t) - V(t, S_{t^-}, \sigma_t) - \gamma_t S_t \partial_S V(t, S_{t^-}, \sigma_t)$$

$$-\phi_t^1(V^1(t, S_{t^-}(1+\gamma_t) - V^1(t, S_{t^-}, \sigma_t) - \gamma_t S_{t-} \partial_S V^1)] dN_t.$$

To hedge the randomness caused by the diffusion part dW_t^1 in the underlying asset price and dW_t^2 in its volatility, we set these two conditions which cancel the dS_t and $d\sigma_t$ coefficients:

$$\frac{\partial V}{\partial S} - \phi_t - \phi_t^1 \frac{\partial V_1}{\partial S} = 0; \quad \frac{\partial V}{\partial \sigma} - \phi_t^1 \frac{\partial V_1}{\partial \sigma} = 0.$$

We assumed that $\frac{\partial V_1}{\partial \sigma} \neq 0$ a.s. so

$$\phi_t^1 = \frac{\frac{\partial V}{\partial \sigma}(t, S_{t-}, \sigma_t)}{\frac{\partial V_1}{\partial \sigma}(t, S_{t-}, \sigma_t)}; \quad \phi_t = \frac{\partial V}{\partial S}(t, S_{t-}, \sigma_t) - \frac{\frac{\partial V}{\partial \sigma}(t, S_{t-}, \sigma_t)}{\frac{\partial V_1}{\partial \sigma}(t, S_{t-}, \sigma_t)} \frac{\partial V_1}{\partial S}(t, S_{t-}, \sigma_t). \tag{4.7}$$

With the choice of function $V^1(t, S, \sigma) = \sigma$, we get $\phi_t^1 = \frac{\partial V}{\partial \sigma}(t, S_t, \sigma_t)$, $\phi_t = \frac{\partial V}{\partial S}(t, S_t, \sigma_t)$ and the associate cost process $dV_t - \partial_S V dS_t - \partial_\sigma V d\sigma_t$. We then use (4.6) and (4.2) to replace dV_t so we get the second expression for the cost process. □

In the second step we will develop another form of the evaluation equation using a mean variance hedging strategy.

5. A mean variance hegding strategy in a jump-diffusion model

In the above we hedged the diffusive part of the random walk for the underlying assets. We now hedge both diffusion and jump process in the presence of information costs as much as we can and our question is the possibility to provide another hedging strategy.

We choose strategy ϕ to minimize the variance of the hedging portfolio. That means to choose a strategy ϕ which minimizes the variance of cost process under an appropriate probability measure:

$$\phi \mapsto E\left[\left(g(S_T) - E[g(S_T)] - \int_0^T \phi_s dS_s\right)^2\right]$$

where $g(S_T)$ is the option to hedge, for instance $(S_T - K)^+$. This approach is called *mean-variance hedging* but actually many authors look at it under a risk neutral probabibility measure ([16, 28, 35, 41, 42] among others), or more exactly under the so called *minimal martingale measure.* These studies are closely related to orthogonal decomposition of the contingent claim (e.g. Ansel and Stricker [2]). In our context, because of stochastic volatility, we aren't sure that such a minimal martingale measure exist. Schweizer [38] provides some necessary and sufficient conditions for the existence of a minimal martingale measure; we translate them in our model: let be $\beta_t := \frac{\lambda_t^S}{S_t(\sigma_t^2+\alpha_t\gamma_t^2)}$, then minimal martingale measure $\widehat{\mathbb{P}}$ exists if and only if the Doléans exponential $\mathcal{E}(-\beta.M)$ is a square integrable martingale, where M is defined as $M_t = 1 + \int_0^t S_s\sigma_s dW_s^1 + \gamma_s S_{s-} d\tilde{N}_s$, and so $\widehat{\mathbb{P}} = \mathcal{E}(-\beta.M).\mathbb{P}$. Under such a minimal martingale measure, the authors look for the so-called *local risk minimizing* ([6, 7, 30, 33, 40, 45] among others). Some of them deal with mixed diffusion, few of them deal with stochastic volatility, none of them deal with cost of information.

Our model is close to Colwell-Elliott' model [9], despite this one is without stochastic volatility nor shadow costs. They relax a little bit minimal martingale measure definition, and look for a minimal martingale measure specific to the contingent claim to be hedged, and not for a general minimal martingale measure concerning general contingent claims. So this idea is a key to solve our problem: once again we choose the risk probability measure $\mathbb{P}^\lambda$ such that the process $\int_0^t \lambda_s^S ds + \int_0^t \sigma_s dW_s^1 = \int_0^t \sigma_s dW_s^*$ where W^* is a $(\mathbb{F}, \mathbb{P}^\lambda)$ Brownian motion, and S is a (at least local) $(\mathbb{F}, \mathbb{P}^\lambda)$-martingale. Under this probability measure $\mathbb{P}^\lambda$, as Colwell-Elliott get it, the contingent claim is decomposed as the sum of two orthogonal $(\mathbb{F}, \mathbb{P}^\lambda)$ martingales: the cost process and $\int \phi_s dS_s$. We obtain the following result:

Theorem 5.1. *We assume*
1. Doléans exponential $G_. := \mathcal{E}(-\int_0^{\cdot} \frac{\lambda_t^S}{\sigma_t}.dW_t^1)$ *is a* $\mathbb{P}$*-martingale and we note* $\mathbb{P}^\lambda := G_T.\mathbb{P}$.
2. There exists a regular function V^* *on* $[0,T] \times \mathbb{R}_*^+ \times \mathbb{R}$ *such that* $E^\lambda[g(S_T)/\mathcal{F}_t] = V^*(t, S_t, \sigma)$.

Then an optimal strategy $\hat{\phi}$ is obtained by minimizing the cost variance:

$$\widehat{\phi_t} = \frac{\sigma_t^2 \partial_S V^* S_{t-} + \rho\sigma_t a_t \partial_\sigma V^* + \alpha_t\gamma_t[V^*(S_{t^-}(1+\gamma_t), \sigma_t, t) - V^*(S_{t^-}, \sigma_t, t)]}{\sigma_t^2 S_{t-} + \alpha_t \gamma_t^2 S_{t-}}, \tag{5.1}$$

V^* being solution to:

$$\begin{aligned}
&\frac{\partial V}{\partial t} + \frac{1}{2}\sigma^2 x^2 \frac{\partial^2 V}{\partial S^2} + \rho\sigma a(t,x,\sigma) x \frac{\partial^2 V}{\partial S \partial \sigma} + \frac{1}{2} a^2(t,x,\sigma)\frac{\partial^2 V}{\partial \sigma^2} + \nu(t,x,\sigma)\frac{\partial V}{\partial \sigma} \\
&\quad + \alpha_t[V(t, x(1+\gamma_t), \sigma) - V(t,x,\sigma) - \frac{\partial V}{\partial S}\gamma_t x] = 0, \qquad (5.2)\\
&\quad V(T,x,\sigma) = g(x), \ (t,x,\sigma) \in [0,T] \times \mathbb{R}^+ \times \mathbb{R}.
\end{aligned}$$

The minimal cost variance value is the $\mathbb{P}^\lambda$-expectation of

$$\begin{aligned}
&\langle C(\widehat{\phi}) \rangle_T \\
&= \int_0^T \left[\alpha_t \sigma_t^2 \frac{[V^*(S_{t^-}(1+\gamma_t), \sigma_t, t) - V^*(S_{t^-}, \sigma_t, t) - \gamma_t S_{t-} \partial_S V^*(S_{t^-}, \sigma_t, t)]^2}{\sigma_t^2 + \alpha_t \gamma_t^2} \right. \\
&+ \rho a_t \sigma_t \partial_\sigma V^* \frac{2\alpha\gamma^2 \partial_S V^* - \rho a_t \sigma_t \partial_\sigma V^* - 2\gamma_t \alpha_t [V^*(S_{t^-}(1+\gamma_t), \sigma_t, t) - V^*(S_{t^-}, \sigma_t, t)]}{\sigma_t^2 + \alpha_t \gamma_t^2} \\
&\left. + a_t^2 (\partial_\sigma V^*)^2 \right] dt.
\end{aligned}$$

Remark 5.2. Here we recover the locally minimizing strategy in Theorems 2.1 and 3.1 [9] in case of non stochastic volatility and Pham's example ([33] pp. 151-152) in case of no jumps.
Similarly we recover Cont and Tankov' result (cf. Proposition 10.6 page 342 [10]) who consider the jump diffusion model $\frac{dS_t}{S_{t^-}} = dZ_t$ where $Z_t = \mu t + \sigma W_t + \sum_{i=1}^{N_t} Y_i$, $(N_t)_{0 \le t \le T}$ being a Poisson process with intensity λ and Y_i i.i.d. random variables; moreover they provide a necessary and sufficient condition for the existence of a minimal martingale measure.
If $\sigma_t = 0$ (pure jump process) we get the hedging stategy:

$$\phi_t = \frac{V^*(t, S_{t^-}(1+\gamma_t)) - V^*(1, S_{t^-})}{\gamma_t S_{t^-}}. \tag{5.3}$$

Proof. : Under $\mathbb{P}^\lambda$, the total cost $C_T(\phi) = g(S_T) - E^\lambda[g(S_T)] - \int_0^T \phi_s dS_s$ is the terminal value of the $\mathbb{P}^\lambda$ martingale $E^\lambda[g(S_T)/\mathcal{F}_t] - E^\lambda[g(S_T)] - \int_0^t \phi_s dS_s$. Using Hypothesis 2, we develop $C_t(\phi)$ according to Itô formula:

$$
\begin{aligned}
dC_t(\phi) = & \left[\frac{\partial V^*}{\partial t} + \frac{1}{2}\sigma_t^2 S_t^2 \frac{\partial^2 V^*}{\partial S^2} + \rho\sigma_t a_t S_t \frac{\partial^2 V^*}{\partial S \partial \sigma} + \frac{1}{2} a_t^2 \frac{\partial^2 V^*}{\partial \sigma^2} + \nu_t \frac{\partial V^*}{\partial \sigma} \right. \\
& \left. + \alpha_t [V^*(t, S_t, \sigma_t) - V^*(t, S_{t-}, \sigma_t) - \frac{\partial V^*}{\partial S} S_{t-}\gamma_t]\right] dt \\
& + (\frac{\partial V^*}{\partial S} - \phi_t) S_t \sigma_t dW_t^* + \frac{\partial V^*}{\partial \sigma} a_t dW_t^2 \\
& + [V^*(t, S_t, \sigma_t) - V^*(t, S_{t-}, \sigma_t) - \phi_t S_{t-}\gamma_t] d\tilde{N}_t.
\end{aligned}
$$

Since $C_.(\phi)$ is a $\mathbb{P}^\lambda$-martingale, we cancel the first term and so prove that V^* is solution to the PIDE (5.2). Secondly, we compute the instantaneous quadratic variation of the cost process: $\langle C(\phi)\rangle_t^{\mathbb{P}^\lambda}$ to be minimized:

$$
\begin{aligned}
\phi \mapsto \frac{d}{dt}\langle C(\phi)\rangle_t^{\mathbb{P}^\lambda} = & (\partial_S V^* - \phi)^2 S_{t-}^2 \sigma_t^2 + (\partial_\sigma V^*)^2 a_t^2 + 2\rho(\partial_S V^* - \phi) S_{t-}\sigma_t \partial_\sigma V^* a_t \\
& + [V^*(S_{t-}(1+\gamma_t), \sigma_t, t) - V^*(S_{t-}, \sigma_t, t) - \phi S_{t-}\gamma_t]^2 \alpha_t.
\end{aligned}
$$

This application is convex with respect to ϕ, optimal $\hat{\phi}$ is solution to $\nabla_\phi \frac{d}{dt}\langle C(\phi)\rangle_t = 0$

$$
\begin{aligned}
& (\partial_S V^* - \phi) S_{t-}^2 \sigma_t^2 + \rho S_{t-}\sigma_t \partial_\sigma V^* a_t \\
& \quad + [V^*(S_{t-}(1+\gamma_t), \sigma_t, t) - V^*(S_{t-}, \sigma_t, t) - \phi S_{t-}\gamma_t] S_{t-}\gamma_t \alpha_t = 0,
\end{aligned}
$$

meaning the solution

$$
\widehat{\phi}_t = \frac{\sigma_t^2 \partial_S V^* S_{t-} + \rho\sigma_t a_t \partial_\sigma V^* + \alpha_t\gamma_t [V^*(S_{t-}(1+\gamma_t), \sigma_t, t) - V^*(S_{t-}, \sigma_t, t)]}{\sigma_t^2 S_{t-} + \alpha_t \gamma_t^2 S_{t-}}. \tag{5.4}
$$

We note that Equation(5.4) exactly means the $\mathbb{P}^\lambda$ orthogonality between processes $(C_t(\hat{\phi}))_{0\le t\le T}$ and $(\int_0^t \hat{\phi}_s dS_s)_{0\le t\le T}$ which yields the orthogonal decomposition of the contingent claim $g(S_T) = E^\lambda[g(S_T)] + \int_0^T \widehat{\phi}_s dS_s + C_T(\widehat{\phi})$.

As a corollary, we get $\frac{d}{dt}\langle C(\hat{\phi})\rangle_t$ minimal value:

$$
\frac{d}{dt}\langle C(\widehat{\phi})\rangle_t^{\mathbb{P}^\lambda} = c - \frac{b^2}{a} \tag{5.5}
$$

with

$$
\begin{aligned}
c = & (\partial_S V^*)^2 S_{t-}^2 \sigma_t^2 + (\partial_\sigma V^*)^2 a_t^2 + 2\rho(\partial_S V^*) S_{t-}\sigma_t \partial_\sigma V^* a_t \\
& + [V^*(S_{t-}(1+\gamma_t), \sigma_t, t) - V^*(S_{t-}, \sigma_t, t)]^2 \alpha_t, \\
b = & -S_{t-}[\partial_S V^* S_{t-}\sigma_t^2 + \rho\sigma_t \partial_\sigma V^* a_t + \gamma_t \alpha_t (V^*(S_{t-}(1+\gamma_t), \sigma_t, t) \\
& - V^*(S_{t-}, \sigma_t, t))], \\
a = & (\sigma_t^2 + \alpha_t \gamma_t^2) S_{t-}^2.
\end{aligned}
$$

After tedious but straightforward cancellations, we get the minimal quadratic variation of cost process under probability measure $\mathbb{P}^\lambda$:

$$\frac{d}{dt}\langle C(\hat{\phi})\rangle_t^{\mathbb{P}^\lambda} = \alpha_t\sigma_t^2 \frac{[V^*(S_{t-}(1+\gamma_t),\sigma_t,t)-V^*(S_{t-},\sigma_t,t)-\gamma_t S_{t-}\partial_S V^*(S_{t-},\sigma_t,t)]^2}{\sigma_t^2+\alpha_t\gamma_t^2}$$

$$+\rho a_t\sigma_t\partial_\sigma V^* \frac{2\alpha\gamma^2\partial_S V^* - \rho a_t\sigma_t\partial_\sigma V^* - 2\gamma_t\alpha_t[V^*(S_{t-}(1+\gamma_t),\sigma_t,t)-V^*(S_{t-},\sigma_t,t)]}{\sigma_t^2+\alpha_t\gamma_t^2}$$

$$+a_t^2(\partial_\sigma V^*)^2$$ □

6. Comparison between these two strategies costs

In this section we compare the strategies found in Sections 4 and 5 in some examples with respect to the $\mathbb{P}^\lambda$ expectation of their quadratic cost process. We first remark that actually, looking at both PIDEs, $V_t = \exp(-\int_t^T \lambda_s^V ds).V_t^*$.

6.1. *Continuous case*

Let be a first example with no jumps, so $\alpha_t = 0$ and recall the first strategy given in Section 4

$$\widehat{\phi_t^1} = (\partial_S V, \partial_\sigma V)$$

which leads to the cost process using (4.4)

$$dC_t(\widehat{\phi^1}) = (\lambda_t^V V_t - \nu_t\partial_\sigma V_t)dt$$

and the quadratic risk is bounded:

$$E^\lambda[(C_T(\widehat{\phi^1})^2] \leq T\int_0^T E^\lambda[(\lambda_t^V V_t - \nu_t\partial_\sigma V_t)^2]dt$$

$$= T\int_0^T e^{-2\int_t^T \lambda_s^V ds} E^\lambda[(\lambda_t^V V_t^* - \nu_t\partial_\sigma V_t^*)^2]dt.$$

In Section 5, the optimal strategy is

$$\widehat{\phi_t^2} = \partial_S V^* + \frac{\rho a_t\partial_\sigma V^*}{\sigma_t S_{t-}}.$$

and the optimal cost process $C(\widehat{\phi^2})$ admits a quadratic risk:

$$E^\lambda[\langle C(\widehat{\phi^2})\rangle_T] = \int_0^T E^\lambda[a_t^2(1-\rho^2)(\partial_\sigma V_t^*)^2]dt.$$

This result shows that in the case of no jumps but stochastic volatility, and $Te^{-2\int_t^T \lambda_s^V ds} E^\lambda[(\lambda_t^V V_t^* - \nu_t \partial_\sigma V_t^*)^2] < E^\lambda[a_t^2(1-\rho^2)(\partial_\sigma V_t^*)^2]$, meaning V information cost high enough, the first strategy is less costly.

6.2. *Non stochastic volatility case*

In a second example, we consider that σ is non stochastic, meaning $\rho = a = 0$. So, we set using again (4.4) then Theorem 5.1:

- $\hat{\phi}_t = \partial_S V_t$; $\hat{\phi}_t^1 = \partial_\sigma V_t$,

$$dC_t^1(\widehat{\phi}) = (\lambda_t^V V_t - \nu_t \partial_\sigma V_t)dt + [V(S_{t^-}(1+\gamma_t), \sigma_t, t) - V(S_{t^-}, \sigma_t, t) - \gamma_t S_{t^-} \partial_S V(S_{t^-}, \sigma_t, t)]d\tilde{N}_t,$$

so

$$E^\lambda[(C_T^1(\widehat{\phi})^2] = E^\lambda[(\int_0^T (\lambda_t^V V_t - \nu_t \partial_\sigma V_t)dt)^2] + E^\lambda[\int_0^T [V(S_{t^-}(1+\gamma_t), \sigma_t, t) - V(S_{t^-}, \sigma_t, t) - \gamma_t S_{t^-} \partial_S V(S_{t^-}, \sigma_t, t)]^2 \alpha_t dt],$$

- while $\widehat{\phi_t^2} = \frac{\sigma_t^2 \partial_S V^* S_{t-} + \alpha_t \gamma_t [V^*(S_{t-}(1+\gamma_t), \sigma_t, t) - V^*(S_{t-}, \sigma_t, t)]}{\sigma_t^2 S_{t-} + \alpha_t \gamma_t^2 S_{t-}}$, and

$$\frac{d}{dt}\langle C_2(\widehat{\phi^2})\rangle_t = \frac{\alpha_t \sigma_t^2 [V^*(S_{t^-}(1+\gamma_t), \sigma_t, t) - V^*(S_{t^-}, \sigma_t, t) - \gamma_t S_{t^-} \partial_S V^*(S_{t^-}, \sigma_t, t)]^2}{\sigma_t^2 + \alpha_t \gamma_t^2}.$$

Since $V_t = \exp(-\int_t^T \lambda_s^V ds).V_t^*$, we can minorate the ratio between both integrands by:

$$\frac{[V(S_{t^-}(1+\gamma_t), \sigma_t, t) - V(S_{t^-}, \sigma_t, t) - \gamma_t S_{t^-} \partial_S V(S_{t^-}, \sigma_t, t)]^2 \alpha_t (\sigma_t^2 + \alpha_t \gamma_t^2)}{\alpha_t \sigma_t^2 [V^*(S_{t^-}(1+\gamma_t), \sigma_t, t) - V^*(S_{t^-}, \sigma_t, t) - \gamma_t S_{t^-} \partial_S V^*(S_{t^-}, \sigma_t, t)]^2}$$
$$= \exp(-\int_t^T \lambda_s^V ds) \frac{\sigma_t^2 + \alpha_t \gamma_t^2}{\sigma_t^2} > 1$$

as soon as $\exp(\int_t^T \lambda_s^V ds) < \frac{\sigma_t^2 + \alpha_t \gamma_t^2}{\sigma_t^2}$ meaning V information cost small enough: then $E^\lambda[(C_T(\widehat{\phi^2})^2] < E^\lambda[(C_T(\widehat{\phi^1})^2]$.

This result shows that in the case of non stochastic volatility, jumps and V information cost small enough case, the second strategy is less costly.

7. A Viscosity solution of the PIDEs

Remark that actually the PIDE (5.2) is formally a particular case of (4.2) with null shadow cost λ^V. After some change of variables and functions, we set a new PIDE to equivalently solve Equation (4.2).

Proposition 7.1. *Let be the U a solution of the following PIDE:*

$$\begin{aligned}
&\partial_t U + \frac{1}{2}\sigma^2(-\partial_y U + \frac{\partial^2 U}{\partial y^2}) + \rho\sigma a(t,x,\sigma)\frac{\partial^2 U}{\partial y \partial \sigma} + \frac{1}{2}a^2(t,x,\sigma)\frac{\partial^2 U}{\partial \sigma^2} \\
&+\nu(t,x,\sigma)\frac{\partial U}{\partial \sigma} + \alpha[U(t,y+\ln(1+\gamma_t),\sigma) - U(t,y,\sigma) - \frac{\partial U}{\partial y}\gamma_t] = 0, \quad (7.1) \\
&U(T,y,\sigma) = g(e^y), \;\; y \in \mathbb{R}, \;\; \sigma \in \mathbb{R}.
\end{aligned}$$

Then, $V(t,x,\sigma) = e^{-\int_t^T \lambda_s^V ds} U(t, \ln x, \sigma)$ *is solution to PIDE:*

$$\begin{aligned}
&\frac{\partial V}{\partial t} + \frac{1}{2}\sigma^2 x^2 \frac{\partial^2 V}{\partial S^2} + \rho\sigma a(t,x,\sigma) x \frac{\partial^2 V}{\partial S \partial \sigma} + \frac{1}{2}a^2(t,x,\sigma)\frac{\partial^2 V}{\partial \sigma^2} + \nu(t,x,\sigma)\frac{\partial V}{\partial \sigma} \\
&+ \alpha_t[V(t,x(1+\gamma_t),\sigma) - V(t,x,\sigma) - \frac{\partial V}{\partial S}\gamma_t x] = \lambda_t^V V(t,x,\sigma), \quad (7.2) \\
&V(T,x,\sigma) = g(x).
\end{aligned}$$

Proof.

1. Define function H such that

$$V(t,x,\sigma) = e^{-\int_t^T \lambda_s^V ds} H(t,x,\sigma)$$

then $\partial_t V(t,x,\sigma) = -\lambda_t^V V(t,x,\sigma) + e^{-\int_t^T \lambda_s^V ds}\partial_t H(t,x,\sigma)$ and H is solution to:

$$\begin{aligned}
&\partial_t H(t,x,\sigma) + \frac{1}{2}\sigma^2 x^2 \frac{\partial^2 H}{\partial S^2} + \rho\sigma a_t x \frac{\partial^2 H}{\partial S \partial \sigma} + \frac{1}{2}a_t^2 \frac{\partial^2 H}{\partial \sigma^2} + \nu_t \frac{\partial H}{\partial \sigma} \\
&+ \alpha_t[H(t,x(1+\gamma_t),\sigma) - H(t,x,\sigma) - \frac{\partial H}{\partial S}\gamma_t x] = 0, \quad (7.3) \\
&H(T,x,\sigma) = g(x), \;\; x > 0, \;\; \sigma \in \mathbb{R}.
\end{aligned}$$

2. Let be the change of variable and function:

$$y = \ln x, \;\; x = e^y, \;\; y \in R, \;\; U(t,y,\sigma) = H(t,x,\sigma).$$

Then:

$$\nabla H = \partial_y U \frac{1}{x}, \;\; D^2 H = -\partial_y U \frac{1}{x^2} + \partial^2_{yy} U \frac{1}{x^2}, \;\; \partial^2_{x,\sigma} H = \partial^2_{y,\sigma} U \frac{1}{x}.$$

Plugging in (7.3) we get a new PIDE:

$$\partial_t U + \frac{1}{2}\sigma^2(-\partial_y U + \frac{\partial^2 U}{\partial y^2}) + \rho\sigma a_t \frac{\partial^2 U}{\partial y \partial\sigma} + \frac{1}{2}a_t^2 \frac{\partial^2 U}{\partial\sigma^2} + \nu_t \frac{\partial U}{\partial\sigma}$$

$$+ \alpha_t[U(t, y + \ln(1+\gamma_t), \sigma) - U(t, y, \sigma) - \frac{\partial U}{\partial y}\gamma_t] = 0,$$

$$U(T, y, \sigma) = g(e^y), \ y \in \mathbb{R}, \ \sigma \in \mathbb{R}.$$

□

Thus, according to Feymann-Kac formula, a candidate to be a solution of partial integro-differential equation (4.2) could be defined as an appropriate conditional expectation of the terminal value $g(e^{Y_T})$ as following.

Proposition 7.2. *We assume that parameters a and ν are Lipschitz functions of σ, and satisfy hypotheses (3.2). Let us assume that f solves equation (7.2). Then necessarily*

$$f(t, y, \sigma) := E^\lambda[g(e^{Y_T})/Y_t = y, \sigma_t = \sigma], \ Y_T = \ln S_T, \tag{7.4}$$

where the diffusion (Y, σ) is the unique solution of the differential system

$$\begin{aligned} dY_t &= [-\frac{1}{2}\sigma_t^2 + \alpha_t(\ln(1+\gamma_t) - \gamma_t)]dt + \sigma_t dW_t^* + \ln(1+\gamma_t)d\tilde{N}_t, \\ d\sigma_t &= \nu(t, \sigma_t)dt + a(t, \sigma_t)dW_t^2, \\ Y_0 &= \ln S_0, \ \sigma_0 \in \mathbb{R}. \end{aligned} \tag{7.5}$$

Proof. First of all remark that $(Y_t, \sigma_t)_{0 \le t \le T}$ is a diffusion the generator of which appears in PIDE (7.2), actually $Y_t = \ln S_t$ acording to (3.1). Moreover the system (7.5) admits a unique strong solution: all the coefficients are locally Lipschitz.

Let us assume that f solves equation (7.2). Then using Itô formula, we get

$$f(T, Y_T, \sigma_T) - f(t, Y_t, \sigma_t)$$

is a local martingale (actually a martingale because $g \in C_b^2$), so

$$f(t, y, \sigma) := E^\lambda[g(e^{Y_T})/Y_t = y, \sigma_t = \sigma].$$

We check that $f(T, Y_T, \sigma_T) = g(e^{Y_T})$. Conditional expectation under $\mathbb{P}^\lambda$ leads to

$$E^\lambda[f(T, Y_T, \sigma_T)/\mathcal{F}_t] = E^\lambda[g(e^{Y_T})/(Y_t, \sigma_t)]$$

since process $(Y_t, \sigma_t)_{0 \le t \le T}$ is a Markov process; if a solution f exists, Doob's theorem implies that it is necessarily a measurable function satisfying:

$$f(t, y, \sigma) = E^\lambda[g(e^{Y_T})/Y_t = y, \sigma_t = \sigma].$$

□

We now mix G. Barles and C. Imbert [3], H. Pham [32], E. Voltchkova [46] definitions of viscosity solution for any function F satisfying standard monotony properties (actually an adaptation of Definition 2 page 572 [3] to parabolic PIDEs).

Below w^* (respect. w_*) denotes the upper (lower)-semicontinuous envelope of w.

Definition 7.3. Let be w function on $[0,T] \times \mathbb{R}^2$ such that there exists C such that $|w(t,x) \leq C(1 + \|x\|)$.
1. Let be $(\bar{t}, \bar{x}) \in [0,T] \times \mathbb{R}^2$, a (test) function $\varphi \in C^{1,2}([0,T] \times \mathbb{R}^2)$, such that $w_*(\bar{t}, \bar{x}) = \varphi(\bar{t}, \bar{x})$, and $(\bar{t}, \bar{x})$ is a minimum of $w_* - \varphi$.

If $F[(\bar{t}, \bar{x}, \varphi(\bar{t}, \bar{x}), \partial_t \varphi(\bar{t}, \bar{x}), \nabla \varphi(\bar{t}, \bar{x}), D^2 \varphi(\bar{t}, \bar{x}), \mathcal{I}(\bar{t}, \bar{x}, \varphi)] \geq 0$, then w is a viscosity super-solution of PIDE:

$$F[t, x, h, \partial_t h, \nabla h, D^2 h, \mathcal{I}(\bar{t}, \bar{x}, h)] = 0.$$

2. Similarly, let be a test function $\varphi \in C^{1,2}([0,T] \times \mathbb{R}^2)$, such that $w^*(\bar{t}, \bar{x}) = \varphi(\bar{t}, \bar{x})$, and $(\bar{t}, \bar{x})$ is a maximum of $w^* - \varphi$.

If $F[(\bar{t}, \bar{x}, \varphi(\bar{t}, \bar{x}), \partial_t \varphi(\bar{t}, \bar{x}), \nabla \varphi(\bar{t}, \bar{x}), D^2 \varphi(\bar{t}, \bar{x}), \mathcal{I}(\bar{t}, \bar{x}, \varphi)] \leq 0$, then w is a viscosity sub-solution of this PIDE.

A function being both a sub and super viscosity solution is named a viscosity solution.

Here the function F of Definition 7.3 is defined with $x = (y, \sigma)$:

$$-F[t, x, U, \partial_t U, \nabla U, D^2 U, \mathcal{I}(U)] = \partial_t U + (r + \lambda_t^S - \frac{1}{2}\sigma^2)\partial_y U \tag{7.6}$$

$$+\nu(t,\sigma)\frac{\partial U}{\partial \sigma} + \frac{1}{2}\sigma^2 \frac{\partial^2 U}{\partial y^2} + \rho\sigma a(t,\sigma)\frac{\partial^2 U}{\partial y \partial \sigma} + \frac{1}{2}a^2(t,\sigma)\frac{\partial^2 U}{\partial \sigma^2} + \mathcal{I}(\bar{t}, \bar{x}, h)$$

where $\mathcal{I}(t, x, h) = \alpha_t[h(t, y + \ln(1 + \gamma_t), \sigma_t) - h(t, y, \sigma_t) - \gamma_t \partial_y h(t, y, \sigma)]$.
For short we note:

$$-\mathcal{A}(U)(t,x) = \partial_t U + (r + \lambda_t^S - \frac{1}{2}\sigma^2)\partial_y U + \nu(t,\sigma)\frac{\partial U}{\partial \sigma} + \frac{1}{2}\sigma^2 \frac{\partial^2 U}{\partial y^2} \tag{7.7}$$

$$+\rho\sigma a(t,\sigma)\frac{\partial^2 U}{\partial y \partial \sigma} + \frac{1}{2}a^2(t,\sigma)\frac{\partial^2 U}{\partial \sigma^2} + \mathcal{I}(t, x, U).$$

Proposition 7.4. *We assume that the function f defined in (7.4) is continuous on $[0,T] \times \mathbb{R}^2$, then it is a viscosity solution of the PIDE (7.2) according to Definition 7.3.*

Proposition 7.6 below proves that f is a continuous function on $[0,T]\times\mathbb{R}^2$ under some appropriate hypotheses.

Proof. : First of all, function F satisfies the usual ellipticity and parabolicity conditions: function F defined in (7.6) is decreasing with respect to $\mathcal{I}(t,x,U)$, $\partial_t u$, D^2u since matrix $-\frac{1}{2}\begin{pmatrix}\sigma^2 & \rho\sigma a\\ \rho\sigma a & a^2\end{pmatrix}$ is negative ($\rho^2\leq 1$). Moreover, the assumption is continuity of f, so f is both upper and lower continuous.

We use Pham's method ([34] p. 38 et sq.): Let be $(\bar{t},\bar{y},\bar{\sigma})\in[0,T]\times\mathbb{R}^2$, a test function $\varphi\in C_b^{1,2}$, $0=\phi(\bar{t},\bar{y},\bar{\sigma})-f(\bar{t},\bar{y},\bar{\sigma})=\min_{(t,y,\sigma)}(\phi(t,y,\sigma)-f(t,y,\sigma))$. There exists a sequence $(t_n,y_n,\sigma_n)_n$ in $[0,T]\times\mathbb{R}^2$ such that $(t_n,y_n,\sigma_n)\to(\bar{t},\bar{y},\bar{\sigma})$ and $f(t_n,y_n,\sigma_n)\to f(\bar{t},\bar{y},\bar{\sigma})$.

To be shorter, we put $x_n=(y_n,\sigma_n)$ and similarly, from now on, the diffusion X is the diffusion (Y,σ). According to the continuity of function f at x_n, we introduce η_n such that in the neighborhood $B(x_n,\eta_n)$, $|f(t_n,x_n)-f(t_n,y)|\leq\varepsilon$. Then let us introduce for any n the stopping time

$$\tau_n=\inf\{s>t_n:\ |X_s^{t_n,x_n}-x_n|\geq\eta_n\},$$

which is finite since the diffusion X is right continuous. Let be a sequence (h_k) decreasing to 0 and define

$$\theta_{n,k}=\tau_n\wedge(t_n+k_k)\in[t_n,t_n+h_k].$$

Apply Itô's formula to φ between t_n and $\theta_{n,k}$:

$$\varphi(\theta_{n,k},X_{\theta_{n,k}}^{t_n,x_n})=\varphi(t_n,x_n)-\int_{t_n}^{\theta_{n,k}}\mathcal{A}(\varphi)(s,X_s^{t_n,x_n})ds+\text{ martingale },$$

X^{t_n,x_n} being the diffusion starting from (t_n,x_n) and $\varphi(\theta_{n,k},X_{\theta_{n,k}}^{t_n,x_n})\geq f(\theta_{n,k},X_{\theta_{n,k}}^{t_n,x_n})$.

Then, for any n and any k, the $\mathcal{F}_{t_n}$ conditional expectation yields:

$$\varphi(t_n,x_n)-E[\int_{t_n}^{\theta_{n,k}}\mathcal{A}(\varphi)(s,X_s^{t_n,x_n})ds/\mathcal{F}_{t_n}]\geq E[f(\theta_{n,k},X_{\theta_{n,k}}^{t_n,x_n})/\mathcal{F}_{t_n}].$$

Let k go to the infinity, so h_k goes to 0, $\theta_{n,k}$ goes to t_n, X right continuity, $\mathcal{A}\varphi$ and f continuity get

$$\varphi(t_n,x_n)-\mathcal{A}(\varphi)(t_n,x_n)\geq f(t_n,x_n).$$

By definition of sequence (t_n), we get $\lim_n(\varphi(t_n,x_n)-f(t_n,x_n))=\varphi(\bar{t},\bar{y},\bar{\sigma})-f(\bar{t},\bar{y},\bar{\sigma})=0$, and using once again $\mathcal{A}(\varphi)$ continuity and n going to infinity,

$$\mathcal{A}(\varphi)(\bar{t},\bar{y},\bar{\sigma})=F[t,x,\varphi,\partial_t\varphi,\nabla\varphi,D^2\varphi,\mathcal{I}(\varphi)]\leq 0.$$

Thus function f is an upper solution.
Symmetrically, f is also a sub solution, and finally a viscosity solution. □

Corollary 7.5. *Assume the function f defined in (7.4) is continuous on $[0,T] \times \mathbb{R}^2$, $g \in C_b^2(\mathbb{R})$, coefficients of diffusion σ satisfy $a_t = a(t, \sigma_t) \in L^2(\Omega \times dt)$, $\nu_t = \nu(t, \sigma_t) \in L^1(\Omega \times dt)$, a and ν Lipschitz with respect to σ. Then the value functions are*

$$V(t,x,\sigma) = e^{-\int_t^T \lambda_s^V ds} f(t, \ln x, \sigma), \; V^*(t,x,\sigma) = f(t, \ln x, \sigma).$$

Proposition 7.6. *Assume $g \in C_b^2(\mathbb{R})$, coefficients of diffusion σ satisfy $a_t = a(t,\sigma_t) \in L^2(\Omega \times dt)$, $\nu_t = \nu(t,\sigma_t) \in L^1(\Omega \times dt)$, a and ν Lipschitz with respect to σ. Moreover, we suppose that the diffusion (Y, σ) is such that the law of Y given $\mathcal{F}_t$ is continuous in time (it could be the case if the Brownian part and the compound Poisson part of Y are independent). Then function f defined by (7.4) satisfies*
1. $f(t,.)$ is continuous on $\mathbb{R}^2$,
2. $f(.,y,\sigma)$ is continuous on $[0,T]$.

Proof. Proposition hypotheses show that the diffusion $(\sigma_s, s \geq t)$ is continuous with respect to initial conditions (t, σ) (cf. Protter [37] pp.307-317). Then

$$Y_T(t,y,\sigma) = y + \int_t^T [-\frac{1}{2}(\sigma_u^\sigma)^2 - \alpha_u \gamma_u] du + \int_t^T \sigma_u^\sigma dW_u^* + \int_t^T \ln(1+\gamma_u) dN_u$$

and using dominated Lebesgue theorem and L^2- continuity of $\sigma \mapsto \int_t^T \sigma_u^\sigma dW_u^*$, we deduce L^1- continuity of

$$(y, \sigma) \mapsto Y_T(t,y,\sigma).$$

Boundness of g, g', g" leads to continuity of $(y,\sigma) \mapsto g(Y_T(t,y,\sigma))$ and the one of $(y,\sigma) \mapsto E[g(Y_T(t,y,\sigma))/Y_t = y, \sigma_t = \sigma] = f(t,y,\sigma)$.

Part 2 of the proposition is proved using the conditional law of diffusion Y as a convolution of the laws of two independent random variables, namely on the event $N_{T-t} = k$

$$Y_T = y + \int_t^T [-\frac{1}{2}(\sigma_u^\sigma)^2] du + \int_t^T \sigma_u^\sigma dW_u^* + \sum_{j=1,k} Y_j. \qquad \square$$

8. Conclusion

This paper concerns the option valuation in a model with stochastic volatility, jumps and information costs. We give two approaches to hedge both jumps and stochastic volatility in presence of 'shadow costs'. This 'shadow costs' modelisation is inspired from Merton's discrete model [27]. The first approach consists in hedging the Brownian part of the randomness, but the inconvenient in this case is to consider the stochastic volatility as an assets that could be traded. The more classical second approach is based on mean variance hedging. In both situations, optimal strategies are provided, the value functions are solution of partial integro-differential equations. Moreover, under some convenient assumptions, the unique solution of these PIDE is exhibited as a conditional expectation of the contingent claim. In two simple examples, we compare the quadratic expectation of the hedging error coming from the chosen strategy; the discussion shows that the choice depends on the higher or lower level of the information cost on the claim to be hedged.

Acknowledgments. The authors would like to thank our presentations listeners and especially Monique Jeanblanc and Caroline Hillairet for their valuable advices and comments.

References

[1] J. Amendinger, P. Imkeller, and M. Schweizer. Additional logarithmic utility of an insider. *Stochastic Processes and Their Applications*, 75:263–286, 1998.

[2] Ansel J.P. and Stricker C., Décomposition de Kunita-Watanabé, *Séminaire de Probabilités XXVII, Springer,* L.N. in Maths 1557, 30-32, 1993.

[3] Barles G. and Imbert C., Second order elliptic integro-differential equations: viscosity solutions' theory revisited, *Annales de l'IHP, AN* 25: 567-585, 2008.

[4] Bellalah M., Jacquillat B., Option Valuation with Information Costs:Theory and Tests, *The Financial Review*, Vol 30, N 3: 617-635,1995.

[5] Bellalah M. , Mahfoudh S., Option Pricing Under Stochastic Volatility with Incomplete Information, *Wilmott Magazine* , March: 50-58, 2004.

[6] Biagini F. and Pratelli M., Local Risk minimization and numéraire, *Journal of Applied Probability* 36(4) 1126-1139, 1999.

[7] Biagini F., Cretarola A. Platen E., Local Risk minimization under the benchmark approach, *preprint*, 2011.

[8] Black F. and Scholes M., The Pricing of Options and Corporate Liabilities, *Journal of Political Economy* 81 May-June: 637-659, .

[9] Colwell D.B. and Elliott R.J., Discontinuous Asset Prices And Non-Attainable Contingent Claims, *Mathematical Finance* Vol 3, No. 3: 295-308, 1993.

[10] Cont R. and Tankov P., Financial Modelling With Jump Processes, *Chapman and Hal, CRC Financial Mathematics series* 2004.

[11] Cox J. C. and Ross S. A., A survey of Some New Results In financial Option Pricing Theory , *Journal of Finance* Vol 31(2): 383-402, 1976.

[12] Cox J.C., Ross S.A., The Valuation of Options for Alternative Stochastic Processes, *Journal of Financial Economics* Vol 3: 145-166, 1976.

[13] Cox J.C. and Ross S.A. and Rubinstein M., Option Pricing: A Simplified Approach, *Journal of Financial Economics* 7: 229-263, 1979.

[14] Engle R., Autoregressive Conditional Hetero-elasticity with Estimates of the Variance of U.K.Inflation, *Econometrica* 50: 987-1008, 1982.

[15] Föllmer H. and Schweizer M., Hedging of contingent claims under Incomplete Information, *in Davis M.H. Eliott eds, Applied Stochastic Analysis, Stochastics monographs* Vol 5: 389-414, 1991.

[16] Föllmer H. and Sondermann D., Hedging of non-redundant contingent claims, *Contributions to Mathematical Economics*, Amsterdam: North Holland, 205-223, 1986.

[17] Geske R. The pricing of options with a stochastic dividend yield, *Journal of Finance* 33: 617-625, 1978.

[18] Harrison J. M. and Pliska S. R. , Martingales and Stochastic and Integrals in The Theory of Continuous Trading, *Stochastic Processes and their Applications* Vol 11: 215-260, 1981.

[19] Harrison J. M. and Pliska S. R. : A Stochastic Calculus Model of Continuous Trading: Complete Markets, *Stochastic Processes and their ApplicationsC*, Vol 15: 313-316, 1983.

[20] Heston S., A closed-Form Solution for Options with Stochastic Volatility with Applications to Bond and Currency Options, *The Review of Financial Studies* Vol 6: 327-343, 1993.

[21] Hull J., White A., The Pricing of Options on Assets with Stochastic Volatilities, *Journal of Finance*, June: 281-320, 1987.

[22] Hull J., White A., An Analysis of The Bias in Option Pricing Caused by a Stochastic Volatility, *Advances in Futures and Options Research*, Vol 3: 29-61, 1988.

[23] Johnson H., Shanno D., Option Pricing When The Variance is Changing, *Journal of Financial Quant. Anal*, Vol 22: 143-151, 1987.

[24] Karatzas I, Optimization Problems In The Theory of Continuous Trading, *SIAM Journal of Control and Optim.* 27: 1221-1259, 1989.

[25] Merton R.C., Option Pricing When Underlying Stock Returns are Discontinuous, *Journal of Financial Economics* 3: 125-144, 1976.

[26] Merton R.C., The Impact on Option Pricing Of Specification Error in the Underlying Stock Price Returns, *Journal of Finance* 31: 333-350, 1976.

[27] Merton R.C., A Simple Model of Capital Market Equilibrium with Incomplete Information, *Journal of Finance*, 3: 483-509, 1987.

[28] Moller T., Risk-minimizing hedging strategies for insurance payment processes, *Finance Stochastic*, 5: 419-446, 2001.

[29] Moller T., Risk-minimization, *prepared for the Encyclopedia of Actuarial Sciences*, www.math.ku.dk/"tmoller, 2003.

[30] Menoukeu-Pamen O. and Momeya R., Local risk-minimization under a partially observed Martkov-modulated exponential Lévy model, *preprint* March, 2011.

[31] Naik, V., M. Lee, General equilibrium pricing of options on the market portfolio with discontinuous returns, *Rev. Financial Studies* 3: 493-521, 1990.

[32] Pham H., Optimisation et Contrle Stochastique Appliqués à la Finance, *Springer-Verlag*, 2007.

[33] Pham H., On quadratic hedging in continuous time, *Mathematical Methods of Operations Research* 51: 315-339, 2000.

[34] Pham H., Lectures on Stochastic Control and Applications in Finance, *Autumn school: Stochastic control problems for FBSDEs and applications, Marrakech*, December 2011, http://www.proba.jussieu.fr/pageperso/pham/pham/html, 2011

[35] Poulsen R., Schenk-Hoppé K.R., Ewald C.O., Risk-minimization in stochastic volatility models: model-risk and empirical performance, *preprint* 21 Sept 2007.

[36] Romano M. and Touzi N., Contingent Claims and Market Completeness In a Stochastic Volatility Model, *Mathematical Finance* Vol 7, No. 4: 399-412, 1997.

[37] Protter P., Stochastic Integration and Differential Equations, *Springer, Berlin Heidelberg New York*, 2005.

[38] Schweizer M., Some remarks on hedging under incomplete information *personal comunication* March 1990.

[39] Schweizer M., Risk miniminality and orthogonality of martingales, *Stochastics* 30: 123-131, 1990.

[40] Schweizer M., Option hedging for semi-martingales, *Stochastics processes and their Applications* 37: 339-363, 1991.

[41] Schweizer M., Approximating random variables by stochastic integrals, *The Annals of Probability* 22(3): 1536-1575, 1994.

[42] Schweizer M. Mean-Variance Hedging for General Claims, *Annals of Applied Probability* Vol 2: 171-179, 1992.

[43] Schweizer M., A guided tour through quadratic hedging approaches, *in Jouini, Cvitanic, Musiela (eds), Option Pricing, Interest Rates and Risk Management* Cambridge Univ. press, 538-574, 2001.

[44] Stein E. M., Stein J. C., Stock Price Distributions with Stochastic Volatility: An Analytic Approach, *The Review of Financial Studies* 4: 727-752, 1991.

[45] Vandaele N. and Vanmaele M., A locally risk-minimizing hedging strategy for unit-linked life insurance contracts in a Lévy process financial market, *Insurance: Mathematics and Economics* 42: 1128-1137, 2008.

[46] Voltchkova E. Thesis, *Université Toulouse 2*, 2005.

Printed in the United States
By Bookmasters